Theory and Method of Fisheries Forecasting

Xinjun Chen
Editor

Theory and Method of Fisheries Forecasting

 Springer

Editor
Xinjun Chen
College of Marine Science
Shanghai Ocean University
Shanghai, China

ISBN 978-981-19-2958-8   ISBN 978-981-19-2956-4   (eBook)
https://doi.org/10.1007/978-981-19-2956-4

This Springer imprint is published by the registered company Springer Nature Singapore Pte Ltd.
The registered company address is: 152 Beach Road, #21-01/04 Gateway East, Singapore 189721, Singapore

Preface

Marine biological resources are not only a part of important source of high-quality protein for mankind, but also provide the material basis for the sustainable development of human society. The rational development and utilization of marine biological resources need the support of fisheries theories and methods. With the development of theories and technologies such as fisheries oceanography, marine remote sensing, and geographic information system, as well as the increasing demand of marine fishery industry, the interdisciplinary discipline of theory and method of fisheries forecasting began to appear in the 1970s and have developed and matured in recent decades. It has become an important research field of marine fishery industry to understand the spatial distribution and change law of marine biological resources, and its relationship with climate and marine environment, which will contribute to the sustainable development and scientific utilization of marine biological resources.

Fisheries forecasting is the main content of fisheries oceanography, and it is also the comprehensive application of the basic principles and methods of fisheries oceanography in marine fisheries. Its research content is to predict the elements of fishery resources in a certain period and within a certain water area in the future, such as fishing season, fishing ground, fish population, and possible catch. With the development and application of information technology (Geographic Information System), space technology (Marine Remote Sensing), and expert system, the means and methods of fisheries forecasting have been deepened and developed continuously. From the traditional single species and single factor to the current marine ecosystem, the accuracy of fisheries forecasting has also been greatly improved and will be improved with the development of artificial intelligence and other technologies. The theory and method of fisheries forecasting have also been further improved and developed in the future.

This book is divided into four parts and seven chapters. The first part is the introduction, which mainly introduces the concept of fisheries forecasting and its relationship with other disciplines and reviews the research status at home and abroad. The second part is the theoretical basis of fisheries forecasting, mainly including description of the global marine environment, the shoaling, and migration of fish and its relationship with marine environment, the basic theory of fishing ground formation, etc. It focuses on the general laws of environmental factors such as surface temperature affecting fish behavior and distribution, as well as the types and basic principles of five kinds of excellent fishing ground. The third part mainly introduces the basic theories and methods of fisheries forecasting, as well as the application of marine remote sensing, geographic information system, habitat theory, and artificial intelligence in the fisheries forecasting. The fourth part is the case and application of fisheries forecasting; some important fishery species are taken as studying cases in forecasting the central fishing ground, fishing season, catch, resource abundance, and so on. At the same time, it summarizes and analyzes the impact of climate change on resources and fishing grounds of important economic species.

This book will provide a reference material for those who want to engage in fisheries or fishery-related work, which is applicable to undergraduate, postgraduate, training education, and fishery-related personnel. Meanwhile, due to the limitations of length and reference materials, as well as the limited level of authors, there are still many inappropriate points in this book. Readers are requested to make corrections and suggestions.

Shanghai, China   Xinjun Chen

Contents

About the Editor

Xinjun Chen is Professor, Dean of College of Marine Sciences, Shanghai Ocean University. He is also Director of National Engineering Research Center for Oceanic Fisheries, Director of Key Laboratory of Sustainable Exploitation of Oceanic Fisheries Resources, Director of China Oceanic Fisheries Data Center, and Head of Squid Fishing Technology Group of China Oceanic Fisheries Association

Contributors

Xinjun Chen College of Marine Sciences, Shanghai Ocean University, Shanghai, China

Jintao Wang College of Marine Sciences, Shanghai Ocean University, Shanghai, China

Wei Yu College of Marine Sciences, Shanghai Ocean University, Shanghai, China

Abbreviations

$\bar{\delta}$	The average density of the floating substances in time T
ξ	The disaster threshold
$\hat{a}$	The gray parameter
$\triangle$	The osmotic pressure
$\triangle S$	The spatial gradient distribution of environmental factors
$\triangle T$	The water temperature difference between points A and B
$\triangle T/\triangle Z$	The intensity of the thermocline
$\triangle T_1$	Sea temperature difference from 0 to 100 m
$\triangle T_2$	Sea temperature difference from 0 to 50 m
$\triangle WS$	The amount of change in salinity
$\triangle WT$	The amount of change in temperature
$\triangle X$	The horizontal distance of sea area
$\triangle Z$	The thickness of the thermocline
AIC	Akaike information criterion
AKFIN	Alaska Fisheries Information Network
AMM	Arithmetic mean model
ANNs	Artificial neural networks
ANOVA	Analysis of variance
APT	Automatic picture transmission
ATA	Air temperature anomaly
AVHRR	Advanced Very High Resolution Radiometer
B	The cumulative matrix
BP	Back propagation
c	The posteriori error ratio
C_0	MSE ratio value for determining precision testing grades
CBR	Case-based reasoning
CCAMLR	The Commission for the Conservation of Antarctic Marine Living Resources
CDOM	Colored dissolved organic matter
Chl-a	Chlorophyll-a
C_i	the yield of fishing area i
CLS	Collecte Localisation Satellites
CPM	Continued product model
CPUE	Catch per unit effort
CZCS	Coastal Zone Color Scanner
D	The shoaling of fish

DMSP	Defense Meteorological Satellite Program
EBP	Error backpropagation
EKE	Eddy kinetic energy
ENSO	El Niño-Southern Oscillation
ERS	European Remote-Sensing Satellite
ε	The error term
FAO	Food and Agriculture Organization
FG_{Lat}	The latitude of the gravity center related to the distribution of *O. bartramii*
FG_{Long}	The longitude of the gravity center related to the distribution of *O. bartramii*
FNN	Fuzzy neural network
FVCOM	Finite Volume Coastal Ocean Model
GAM	Generalized additive model
GEO	Geosynchronous
GEOSAT	Geodetic Satellite
GIS	Geographic Information System
GLMs	Generalized linear models
GM	Gray model
GMM	Geometric mean model
GOES	Geostationary Operational Environmental Satellite
GPS	Global Positioning System
HEP	Habitat evaluation procedures
HNLC	High-nutrient, low-chlorophyll
HSI	Habitat suitability index
HY-1B	Hai Yang-1B
IBM	Individual-based model
JAFIC	Japan Fisheries Information Service Center
K	The degree of convergence
KOMPSAT	Korean Multi-Purpose Satellite
K-S	Kolmogorov-Smirnov
L	The distance between the two water systems
LA-ICP-MS	Laser ablation-inductively coupled plasma mass spectrometry
Lat_{155}	The latitude value for the 20°C isotherm on the 155°E latitude line
Lat_{160}	The latitude value for the 20°C isotherm on the 160°E latitude line
MAXM	Maximum model
MINM	Minimum model
MLD	The mixed layer depth
MODIS	Moderate Resolution Imaging Spectroradiometer
MSE	Mean square error
N	The resource quantity of the tagged fish species
M	The total number in the fishing area
NAO	North Atlantic Oscillation
NMFS	National Marine Fisheries Service
NOAA	National Oceanic and Atmospheric Administration

NPI	North Pacific Index
NRC	National Research Council
p	The small error probability
p_0	Small error probability for determining precision testing grades
PacFIN	Pacific Fisheries Information Network
PDO	Pacific Decadal Oscillation
PFRP	Pelagic Fisheries Research Program
P_S	The ratio of the range occupied by the optimal SST to the total area of the spawning grounds
PSAT	Pop-up satellite archival tags
QR	Quantile regression
R	Multiple correlation coefficient
Radarsat	Radar Satellite
R_C	The resolution coefficient
RecFIN	Recreational Fisheries Information Network
R_{Long} and R_{Lat}	The correlation coefficients
ROCSAT	Republic of China Satellite
RS	Remote sensing
RSF	Resource selection function
S	A group of environmental factors
S'	The rate of change of environmental factor
s_1	The deviation of the observed data
s_2	The deviation of the residual error:
SD	Residual standard deviation
SDMs	Data-based species distribution models
SeaWIFS	Sea-viewing Wide Field-of-view Sensor
SOI	Southern Oscillation Index
SPC	Secretariat of the Pacific Community (formerly the South Pacific Commission)
SSC	Sea surface chlorophyll
SSH	Sea surface height
SSHA	Sea surface height anomaly
SSS	Sea surface salinity
SST	Sea surface temperature
SSTA	Sea surface temperature anomaly
T	The month
T_1	Mean SST anomaly value for spawning grounds (20-30°N, 140-170°E)
T_2	Mean SST anomaly value for feeding grounds (39-45°N, west of 150°E)
$T_3\ T_1 \times T_2$	Is the sympathetic effect between the spawning grounds and the feeding grounds
TIROS	Television Infrared Observation Satellite
TOPEX	Ocean Topography Experiment
u_1 and u_2	The velocity of the two water systems, respectively
URI	University of Rhode Island

WGM	Weighted geometric mean
WGMM	Weighted geometric mean model
WMM	Weighted mean model
X	The number of the tagged fish
$X_1, X_2, \ldots,$ X_h	The original data sequence $x_i^{(0)} = \{x_i^{(0)}(1), x_i^{(0)}(2), \ldots, x_i^{(0)}(n)\}$ $(i=1, 2, \ldots, h)$
X_{effort}	Is the total fishing effort during the winter fishing season
X_i and Y_i	The longitudinal and latitudinal positions of the center of fishing area i, respectively
X_{long} and Y_{lat}	The longitude and latitude of the position of the gravity center, respectively
X_{NB}	Is the *Trichiurus haumela* resource index from May to August from the Ningbo Fishery Company
X_{RE}	Is the corrected number for the relative resource of *Trichiurus haumela* in September
X_{s1}	Represents the relative shrimp yield in the Bohai Sea
$X_{s2}, X_{s3},$ and X_{s4}	Represent the relative quantity of juvenile shrimp in Bohai Bay, Laizhou Bay, and Liaodong Bay, respectively
X_{SH}	The *Trichiurus haumela* resource index from May to September from Shanghai Fisheries company
X_{SSTA}	The SSTA for the Nino 3.4 region
X_{YR}	The average runoff volume of the Yangtze River (September)
X_{ZJ} and X_{SS}'	The total fishing effort invested in the inshore area of Zhejiang and in the Shengshan fishing ground, respectively, in all fishing seasons that year
Y	The number of the recaptured tagged fish
Y_{catch}	The *Trichiurus haumela* catch during the winter fishing season
Y_{date}	The fishing season date (April y)
Y_{E}	The longitude of quarterly production gravity center
Y_N	The constant term vector
Y_{ZJ} and Y_{SS}	The total *Trichiurus haumela* yield in all fishing seasons in the inshore area of Zhejiang and the total *Trichiurus haumela* yield in all fishing seasons in the Shengshan fishing ground, respectively
Z	The total number of fish captured during the fishing season
Z_a	The depth at point A is the upper boundary depth of the thermocline
Z_b	The depth at point B is the lower boundary depth of the thermocline
α_0	Relative error for determining precision testing grades
$\beta 0, \beta 1, \beta 2,$ and so on	Regression coefficients
ε_0	Correlation rate for determining precision testing grades
μ and ν	The components of the horizonal current in the directions of x and y
ρ_1 and ρ_2	The density of the two water systems, respectively

Introduction

1

Xinjun Chen

Abstract

Marine fishery is an important part of the world fishery. For a long time, it not only provides high-quality animal protein for human but also provides employment opportunity, social welfare, and economic benefit. The sustainable development and use of marine fisheries depend on an understanding of the dynamics of fishery resources and the distribution mechanisms of fisheries, as well as on projections of resources and fishing ground. Therefore, with the progress of science and technology, as well as the increasing demand for the exploitation of marine fishery resources, fisheries forecasting began to emerge in the 1970s and has gradually become an emerging interdisciplinary subject; it covers fishery oceanography, oceanography, marine biology, fish behavior, satellite ocean remote sensing, and geographic information system. Fisheries forecasting is an applied subject, which studies the relationship between the action state of fish and other marine organisms and the surrounding environment, and grasps the laws of the quantity change of fishery resources and the distribution of fishing grounds, and is able to forecast the amount of catch. The forecast of fishery situation refers to the forecast of fishery resources in a certain period and a certain range of waters in the future, such as fishing period, fishing ground, fish quantity and quality, the possible catch, etc. Its forecast is based on the relationship between fish movement and biological features and environmental conditions, as well as the catch, resource status, marine environment, and other information from near-real-time pre-fishing season survey. Accurate forecasting of fishing conditions can provide scientific bases for fishery authorities and production units regarding how to carry out production deployment and production management during the fishing season, and furthermore, it can also provide a basis for fishery management departments regarding predicting the amounts of available resource. With the development of information technology and space technology as well as artificial intelligence and big data technology, the means and tools of fisheries forecasting have developed continuously, and the accuracy of fisheries forecasting has also improved. After undergoing decades of development, a comparatively complete fisheries forecasting discipline and technical system have initially been established.

X. Chen (✉)
College of marine sciences, Shanghai Ocean University,
Lingang Newcity, Shanghai, China
e-mail: xjchen@shou.edu.cn

© The Author(s), under exclusive license to Springer Nature Singapore Pte Ltd. 2022
X. Chen (ed.), *Theory and Method of Fisheries Forecasting*,
https://doi.org/10.1007/978-981-19-2956-4_1

Keywords

Fisheries forecasting · Marine fishery ·
Geographic information system · Ocean
remote sensing

1.1 The Concept of Fisheries Forecasting

Fisheries forecasting is the primary content of fishery oceanography research, and it involves the comprehensive application of the basic principles and methods of fishery oceanography in marine fisheries, representing the main tasks of scientific production and management in marine fisheries. Fisheries forecasting refers to making forecasts for each element of fishery resource conditions within a certain period in the future and for a certain range of waters, for example, fishing season, fishing grounds, number and quality of fish stocks, fishing catch, and so on. The bases of the forecasting are the relationship and laws between the movement and the biological conditions of fish and the environmental conditions, as well as various data on fishery oceanographic conditions, such as catch, abundance index, and marine environment, obtained in various near-real-time pre-fishing season surveys (Chen 2014). The main task of fisheries forecasting is to predict the central fishing ground, fishing season, amounts of available resource or resource abundance, and possible fish catch, that is, to answer questions regarding at what time and in what place, which fish to catch, how long the operation can last, when the beginning and end of the fishing season and the peak fishing period occur, where central fishing grounds are located, what the possible fish catch is for the entire fishing season, and what the amount of available resource or resource abundance is for the year. Accurate forecasting of fishing conditions can provide scientific bases for fishery authorities and production units regarding how to carry out production deployment and production management during the fishing season, and furthermore, it can also provide a basis for fishery management departments regarding predicting the amounts of available resource.

Since the 1950s, with the development of inshore fishery resources in China, each aquatic research unit in China has used fishery oceanography theories, traditional statistics, and other methods to conduct research on fisheries forecasting for the main inshore traditional commercial fish species, with an emphasis on forecasting central fishing grounds and predicting amounts of available resource and fish catch. Some achievements were attained, with an accumulation of valuable experience, making contributions to the research and development of fisheries forecasting. Beginning in the 1990s, with the development of offshore fisheries in China, research institutions such as Shanghai Ocean University have also started to utilize ocean remote sensing and a geographic information system (GIS) as well as an expert system, among other methods, to carry out research related to fisheries forecasting for the main fish species of offshore fisheries, which mainly include squid, tuna, and mackerel.

Internationally, since the 1970s, with the continuous development of ocean remote sensing technology, Japan, the United States, and France, among other countries, have also started to utilize sea data obtained by ocean remote sensing (Yamanaka et al. 1988), for example, surface temperature and chlorophyll concentration, to carry out fisheries forecasting for commercial capture fisheries and recreational fisheries. Additionally, fisheries forecasting technology has continuously developed. For example, Japan has established research centers for fisheries forecasting. With the development of information technology (i.e., GIS) and space technology (i.e., ocean remote sensing) as well as artificial intelligence and big data technology, the means and tools of fisheries forecasting have developed continuously, and the accuracy of fisheries forecasting has also improved. After undergoing decades of development, a comparatively complete fisheries forecasting discipline and technical system have initially been established.

1.2 Nature and Research Content of Fisheries Forecasting Discipline

1.2.1 Nature and Status of the Discipline

The fisheries forecasting discipline is a comprehensive applied science that studies the interrelation between the state of movement of marine organisms, such as fish, and the surrounding environment, masters the laws of change in the amount of fishery resources and the laws of distribution for fishing grounds, and carries out forecasting and prediction.

The content in this field includes professional basic theories and basic skills that scientific and technical personnel must possess for marine fishery production, management, and research. Through fisheries forecasting education, one can basically master basic knowledge regarding the marine fishery environment, understand the basic relationship between the movement of fish and the marine environment, master the basic principles regarding the types and formation of fishery oceanography, master the basic principles and methods of fisheries forecasting, and understand the application of new technology and new methods in fisheries forecasting, laying a foundation for future marine fishery production, fishery resource management, and teaching and scientific research while providing scientific methods and means for fishery production, fishery resource management, and the sustainable use of fisheries.

1.2.2 Research Content of the Discipline

In the ocean, commercial fish species, followed by cephalopods, shrimps, crabs, and so on, are the primary targets, and these are collectively referred to as aquatic commercial animals. To sustainably develop marine fishery resources, one must be familiar with the reserves of fishing targets in the waters, the migratory distribution of fisheries, and the mechanism and conditions of fishing

ground formation, among other issues, which are extremely important topics of research in the study of fisheries forecasting. The objectives and tasks in the study of fisheries forecasting technology are to learn the basic methods for studying and forecasting the distribution of central fishing grounds and predicting the resource amounts in order to master changes in the number of fishery resources and to provide scientific means of ensuring the sustainable use of fishery resources. The main research content includes the following (Chen 2016):

1. Analyze and master the relationship between the marine environment and the movement of fish. This includes understanding the current distribution of various oceans in the world and their general laws, the relationship between various marine environments (living and nonliving things) and the movement of fish, the migratory distribution of fisheries, the effects of global climate change on marine fishery resources, and so on.
2. Master the basic theories and laws of fishing ground formation. This includes an introduction to the basic concepts of fishing grounds and fishing seasons and the types of fishing grounds, the general principle of good fishing ground formation, general methods for finding central fishing grounds, and so on.
3. Master the basic theories and methods of fisheries forecasting. This includes an introduction to the concepts and types of fisheries forecasting, research methods, enumeration of typical fisheries forecasting cases, and an introduction to the application of advanced and new technology, such as ocean remote sensing and GIS, in fisheries forecasting.

1.3 Relationship Between Fisheries Forecasting and Other Disciplines

Fisheries forecasting is an applied interdisciplinary science involving fishery science and marine science, among other fields. It has a very close relationship with many other relevant disciplines, and these disciplines have enriched the research content, research means, and research methods of fisheries forecasting, jointly promoting the development of fisheries forecasting technology. The following are the primary relevant disciplines (Chen 2016):

1. Fishery oceanography. Fishery oceanography investigates the interrelation between the movement of biological resources in fisheries (clustering, distribution, migratory movement, etc.) and the surrounding environment (biotic environment and abiotic environment) and ascertains the laws of change in fishing conditions and the principles of fishing ground formation. It is based on fields such as fishery resource biology, oceanography, fish ethology, etc., and it has a close relationship with fields such as fishing gear, fishing laws, and ocean remote sensing. Fishery oceanography is a comprehensive applied science and is the basis for the study of fisheries forecasting.
2. Oceanography. Oceanography investigates the changes in and laws of interaction between hydrology, chemistry, and other inorganic and organic environmental factors in the ocean; therefore, the marine water environment, as the carrier of the object of study, and ichthyology are the bases of this field.
3. Marine biology. Marine biology investigates marine plankton and benthic organisms. Because marine plankton, benthic organisms, etc. are closely related to the object of study in fishery resources and fishery oceanography and provide sufficient food for the growth of fish, they are the bases of this field.
4. Fish ethology. Fish ethology investigates the interrelation between the movement of fish and environmental conditions, especially the relationships among water temperature, salinity, ocean current, light, and other conditions, and the movement of fish. Fish ethology lays a foundation for the development of and research related to fisheries forecasting.
5. Satellite ocean remote sensing. Satellite ocean remote sensing, referred to as space oceanography, is a subdiscipline that utilizes the principle of interaction of electromagnetic waves with the atmosphere and the ocean to observe and study the ocean from a satellite platform. It is an emerging discipline that crosses multiple disciplines, and its content includes physics, marine, and information disciplines; it is closely related to space technology, optoelectronic technology, microwave technology, computer technology, and communications technology. The marine information that satellite ocean remote sensing can provide includes sea surface temperature, sea surface height, chlorophyll concentration, sea wind, sea waves, sea ice, seafloor topography, storm surge, water vapors, rainfall, and so on. The development of satellite ocean remote sensing has provided abundant marine information for fisheries forecasting research and business operations and created conditions for employment in fisheries forecasting.
6. Geographic information system (GIS). GIS is a specific and very important spatial information system. It is a technical system that collects, stores, manages, calculates, analyzes, displays, and describes data related to geographical distribution in all or part of the Earth's surface space under the support of computer hardware and software systems. GIS is a computer-based tool; it can analyze and process spatial information, and the technology can be applied to scientific investigations, resource management, mapping, and other fields. The development of GIS has provided a powerful tool for research and development related to fisheries forecasting and has promoted the development of the study of fisheries forecasting.

1.4 General Situation of Foreign and Domestic Fisheries Forecasting Research

1.4.1 United States

Fish stocks are closely related to the environmental conditions of fishing grounds. The detection of environmental factors and parameters of fishing grounds by scientific methods and using the obtained information to analyze and guide fishery production are subsequent to the emergence of the use of aircraft and satellite ocean remote sensing for the detection of marine environmental conditions. Because the traditional basic and conventional approach involves inputting hydrological parameters forecasted by individual hydrologic stations (observation stations) and ships into a distribution map of marine parameters, this method is neither accurate nor timely. The utilization of aircraft and satellites to assess marine environmental parameters (such as water temperature and water color) has been very successful, and their utilization for fisheries is very convenient, providing timely data. The age of space technology has brought new prospects to fishery remote sensing. Humankind has attained the ability to observe the entire ocean area and sea area within minutes, enabling the use of parameters of vast marine environmental characteristics for fishery resource surveys and fishing ground analysis and forecasting.

Laurs (1971) described the first use of satellite data in the operational support to the US tropical Pacific tuna fleet. The satellite data received by automatic picture transmission (APT) were used in conjunction with other in situ oceanographic and weather information to produce fishery advisory charts, in which were included the location of oceanic thermal fronts as observed from satellites. The charts were transmitted daily to the fishing vessels via radio facsimile broadcast. Starting in 1975, satellite data began to be applied to fishing operations along the Pacific coast. At that time, satellite infrared images were utilized to obtain drawings that represented the positions of oceanic thermal boundaries, and these drawings (through telephone, telex, and mail) were provided to commercial and recreational fishermen for use in fish production areas with real potential. After 1980, radio facsimile was also used to send these drawings to fishing vessels at sea. These drawings were created one to three times per week and were mainly broadcasted via radio facsimile by the US Coast Guard. Fishers used these drawings to save time when finding fish production areas among oceanic fronts. On

the East Coast and in the Gulf of Mexico, the US National Weather Service, the National Marine Fisheries Service, and the National Environmental Satellite, Data, and Information Service often cooperate in the use of satellite infrared images and ship forecasting to create drawings to provide to fishers that indicate oceanic fronts, warm eddies, and sea surface temperature distribution. Driven by the United States, the United Kingdom, France, Japan, Finland, South Africa, and the United Nations Food and Agriculture Organization have successively organized various fishery remote sensing application studies and experiments, and some countries have also established corresponding service agencies. From 1993 to 1998, the US Pelagic Fisheries Research Program (PFRP) measured sea surface height through the Ocean Topography Experiment (TOPEX)/Poseidon satellite; the data revealed the relationship between the strength of subtropical fronts and the Hawaiian longline fishing ground for swordfish. During the period, 75% of the changes in catch per unit effort (CPUE) of the swordfish fisheries from January to June each year can be explained by the RS data (Lei 2016).

The National Marine Fisheries Service (NMFS) of the US National Oceanic and Atmospheric Administration (NOAA) applied ocean remote sensing and GIS to investigate marine fishery resources and fishing conditions and developed a series of fishery information systems, including the Alaska Fisheries Information Network (AKFIN), the Pacific Fisheries Information Network (PacFIN), the Recreational Fisheries Information Network (RecFIN), and other fisheries information system.

1.4.2 Japan

Japan's marine fisheries are more developed, and research and forecasting work related to the fishing conditions of important inshore commercial fish species were launched in the 1930s and 1940s. Through the early development of ocean remote sensing technology, Japan began applying and researching fishery remote sensing in the 1970s. In 1977, the Japan Science and

Technology Agency and the Fisheries Agency formally launched oceanic and fishery remote sensing experiments. Japan's Fisheries Agency established the "Remote Sensing Technology Center of Japan" in 1980; the purpose was to apply the remote sensing technology of artificial satellites to fisheries. The satellite utilization survey and review project that the Fisheries Agency entrusted to the "Fisheries Information Service Center" was divided into two stages. The first stage occurred from 1977 to 1981, and the main research content involved collecting and interpreting information from the artificial satellites and drawing isothermal diagrams of the sea surface temperature at 1 °C intervals; the second stage involved the transmission of such processed images to fishermen by two means, printed material and fax; and these products mainly included sea-state charts (water temperature charts) and fishing ground model forecasting. In October 1982, Japan's Fisheries Agency announced that it had utilized artificial satellites and computers to successfully search for fish stocks, such as sauries and tunas. Since then, the fishing ground and fishing condition chart (satellite interpretation map) has become a routine service product of Japan's Fisheries Information Service Center. In the early 1980s, Japan had approximately 900 fishing vessels that were equipped with fax machines able to receive images, and the "Fisheries Information Service Center" was thus established accordingly, creating a fishery information service system that included satellites, special survey aircraft, survey ships, fishing vessels, fishery communication networks, and the Fisheries Information Service Center. The Fisheries Information Service Center is responsible for the collection, analysis, archiving, and distribution of data. It uses a certain frequency at a set time every day to release, to domestic production fishing vessels, scientific research units, fishery companies, and so on, quick reports and charts related to fishing and sea conditions, providing more than ten items of information pertaining to the fishing ground environment, such as sea temperature, flow rate, flow direction, eddy current, water color, central fishing grounds, wind force, wind direction, air

temperature, and fishing conditions, thus playing an important role in maintaining Japan's status as an advanced country regarding fisheries. The center effectively utilizes remote sensing data from NOAA satellites to compile fishery forecasts and can obtain a large amount of marine environmental data in a short period of time, for example, hydrology, turbidity, and water color, greatly improving the effect and accuracy of fisheries forecasting. At present, Japan's Fisheries Information Service Center has already expanded its scope of forecasting and service to the sea areas of the three major oceans and directly provides information to Japan's offshore fishing vessels.

The sea areas for which Japan's Fisheries Information Service Center carries out fisheries forecasting include the waters of the Southwest Pacific Ocean, the Southeast Pacific Ocean, the North Atlantic Ocean, the South Atlantic Ocean, and the Indian Ocean, and the information includes quick reports on the inshore and open sea fishing and sea conditions of the Pacific Ocean, quick reports on the sea and fishing conditions of the Sea of Japan, quick reports on the sea and fishing conditions of the East China Sea, quick reports on the sea and fishing conditions in the eastern sea area of the Pacific Ocean corridor, quick reports on the sea and fishing conditions in the northeastern area of the Sea of Japan, quick reports on the sea and fishing conditions in the central and western sea areas of the Sea of Japan, quick reports on the sea conditions in the entire sea area of the North Pacific Ocean, quick reports on the sea conditions in the sea area of the East Pacific Ocean, quick reports on the sea conditions in the sea area of the Southeast Pacific Ocean, quick reports on the sea conditions in the sea area of the Southwest Pacific Ocean, quick reports on the sea conditions in the sea area of the Indian Ocean, quick reports on the sea conditions in the sea area of the South Atlantic Ocean, and quick reports on the sea conditions in the sea area of the North Atlantic Ocean, among other services. The fish species for which fishery forecasts are provided are the main fishery species distributed within the inshore water of Japan, for example, sei whale (Balaenoptera), Pacific saury, skipjack, Japanese common squid

(*Todarodes pacificus*), neon flying squid (*Ommastrephes bartramii*), Japanese mackerel (*Scomber japonicus*), horse mackerel (*Trachurus japonicus*), Japanese amberjack (*Seriola quinqueradiata*), and tuna.

1.4.3 China

Compared with some developed fishery countries and regions in the world, China started comparatively early in terms of fisheries forecasting research. In the 1950s and 1960s, under the influence of the Soviet Union and Japan, China's fisheries forecasting emphasized predicting fishing grounds and forecasting the fishing season. Primarily, the purpose of fisheries forecasting was to compile and draw fishing charts based on the relationship among biological data, such as water temperature, salinity, quantity and distribution of food organisms, the stock composition of the population, and sexual maturity, obtained from environmental surveys of fishing grounds, the distribution of population migration, and the external environment and to regularly issue various forecasts to fishery authorities and fishers. With the development of ocean remote sensing technology, satellite remote sensing has replaced large-scale fishing ground surveys (Chen 2016).

In China, technology such as ocean remote sensing and GIS in terms of fisheries forecasting was applied relatively late. During the "Seventh Five-Year Plan" period, research on the application of satellite remote sensing for fisheries was comparatively active, and the projects launched were primarily service oriented. The Fujian Provincial Department of Ocean and Fisheries (1986–1987) utilized a combination of satellite and hydrological data combined with the "Sea and Fishing Conditions Report" issued for the coastal sea area of Fujian, the "Quick Reports and Charts for the Fishing and Sea Conditions of the East China Sea and Yellow Sea" issued by way of radio facsimile by the Second Institute of Oceanography of the Oceanic Administration (1987–1988) on the basis of satellite images, the "Sea conditions of the Winter Fishing Season from satellite in the Tsushima strait" issued by

the Fishery Machinery and Instrument Research Institute (1988–1989), the "Satellite Remote Sensing Map of the Fishing Ground Environment" from the Institute of Oceanography of the Chinese Academy of Sciences, and the "Quick Report on the Fishing and Sea Conditions of the East China Sea and Yellow Sea" issued by the East China Sea Fisheries Research Institute. The aforementioned information is roughly divided into two types: (1) quick satellite reports and charts drawn up and issued mainly based on satellite images and (2) conventional hydrological measurement information that is sometimes combined with the distribution of satellite image information.

During the "Eighth Five-Year Plan" period, relevant scientific research institutions in China developed and applied "remote sensing (RS)" technology and "Global Positioning System (GPS)" technology; utilized NOAA satellite information; obtained dynamic change maps of the ocean temperature field, oceanic fronts, and cold and warm water masses through image processing technology; carried out research on the correlation between satellite information and fishing grounds; and explored the establishment of a measurement and forecast operation system for sea and fishing conditions.

During the "Ninth Five-Year Plan" period, the topic "GIS Technology for Marine Fishery Service" and the special topic "Marine Fishery Remote Sensing Service System" were proposed, in accordance with the goals of obtaining quick fishery forecasts and producing information to serve the three commercial fish species in the East China Sea area (largehead hairtail, black scraper, and mackerel). Based on improving the geographic information support software for marine fishery services, a desktop GIS system with marine fishery applications and a fishery stock assessment and resource assessment system were developed, composing the foundation of the Marine Fishery Geographic Information Application System.

At the end of the "Ninth Five-Year Plan" period, with funding from the Ministry of Science and Technology of China, the Squid Fishery Information Service System in the North Pacific based on GIS and ocean remote sensing technology were launched, initially establishing the Offshore Fishery and Fishing Condition Information Service Center. On the basis of key technologies, such as GIS-based temporal and spatial correlation analysis of central fishing grounds and environmental elements, the Rapid Forecasting System for Squid Fishing Conditions in the North Pacific and the Dynamic Management System for Offshore Fishery Production were developed, providing quick reports on fishing conditions and forecast information service products for squid production in the North Pacific and providing decision-making support for managing offshore fishery production.

During the "Tenth Five-Year Plan" period, the Environmental Information Application Service System for Oceanic Fishery Resource Development was launched, establishing the Environmental Information Acquisition System for Oceanic Fishing Grounds of tuna. Since the "Eleventh Five-Year Plan," the detection capabilities of self-developed ocean satellites and polar and shipborne RS data receiving systems as well as field monitoring by ocean fishing vessels have been utilized to establish an acquisition system for RS environmental information and field information for global fishing grounds in China. Quantitative processing technologies for various satellite RS data have been launched, emphasizing the acquisition of environmental factors such as sea temperature, water color, and sea surface height, and a comprehensive processing system for environmental information in global ocean fishing grounds with independent intellectual property rights has been established; on this basis, a global key fishing ground environment and fishing condition system was established, forming the technical platform for China's ocean fishery environmental monitoring and information service.

Since 1996, the Squid-jigging Technology Group at Shanghai Ocean University has been providing quick reports on the fishing and sea conditions for squid in the North Pacific Ocean; the reports are issued once a week, generating good results. Since 2008, Shanghai Ocean University and the National Satellite Ocean Application Service have collaborated in the Hai Yang-1B (HY-1B) satellite ground system to carry out fisheries forecasting research in connection with the main species in the three major oceans, for example, the purse seine mackerel and mackerel fisheries in the East China Sea, neon flying squid in the Northwest Pacific, jumbo flying squid in the Southeast Pacific Ocean, the Argentine shortfin squid in the Southwestern Atlantic Ocean, the Chilean jack mackerel in the Southeast Pacific Ocean, and the tuna in the central western Pacific; obtained various marine fishery environmental data, such as sea surface temperature, chlorophyll concentration, fronts, and eddy currents; and developed the corresponding software systems, thus achieving business-oriented operations and obtaining comparatively good economic benefits and ecological benefits (Chen 2016).

References

Chen XJ (2014) Fisheries resources and fisheries oceanography. Ocean Press. (In Chinese)

Chen XJ (2016) Theory and method of fisheries forecasting. Ocean Press. (In Chinese)

Laurs RM (1971) Fishery-advisory information available to tropical Pacific tuna fleet via radio facsimile broadcast. Commun Fish Rev 33(4):40±42

Lei L (2016) Remote sensing of marine fisheries. Ocean Press. (In Chinese)

Yamanaka I, Ito S, Niwa K, et al (1988) The fisheries forecasting system in Japan for coastal offshore fish. FAO Fisheries Technical Paper. No.301. Rome. FAO

Overview of the Marine Environment

2

Xinjun Chen

Abstract

Fish, cephalopod, shrimp, and other economic marine species are widely distributed in the ocean, but they are not evenly distributed in the ocean. This means that these marine economic species are not concentrated in all areas, nor can they form commercial fishing grounds in any sea area. A large number of scientific investigations and production practices have shown that the spatial distribution, life history, and resources of fish and other marine economic species are closely related to the marine environment, they are usually concentrated in the area around the front of the two currents and near waters of eddy, so it is very important to understand and master the state of the ocean environment, such as the shape of the ocean, the distribution of the ocean current, the distribution of the water temperature, and so on. This chapter briefly describes the division of the world's oceans, the topography of the seabed and the ocean sediments, the concept of ocean circulation and its causes, the generation of upwelling and downwelling currents, and the distribution of the world's ocean circulation; the distribution and characteristics of main currents in the Pacific Ocean, Atlantic Ocean, Indian Ocean, and Antarctic Ocean are described. Water temperature, nutrient salt, and primary productivity are the important environmental factors that affect the distribution of marine economic species, such as fish and the formation of fishing grounds, and are also the environmental factors that have been applied in fisheries forecasting. Therefore, this chapter briefly describes the distribution of water temperature, nutrient salt, and primary productivity in different sea areas of the world. To master the distribution law of ocean current, water temperature and other important environmental factors will help us to correctly carry out the research of fisheries forecasting and provide the theoretical basis for establishing scientific models of fisheries forecasting.

Keywords

Ocean current · Water temperature · Primary productivity · Global ocean

2.1 The Marine Morphology

2.1.1 Marine Area and Partitioning

The marine area on the earth is 316 million km^2, accounting for approximately 70.8% of the total area of earth. Oceans are distributed unevenly in the northern and southern hemispheres. In the northern hemisphere, oceans account for 60.7%

X. Chen (✉)
College of Marine Sciences, Shanghai Ocean University, Lingang Newcity, Shanghai, China
e-mail: xjchen@shou.edu.cn

© The Author(s), under exclusive license to Springer Nature Singapore Pte Ltd. 2022
X. Chen (ed.), *Theory and Method of Fisheries Forecasting*,
https://doi.org/10.1007/978-981-19-2956-4_2

of the total area, while land accounts for 39.3%; in the southern hemisphere, oceans account for 80.9%, with land accounting for only 19.1%. Moreover, earth can also be divided into the water hemisphere, which contains mostly surface water, accounting for approximately 91% of the area, and the land hemisphere, which contains most of the land but still only accounts for 47% of the area (Fig. 2.1).

Based on marine elements and morphological characteristics, marine waters can be divided into main parts and their subsidiary parts. The main parts are oceans, and the subsidiary parts are seas, bays, and straits.

Oceans

Oceans refer to waters offshore of continents that have a depth of more than 2000–3000 meters. Their area accounts for approximately 89% of the total marine area. Marine factors such as salinity and temperature, among others, are unaffected by the continents. The average salinity of oceanic water is 35‰, with little annual change; oceans have high water color with great transparency, and they have their own independent tidal and ocean current systems.

Based on the aforementioned characteristics, the world's oceanic water can be divided into three oceans, that is, the Pacific Ocean, the Atlantic Ocean, and the Indian Ocean. The boundary between the Pacific Ocean and the Atlantic Ocean is at Cape Horn, the apex of South America, at 70°W longitude, the boundary between the Atlantic Ocean and the Indian Ocean is the Cape of Good Hope (20°E longitude), and the boundary between the Pacific Ocean and the Indian Ocean is from the Malay Peninsula, Sumatra, Java, and East Timor via Cape Londonderry in Australia to Tasmania to the South Pole (147 °E longitude as the boundary). Some people also refer to the ocean surrounding the Antarctic continent as the Southern Ocean or Antarctic Ocean, and some people also refer to the Arctic Sea as the Arctic Ocean.

Seas

Seas refer to waters with a comparatively shallow depth, generally 200 to 300 m. The total area of seas is comparatively small, accounting for only 11% of the total marine area. The temperature of seas is greatly affected by the continents, and there are significant seasonal changes. The salinity is higher in inland sea areas, where there is no inflow of fresh water, and evaporation is strong but lower in sea areas with abundant river water inflow and little evaporation capacity, generally below 32‰. The water color is low, and there is little transparency. There are almost no independent tidal and ocean current systems, and they are mainly affected by the oceans into which they drain. Seas can be divided into two types: intercontinental seas and marginal seas. Intercontinental seas lie between continents or extend into the interior of continents, such as the Mediterranean Sea in Europe, Baltic Sea, South China Sea, Gulf of Mexico, Persian Gulf, and Red Sea, among others. Marginal seas are located on the margins of continents, such as the North Sea, Sea of Japan, East China Sea, and Yellow Sea, among others.

Bays

Bays refer to waters that are a part of oceans or seas that extend into continents, and their depths gradually decrease. In general, the connecting line between capes at the mouth of a bay or the isobath at the mouth of a bay is used as the boundary with the ocean or sea. The properties of the sea water in a bay are therefore very similar to the marine conditions of the connecting ocean or sea due to free water interchange with the adjacent sea or ocean. The maximum tidal range often occurs in a bay because of the continuous decrease in depth and width.

Straits

Straits refer to waterways that are comparatively narrow in width between adjacent sea areas in a sea or ocean. The main marine condition in straits is fast water flow, especially a very fast tidal flow rate, with mostly rock or sand and gravel as the substrate and very little fine sediments, a property that is related to the greater flow rate. Some ocean currents flow in or flow out of the upper or lower layer, such as the Strait of Gibraltar; some currents flow in or flow out from one side or the other, such as the Bohai Strait. Because there are

Fig. 2.1 Land hemisphere and water hemisphere (Chen 2014, 2016)

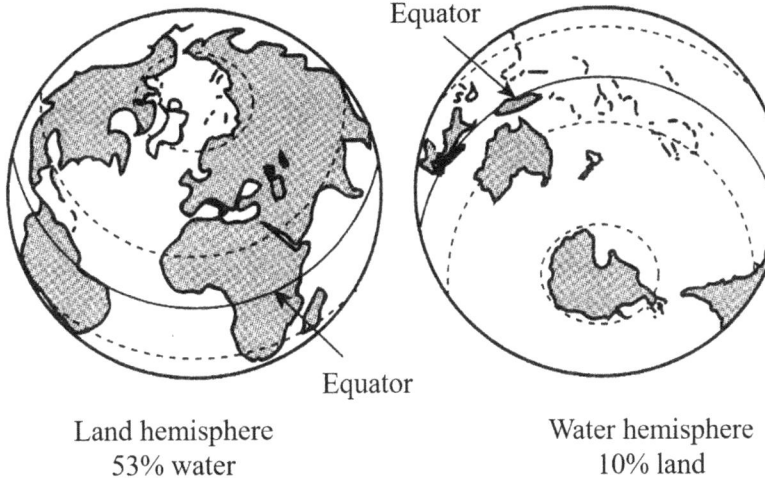

Land hemisphere
53% water

Water hemisphere
10% land

two types of water masses from different sea areas in a strait, the marine environmental conditions are quite different.

2.1.2   Seabed Morphology

Oceans are a continuous whole, but in different areas of oceans, environmental factors still differ greatly. Different species of organisms inhabit different habitats, and no organism can live in all of the environments of oceans.

Oceans can be divided into the water layer and the seabed. The former refers to the entire body of water of an ocean, and the latter refers to the entire seabed, and they can each be divided into different environmental zones. There are two major categories for marine organisms: species that float or swim in the water layer and species that inhabit the bottom of the ocean (on the bottom or in the bottom).

Seabed topography is an important factor in the formation of fishing grounds, for example, flat continental shelf fishing grounds, raised seabeds, and so on. Seabed topography is generally divided into the continental shelf, continental slope, ocean floor (deep-ocean basin), ocean trench, etc. In addition, there are also shoals, banks, reef piles, and so on, which all have a relationship with the formation of fishing grounds. Protruding topography, such as sea

rises or rises, ocean ridges or submarine ridges, submarine plateaus, banks, crests, reefs, shoals, and so on, are all related to the formation of fishing grounds and the schooling of fish. Seabed morphology can be roughly divided as follows: coast, continental margins (including the continental shelf, continental slope, and continental rise), and deep-ocean basin (including abyssal plains, various submarine highlands, marshlands, and so on).

Coast

The coast is the boundary between the sea and the land and refers to those regions where interactions of sea and land processes occur, namely, regions that are submerged when the water level rises (due to the increase in water caused by the tides, wind, and other factors) and exposed when the water level lowers.

Because the coast is an area where the land and the sea or ocean interacts, changes in the contour of the coast, changes in the topography of the seabed, and the displacement of marine sediments occur rapidly. The coastline refers to the dividing line between land and sea, and it is not fixed to a certain extent. Due to the rise and fall of tidal levels and the effect of the increase or decrease in water caused by wind, coastlines can shift; the range of the sea surface in the vertical direction can reach 10–15 m, but advancement and retreat in the horizontal direction can sometimes reach tens of kilometers. In coastal zones, areas affected by the rise and fall of tides are referred to as intertidal zones. Intertidal zones have a certain importance in fisheries production and scientific research.

Continental Margins

Continental margins specifically include the continental shelf, continental slope, and continental rise (Fig. 2.2).

Continental Shelf (Referred to as the Shelf)

The continental shelf is also named as the shelf. According to the *Convention on the Continental Shelf* passed at the International Conference on the Law of the Sea in 1958 (United nations 2005), the continental shelf is defined as "the seabed and subsoil of the submarine areas adjacent to the coast but outside the area of the territorial sea, to a depth of 200 meters or, beyond that limit, to where the depth of the super adjacent waters admits the exploitation of the natural resources of the said areas," as well as "the seabed and subsoil of similar submarine areas adjacent to the coasts of islands." Based on the viewpoint of natural science, the continental shelf is the shallow water zone submerged by sea water around a continent and a natural extension of the continent toward the bottom of the sea or ocean. Its scope starts from the low tide line, with an extremely gentle slope, and extends to the location where the slope increases suddenly. Although this zone is submerged in sea water, it is still a part of the continent. The depth of the continental shelf generally does not exceed 200 meters, and the depth of some individual areas is greater than 800 meters or less than 130 m. The average depth is approximately 130 m.

The shelf has a gentle slope, with an average slope of 7′; most shelves are a continuation of the submerged coastal plains. Near rocky coasts, the slope of the shelf is greater, but it still does not exceed 1°–2° in ordinary circumstances.

The width and depth of the shelf vary greatly and are closely related to the land topography. Adjacent to coasts with high mountains and steep hills, the shelf is narrow; conversely, the shelf is very wide adjacent to coasts formed by glaciers, broad plain, or estuaries of large rivers. In terms of the world, the average width of the shelf is approximately 70 km, but the range can vary from 0 to 700 km. The shelf along the coast of northern Europe and Siberia is very wide, reaching 600–800 km, and the shelf along the coast of China is very wide, with the shelf accounting for approximately 7.6% of the entire seabed area.

Many marine phenomena in the shelf area undergo significant seasonal changes, and the actions of tides, waves, and ocean currents are more intense; therefore, vertical mixing between the water layers is very active, and the bottom layer of sea water is constantly renewed, thereby generating a large amount of dissolved oxygen and various nutrient salts in sea water. As a result, the shelf, particularly estuarine zones, is an important place for fisheries and cultivation enterprises.

The shelf floor is mainly composed of terrigenous detrital sediments, including large rocks, gravel, pebbles, sand, fine mud, and so on, brought by rivers and streams of the continent and formed by the action of wave erosion. The distribution of these sediments on the seabed is regular; the farther away from the coast, pebbles and sand are gradually replaced by fine sand and mud.

Continental Slope

The continental slope refers to the steeper area at the outer margin of the shelf; it actually refers to the area within the margin of continental tectonics and is located above the transition zone from thick continental crust to thin oceanic crust. The slope ranges from 4° to 7°, sometimes reaching 13°–14°, such as in the Bay of Biscay. However, the dip angle near the coast of volcanic islands can be particularly large; the maximum can reach 40°, and sometimes it is almost vertical.

The slope of the continental slope varies with the properties of the coast. The average slope of the continental slope along the coast of mountainous areas is 3°33′, but the average slope of the continental slope outside the coastal plain is only 2°. The depth to which the continental slope can

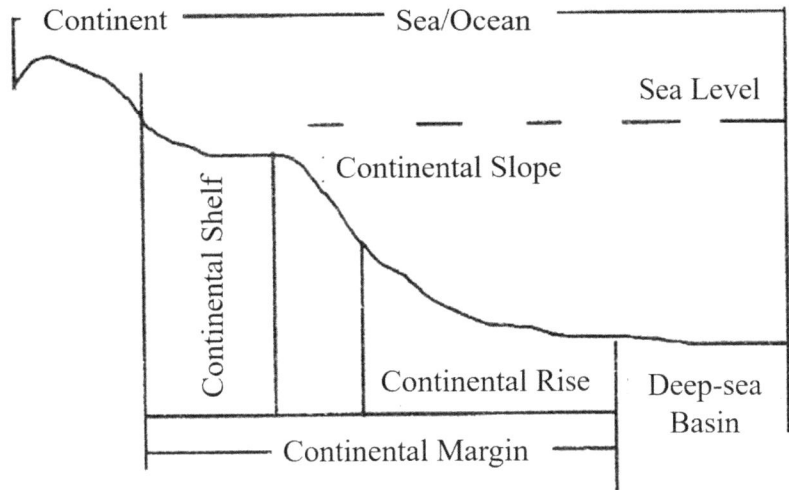

Fig. 2.2 Schematic diagram of seabed morphology (Chen 2014, 2016)

extend is inconsistent, and most researchers think it should include a depth of 200–2500 m.

The sea areas located on the continental slope are less affected by the continent due to being farther from the continent; therefore, the marine conditions of the continental slope are more stable than those of the sea areas associated with the continental shelf. Daily changes in marine elements cannot affect the bottom water layer, and even annual changes are very weak. The movement of the bottom sea water layer is derived mainly from the action of ocean currents and tides, and the effects of wind and waves have already dissipated. The deposition of the seabed also differs from that at the continental shelf; it is mainly terrigenous detrital ooze. Solar energy is absorbed and scattered by the upper sea water layer, and the energy that reaches the bottom is extremely weak or has completely dissipated; therefore, there are basically no plants in deep oceans and ocean bottoms, and plant-feeding animals become less abundant than mud-feeding animals. The skeletons of these animals form biogenic ooze that mixes with the terrigenous detrital ooze. Landslides often occur in seabeds with the greatest inclination, making loose sediment slide deeper along the surface of the slope; therefore, the seabed in these areas is often a rocky reef.

The most special topography on the continental slope is submarine canyons; they have cliffs, have a V shape, and can be tens of kilometers to hundreds of kilometers long. It is believed that most submarine canyons are generated by changes in the stratigraphic configuration. The continental slope is an active zone of the earth's crust, and crustal fractures acting on the continental slope give rise to some very large fissures. Submarine canyons are formed under the actions of powerful submarine turbidity, currents, and ice. There are submarine canyons along the coast of the Sea of Japan and along the west coast of North America, India, Africa, South America, and other areas.

Continental Rise

If the continental slope becomes flat before reaching the deep seabed, then the lower part is referred to as the continental rise or the continental apron. It is a sedimentary apron that slowly inclines from the base of the continental slope toward the depths of the sea or ocean. The water depth at these rises ranges from 2500 to 4000 meters, and the rises can traverse the ocean floor and extend as much as 1000 kilometers. Continental rises total approximately 1900 km², accounting for approximately 5% of the entire ocean floor. Continental rises are particularly broad in the vicinity of large deltas, such as the deltas of the Indus River, Ganges River, Amazon River, Zambia River, Congo River, and Mississippi River.

Deep-Ocean Basin (Ocean Floor)

The deep-ocean basin is the main part of an ocean. The topography is broad and flat and accounts for 72% or more of the marine area. The degree of inclination is small, at approximately $0°20'–0°40'$. The depth can continuously extend to approximately 6000 m from the continental rise. Based on the topographical properties, the ocean floor is a flat plain that adapts to the curvature of the earth and is slightly arched. There are many horizontal and vertical ocean ridges that are interlaced and stretched, dividing the seabed into a series of sea basins. In the ocean, there are also individual highlands formed by coral islands and volcanic islands that rise from the seabed to a height of 5000–9000 m and sunken zones deeper than 6000 m. The following types of topography are commonly seen in deep-ocean basins:

1. Ocean trenches – long and narrow deep depressions in the deep-sea seabed, with two comparatively steep walls.
2. Ocean troughs – long and wide submarine depressions in the deep-sea seabed, with gentle slopes on both sides.
3. Oceanic basins – depressions with a very large area and more or less basin-like shape.
4. Submarine ridges – narrow and long highlands on the deep-sea bottom, with steeper margins and irregular topography compared with that of a continental rise.
5. Ridged lands (oceanic rises) – long and wide highlands on the deep-sea bottom, with a gradual protrusion.
6. Seamounts and table mountains – isolated or relatively isolated highlands at the deep-sea bottom (nearly 1000 m or greater) are called seamounts; seamounts at a depth of greater than 1200 m with roughly flat plateaus at their top are referred to as table mountains. When seamounts and table mountains are arranged in a line or are densely grouped within a range, they are referred to as a seamount chain.
7. Abyssal plains – wide and unobvious highlands on the deep-sea bottom with tops that can vary widely due to smaller fluctuations.

Due to the lack of light and very low temperatures, there are sparse submarine fauna in the deep ocean, and therefore, significant deposition cannot be formed. All sediments in deep-ocean basins are formed from the deposition of the calcareous and siliceous skeletons of plankton propagating in the upper ocean layers. The biogenic ooze in this oceanic area is mainly globigerina ooze, diatomaceous ooze, and radiolarian ooze from rhizopods.

Ocean Trenches

Ocean trenches refer to the long and narrow sunken zones deeper than 6000 meters in the ocean. Ocean trenches and ocean ridges are often linked together, and they usually present an arced shape. Ocean ridges are sometimes exposed, forming islands or archipelagoes. Deep-ocean trenches are generally located on the convex surface of arc-shaped ocean ridges. There are a total of five deep-ocean trenches with a depth of more than 10,000 m, all of which are in the Pacific Ocean; the deepest ocean trench is the Mariana Trench (11,500 m). Pacific Ocean trenches are mostly concentrated in the western Pacific Ocean and along the coast of Asia, extending as a single arc at the intersection between the Pacific Ocean and the Indian Ocean until reaching Australia.

2.1.3   Marine Sediments

Because the relationship between seabed substrate and the distribution of benthic organisms is particularly close, it is important for us to understand the ecology and reasonable exploit of fish, especially fish that feed on benthic organisms. The bottom of the oceans is covered by different substances of various sources and properties, composing marine sediment through physical, chemical, and biological sedimentation.

Based on origin, marine sediments can be divided into two major categories: terrigenous sediments and pelagic sediments. The sediments of the continental margins are terrigenous debris carried into the sea from the mainland or neighboring islands through the action of rivers, wind, glaciers, and so on, including shore and shelf sedimentation and land slope and slope apron sedimentation. Shore and shelf sedimentation refers to the sediments distributed in the intertidal zone and on the continental shelf; their granulometric composition varies greatly but is mainly sand and mud. Land slope and slope apron sedimentations are distributed in the sediments of the flat zone beyond the continental slope and its steep slope; except for being mainly local substances from organisms or volcanoes, the vast majority of the sediments are composed of terrigenous debris, including various types of sand, silt, mud, and so on.

Pelagic sedimentation (also referred to as deep-sea sedimentation) mainly includes red clay ooze, calcareous ooze, and siliceous ooze. Among them, red clay ooze is a sediment formed by submarine weathering of red (brown) clay minerals and some volcanic substances brought from the mainland. Calcareous ooze is mainly composed of foraminifera *Globigerina* and the shells of planktonic Mollusca (pteropoda and heteropoda) and is widely distributed in the Pacific Ocean, Atlantic Ocean, and Indian Ocean, covering approximately 47% of the world's ocean floor area. Siliceous ooze is mainly siliceous sedimentation composed of the cell walls of diatoms and radiolarian bone needles.

The seabed substrate of the continental shelf is mainly sourced to land. In the absence of strong currents, the general rule is that from the shore to the open sea, the grains that appear in the substrate have belt-like distribution from coarse to fine, with coarser sands near the shore; moving away from the shore, the sediments are fine sand, silt, silty mud, sludge, and so on. However, in sea areas with very strong ocean currents, coarse grains will be carried very far, thereby breaking the aforementioned distribution rule.

"Large scale" refers to a large spatial scale, i.e., several hundreds to thousands of kilometers, even a global scope; the meaning of "relatively stable" is that within a longer period, such as 1 month, one season, 1 year, or multiple years, the flow direction, speed, and path are roughly similar.

Ocean currents are generally three-dimensional; that is, there is not only flow in the horizontal direction but also flow in the vertical direction. Because the horizontal scale of the oceans is far greater than the vertical scale, flow in the horizontal direction is far stronger than flow in the vertical direction. Although the latter is relatively weak, it has its special importance in oceanography. Customarily, the horizontal movement of ocean currents is usually referred to as ocean currents, in a narrow sense, and movement in the vertical direction is referred to as upwelling and downwelling.

Ocean circulation generally refers to the relatively independent circulation systems connected end to end formed by ocean currents in a sea area. For all oceans on earth, the temporal and spatial changes in ocean circulation are continuous. Ocean circulation connects the oceans and enables the various hydrological and chemical elements and physical conditions of the oceans to remain relative stable long term.

Although there are many causes of ocean current formation, there are mainly two types. The first is wind power on the sea surface, generating wind-induced ocean currents. The second is changes in sea temperature and salinity. In addition, the densification effect on the sea surface can also directly cause sea water movement in the vertical direction. After ocean currents form, because of the continuity of sea water, upwelling and downwelling will form in places where the sea water generates divergence or convergence.

2.2 Distribution of the Ocean Currents

2.2.1 Concept of Ocean Circulation and Its Genesis

Ocean current refers to the relatively stable flow of sea water on a large scale, and it is one of the important general movement forms for sea water.

2.2.2 Generation of Upwelling and Downwelling

Upwelling refers to the upward surge of sea water from a deep layer, and downwelling refers to the vertical movement of sea water resulting from sinking from the upper layer. The ocean is bounded, and the wind field is not uniform and stable. Therefore, the volumetric transport of wind and ocean currents inevitably leads to the occurrence of sea water divergence or convergence in certain sea areas or shores. Because of continuity, the sea water can rise or sink in these areas, thus changing the structure of the density field and pressure field of the ocean and thereby deriving other flows.

The volumetric transport of infinite deep-sea wind and ocean currents indicates that wind energy parallel to the shore leads to the greatest sea water convergence or divergence by the shore, thereby causing the sinking of surface sea water or a surge of sea water from a lower layer. However, wind that is perpendicular to the shore does not generate the same effect. For shallow seas, for wind that blows at a certain angle to the shoreline, components that are parallel to the shoreline can also cause similar movement. For example, along the coast of Peru and California in the United States, there are strong southeast trade winds and northeast trade winds, respectively, that blow along the coast in the direction of the equator. Because the drifting volumetric transport moves sea water away from the shore, sea water from the lower layer surges to the upper layer, forming world-famous upwelling areas. Another example is along the northwest coast of Africa and along the coast of Somalia (during the southwest monsoon period); for the same reason, there is also upwelling. Upwelling generally originates from a depth of 200 to 300 meters below the sea surface, and the ascending speed is very slow. Although the upwelling speed is very slow, because the upwelling is perennial, it continuously brings nutrient salts to the surface layer, which is conducive to biological reproduction. Therefore, upwelling areas are often well-known fishing grounds; for example, one of the world-famous fishing grounds is near the shore of Peru.

In sea areas in the vicinity of the equator, because the trade winds cross the equator, the volumetric transport of sea water on either side of the equator is opposite in direction and away from the equator, thereby causing divergence in the surface sea water at the equator and forming upwelling. In the ocean, upwelling and downwelling can also be generated by an uneven

wind field. The divergence and convergence of surface sea water have a certain relationship with the horizontal vorticity of wind stress; their relation can be expressed as:

$$Divergence\ (sea\ water\ divergence) = \frac{\partial \tau_y}{\partial x} - \frac{\partial \tau_x}{\partial y}$$

When divergence takes a positive value, the sea water is divergent, which generates upwelling; when the divergence takes a negative value, the sea water is convergent, which generates downwelling.

Cyclones and anticyclones above the ocean can also cause sea water to rise or sink. For example, the "cold wake" observed on the surface layer of the sea over which a typhoon (tropical cyclone) has passed is the cooling caused by the rising of low-temperature water from a lower layer to the sea surface.

In an uneven wind field, due to the uneven drifting volumetric transport, divergence and convergence can be generated (Fig. 2.3). In a cyclonic wind field, upwelling will similarly be generated due to divergence (Fig. 2.3). In the northern hemisphere, surface divergence and convergence in an uneven wind field and upwelling in a cyclonic wind field are the upwelling and the downwelling generated by wind power acting on a coastal area (Fig. 2.4).

2.2.3 Distribution of the Ocean Circulation

The general characteristics of upper ocean circulation on earth can be additionally explained by using the theory of wind-driven circulation (Fig. 2.5). There are similarities in the circulation patterns of the Pacific Ocean and the Atlantic Ocean. A very large anticyclonic circulation corresponding to a subtropical high is present in both the northern and southern hemispheres (clockwise in the northern hemisphere and counterclockwise in the southern hemisphere), and the Equatorial Counter current is between the hemispheres. The western-boundary currents of both oceans in the northern hemisphere (referred

Fig. 2.3 Diagram of divergence and convergence generated in an uneven wind field and a cyclonic wind field (Chen 2014, 2016)

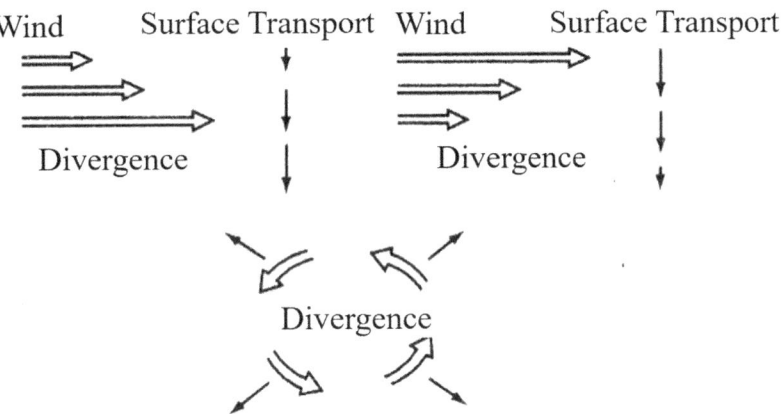

to as the Gulf Stream in the Atlantic Ocean and referred to as the Kuroshio Current in the Pacific Ocean) are very powerful, but the western-boundary currents in the southern hemisphere (the Brazil Current and the East Australian Current) are comparatively weak. There are cold currents from the north along the west side of the ocean basin in both the North Pacific Ocean and the North Atlantic Ocean, and there is a small cyclonic circulation in the northern part of the main vortex.

The differences in the various ocean circulation patterns are caused by their different geometric shapes. The circulation pattern in the southern part of the Indian Ocean is similar, in general characteristics, to the circulation patterns of the South Pacific Ocean and the South Atlantic Ocean, but the northern pattern is a monsoon-type circulation, with circulation in opposite directions for two halves (summer and winter).

In the high-latitude sea areas of the southern hemisphere, there is a powerful circumpolar current from west to east corresponding to westerlies. In addition, there is a circumpolar wind-induced current from east to west along the coast near the Antarctic continent.

Equatorial Currents

Corresponding to the trade-wind zones in the two hemispheres are the westward North Equatorial Current and South Equatorial Current, also referred to as the trade-wind currents (Fig. 2.6). These are two comparatively stable wind-induced drift currents caused by the trade winds, and both are a component of the very large cyclonic circulation in the northern and southern hemispheres. Between the northern and southern trade winds and corresponding to the equatorial calm belt is the Equatorial Countercurrent that moves eastward, with a flow range of approximately 300 to

Fig. 2.4 Schematic diagram of the generation of wind and ocean currents in the northern hemisphere (Chen 2014, 2016)

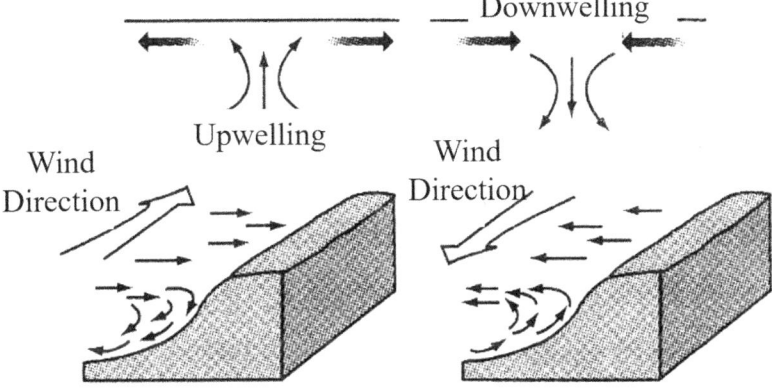

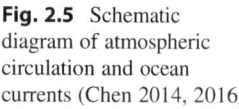

Fig. 2.5 Schematic diagram of atmospheric circulation and ocean currents (Chen 2014, 2016)

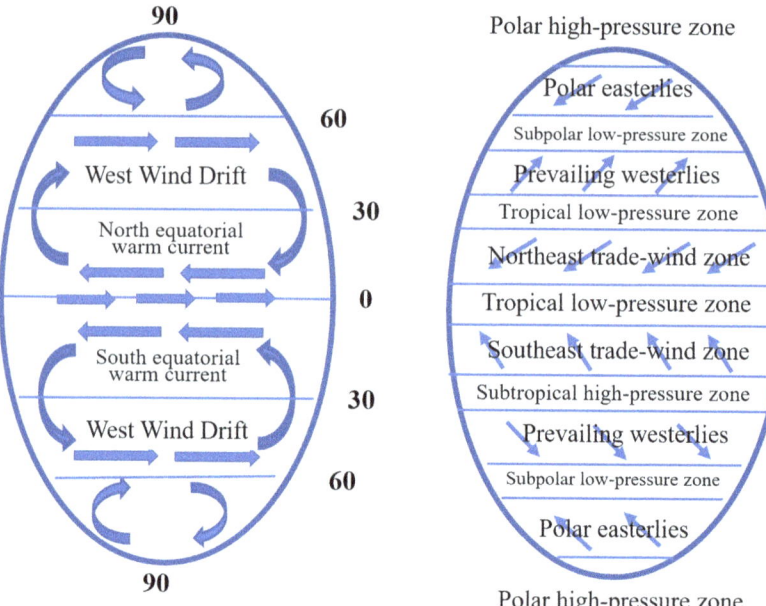

500 kilometers. Because the average position of the equatorial calm belt is between 3 and 10°N, the North Equatorial Current and South Equatorial Current are asymmetric with the equator. In the summer (August), the North Equatorial Current is approximately between 10°N and 20 to 25°N, and the South Equatorial Current is approximately between 3°N and 20°S. They are slightly southerly in the winter.

The equatorial current gradually strengthens from east to west. The equatorial current system is mainly limited to the upper layer 100 to 300 meters below the surface, with an average flow rate of 0.25–0.75 m/s. There is a powerful thermocline in the lower part. Above the thermocline is fully mixed surface water that is warm and has a high salinity content and dissolved oxygen content but a very low nutrient salt content, which is not preferable for plankton reproduction, thus resulting in high transparency and high water color. In short, the equatorial currents are current systems that feature high temperature, high salinity, high water color, and great transparency.

The equatorial current system in the Indian Ocean is mainly controlled by monsoons. The wind direction at the equatorial region is dominated by the meridian direction and changes

in the seasons. The northeast monsoon prevails from November to March of the following year, and the southwest monsoon prevails from May to September. There is a South Equatorial Current all year round south of 5°S, and the Equatorial Countercurrent exists south of the equator all year round. The North Equatorial Current flows westward from November to March of the following year, when the northeast monsoon prevails, and it flows eastward under the influence of the southwest monsoon at other times, becoming confluent with the Equatorial Countercurrent and difficult to distinguish.

There is abundant precipitation in the Equatorial Countercurrent region; therefore, the equatorial current region features high temperatures and low salinity. There is a divergent rising movement of sea water between the Equatorial Countercurrent and the equatorial current that transports sea water with low temperature but high-nutrient salts upward, making the water fertile, which is conducive to the growth of plankton; therefore, water color and transparency are relatively reduced.

There is an Equatorial Undercurrent flowing from west to east, opposite in direction to the equatorial currents in the South Equatorial

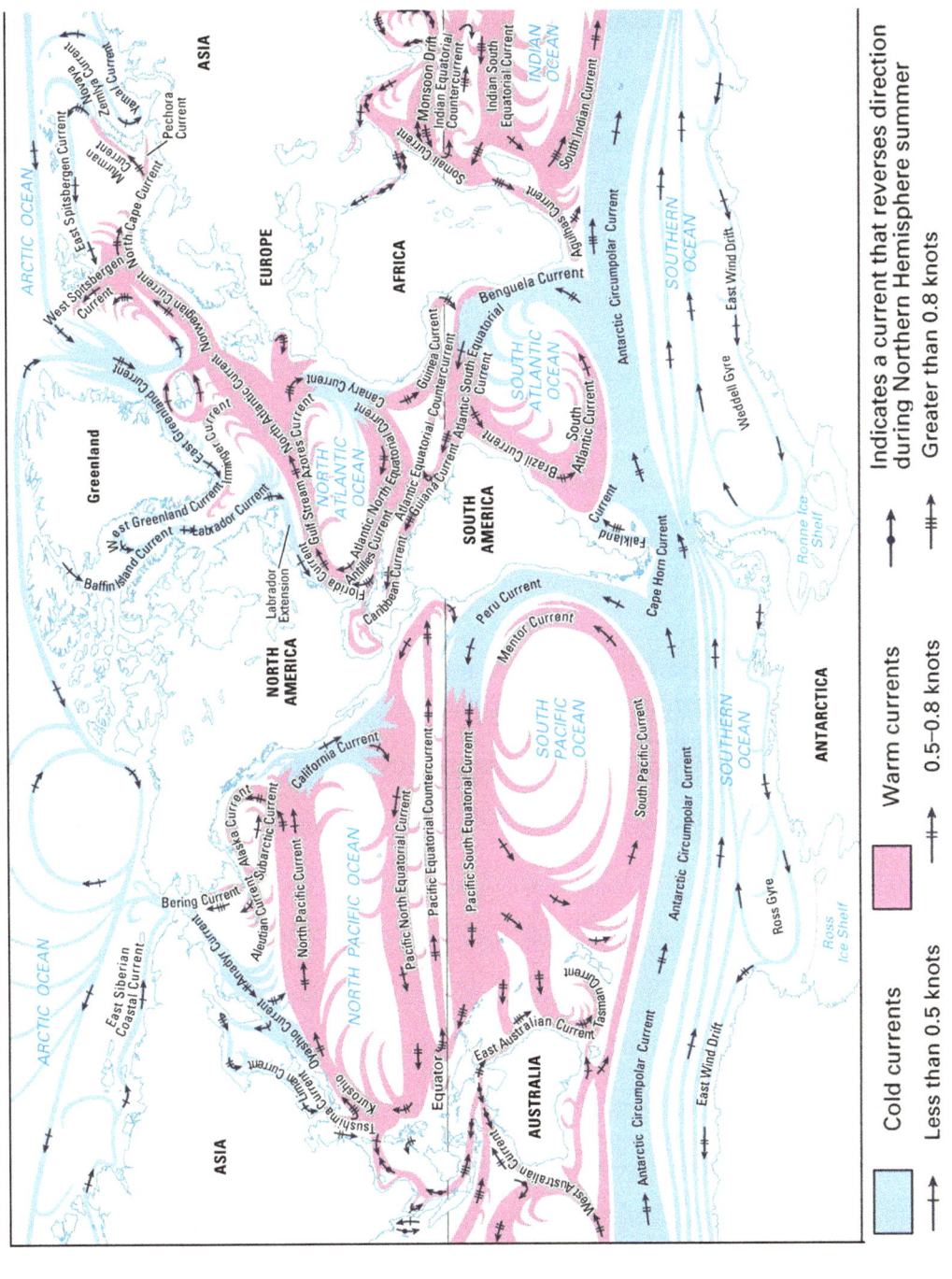

Fig. 2.6 Schematic diagram of surface current in three oceans during the winter of north hemisphere (Cenedese and Gordon 2021)

Current region of the Pacific Ocean. It generally forms a banded distribution with a thickness of approximately 200 m, a width of approximately 300 km, and a maximum flow rate as high as 1.5 m/s. The current axis is often consistent with the thermocline, which is located at a depth of 50 meters or shallower in the eastern part of the ocean and at a depth of 200 m or greater in the western part of the ocean. The generation of the Equatorial Undercurrent is not directly caused by wind, and there are many viewpoints in regard to its formation and maintenance mechanism. This type of undercurrent has been discovered successively in the Atlantic Ocean and the Indian Ocean.

Western-Boundary Currents

The western-boundary currents refer to the ocean currents from low latitudes to high latitudes along the continental slope on the west side of the ocean; these currents include the Kuroshio Current and the East Australian Current in the Pacific Ocean (Fig. 2.6), the Gulf Stream and the Brazil Current in the Atlantic Ocean, and the Mozambique Current in the Indian Ocean, among others. They are a major part of the anti-cyclonic circulation in the northern and southern hemispheres and are also continuations of the North and South Equatorial Currents. Therefore, compared with nearshore sea water, the western-boundary currents have high temperatures, high salinity, high water color, and great transparency.

West Wind Drift

Corresponding to the prevailing westerlies in the northern and southern hemispheres is the powerful West Wind Drift that moves from west to east, that is, the North Pacific Current and the North Atlantic Current in the northern hemisphere and the Antarctic circulation in the southern hemisphere, which are the components of a large anti-cyclonic circulation in the northern and southern hemispheres, respectively (Fig. 2.6). Their boundaries are as follows: the polar side is bounded by the Polar Ice Fields, and the equatorial side stops at the subtropical convergence zone. Their common feature is that there is an obvious temperature gradient in the meridian direction in the West Wind Drift zone, and this area with an observable gradient is referred to as the Ocean Polar Front. The hydrological and climate conditions on both sides of the polar front have obvious differences.

The convergence of polar fronts in the northern hemisphere is not obvious, and only a stronger convergent sinking phenomenon is present in the confluence area between the Kuroshio Current and the Oyashio Current in the northwestern part of the Pacific Ocean and in the confluence area between the Gulf Stream and the Labrador Current in the northwestern part of the Atlantic Ocean, which is generally referred to as the Northwest Convergence Zone. Due to the strong mixing generated by the confluence of cold and warm currents, the marine productivity is high, thereby making the Northwest Convergence Zone a good fishing ground. This is the sea area where the world-famous Hokkaido fishing grounds and Newfoundland fishing grounds are located.

Eastern-Boundary Currents

The eastern-boundary currents are the California Current and the Peru Current in the Pacific Ocean, the Canary Current and the Benguela Current in the Atlantic Ocean, and the West Australian Current in the Indian Ocean. Because they flow from high latitudes to low latitudes, they are all cold currents (Fig. 2.6). Additionally, they are all located on the eastern boundary of the ocean; therefore, they are referred to as the eastern-boundary currents. Compared with western-boundary currents, their flow range is broad, the flow rate is small, and the depth of influence is shallow.

Upwelling is an important marine hydrological feature in sea areas with eastern-boundary currents. This is due to the trade winds that blow along the coast almost all year round; the wind speed is unevenly distributed, that is, it is slower near the shore and faster on the sea surface, which then results in the movement of sea water away from the shore. The upwelling formed in the aforementioned sea areas often produces good fishing grounds.

Polar Circulation (Fig. 2.6)

In the Arctic Ocean, circulation is generated mainly by the Norway Current, which enters from the Atlantic Ocean, and some currents along the coast. There is a very large anticyclonic circulation in the Canada Basin, which reaches the Greenland Sea by passing through the North Pole from the Chukchi Sea at the junction between Asia and America. Part of the circulation turns and flows westward, and part of it merges into the East Greenland Current; together they carry large ice floes into the Atlantic Ocean. Other small cyclonic circulations are also present.

The Antarctic circulation has a very narrow scope at the Antarctic continental margins. Due to the action of the polar easterlies, a small circulation forms around the Antarctic continental margins from east to west, referred to as the East Wind Drift. Between the East Wind Drift and the Antarctic circulation, the Antarctic Divergence Zone is formed due to the dynamic action. Convergence and the sinking of sea water along the continental shelf form between the Antarctic Divergence Zone and the Antarctic continent, that is, the Antarctic continental convergence. These dynamics cause the surface sea water to sink in the Antarctic shelf area.

There are common characteristics of the polar sea areas. They are covered by ice almost all year round or all year round, and the freezing and thawing process leads to comparatively low sea temperature and salinity the whole year, forming low-temperature and low-salinity surface water.

2.2.4   Main Ocean Currents in Various Oceans

Pacific Ocean

In the North Pacific Ocean, the main circulation systems are the North Equatorial Current, the Kuroshio Current, the North Pacific Current, and the California Current, and the ocean currents in the adjacent seas are the Alaska Current, the Oyashio Current, the East Karafuto Current, the Liman Current, the China Coastal Current, the Tsushima Current, and the South China Sea Monsoon Current. In the South Pacific, the main circulation systems are the South Equatorial Current, the East Australian Current, the West Wind Drift, and the Humboldt Current (Peru Current) (Fig. 2.6). The ocean currents in the equatorial Pacific are the Equatorial Countercurrent and the Equatorial Undercurrent (Cromwell Current). The following provides an analysis of the main ocean currents with greater influence on fisheries science.

Kuroshio Current

The North Pacific circulation starts from the North Equatorial Current and flows westward to the western land boundary, where it is divided into two; one part goes south and the other part goes north. The northward branch forms a powerful western-boundary current in the Pacific Ocean, that is, the Kuroshio Current. The southward branch is referred to as the Mindanao Current. The main current in the Kuroshio Current flows eastward along 36°–37°N, passing the southern coast of Honshu, Japan. After leaving Japan, it continues to flow to the east to approximately 170°E; this section is referred to as the Kuroshio Extension, becoming the North Pacific Current after the Extension. As the Kuroshio Current flows in the vicinity of the Ryukyu Islands, there is a branch that ascends northward along the margins of the continental shelf, becoming the Tsushima Warm Current, which flows into the Sea of Japan through the Korea Strait.

In the offshore waters of Sanriku, Japan, the Kuroshio Current meets the Oyashio Current from the north, forming a boundary fishing ground at the intersection of the warm and cold currents that teems with Pacific saury, cetaceans, tuna, etc.

Oyashio Current

The Oyashio Current mainly comes from the Bering Sea and partly comes from the Okhotsk Sea. When the North Pacific Current approaches the North American continent, it divides into north and south branches; the southern branch is the California Current, which ultimately connects with the North Equatorial Current; the northern branch forms the Alaska circulation in the Gulf of Alaska, with a portion flowing between the Aleutian Islands and entering the Bering Sea. The Oyashio Current has high biological productivity and is rich in phytoplankton, and the water color and transparency are both lower than those of the Kuroshio Current.

California Current

The California Current descends southward along the west coast of North America and becomes the eastern-boundary current of the ocean. Its surface flow rate is generally comparatively slow, at approximately 1 km/h. In the summer, under the powerful action of the northerlies, the surface water of the California Current, which descends southward along the coast, flows toward the open sea, and the deep-layer water acts as the compensating current and rises along the coast, becoming the famous California upwelling. Part of the California Current descends southward along the coast of Central America, reaching the low-latitude sea areas of the East Pacific Ocean. In addition, the eastern end of the Equatorial Countercurrent that flows eastward along and in the vicinity of the equator flows and turns north and west in the offshore waters of Mexico and becomes the North Equatorial Current, which flows westward with 10°N as the center. The North Equatorial Current merges with the California Current where it turns to a westward flow and continues to flow westward, becoming part of the large-scale horizontal cycle of the North Pacific Ocean. In the vicinity of this confluence, a purse seine fishing ground for tuna has formed.

Equatorial Currents and Their Undercurrent

There are at least four main ocean currents included in the equatorial current system of the Pacific Ocean, three of which extend to the sea surface, with the other below the sea surface. The three main upper-layer ocean currents are all observable on the surface: the first is the westward North Equatorial Current, which is approximately in the range of 2°–8°N; the second is the westward South Equatorial Current, which is approximately in the range of 3°N–10°S; and

the third is the narrower and eastward-flowing North Equatorial Countercurrent between the aforementioned two ocean currents. In addition, the Equatorial Undercurrent that flows to the east below the sea surface crosses over the equator and occupies a range from 2°N to 2°S; this ocean current can be traced from Panama Bay in the east to the Philippines in the west, with a distance of approximately 15,000 km. In the summer, the Equatorial Countercurrent changes its direction of flow toward the offshore waters of Costa Rica, forming a vortex that rotates counterclockwise, thereby inducing strong upwelling. This upwelling is the Costa Rica Dome, which is an important marine condition for the formation of tuna fishing grounds.

In equatorial sea areas, the surface water of the westward-flowing North and South Equatorial Currents flows northward in the northern hemisphere, and that in the southern hemisphere flows southward. Therefore, an upwelling with a stronger divergent phenomenon is generated in the sea areas of the equator, causing the ascent of deep-layer water rich in nutrient salts, promoting an increase in biological productivity, and forming a cline of sea temperature and dissolved oxygen. The thermocline of the North Equatorial Current watershed gradually becomes shallower from west to east. The depth of the thermocline affects the water layer in which tuna is distributed, having important significance for fisheries.

Peru Current

The Peru Current is equivalent to the cold current portion of the circulation that rotates counterclockwise in the Southeast Pacific Ocean, which originates from the subarctic sea area. The high-latitude West Wind Drift reaches the west coast of South American in the vicinity of 40°S, and the ocean current that flows northward is the Peru Current. The Peru Current near the coast is referred to as the Peru Coastal Current, and the branch in the open sea is referred to as the Peru Ocean Current. These two ocean currents are separated by the irregular Peru Countercurrent, which descends southward, and this Countercurrent is referred to as the Pacific equatorial waters, which is usually a subsurface current 180–500 km away from shore. From November to March, when the Peru Countercurrent is strongest, the current is closer to the surface. The flow potential is weak before November, and the current remains below the sea surface. During this time, the Peru Current becomes a single ocean current, without dividing into the two coastal and ocean branches, and it is the time when the Peru Current is strongest. The southern end of the ocean currents along the coast of Peru is the southern limit of the upwelling area formed along the coast of Chile, and its position is in the vicinity of approximately 36°S.

Atlantic Ocean

In the Atlantic Ocean, the main ocean currents are the Gulf Stream, the North Atlantic Current, the Labrador Current, the Canary Current, the Benguela Current, the Brazil Current, and the Falkland Current, among others (Fig. 2.6). These ocean currents are closely related to fishing grounds.

Gulf Stream

In the Northwest Atlantic, the oceanographic features that are extremely important to fisheries are the result of the warm current system of the Gulf Stream and the cold current system of the Labrador Current. The Gulf Stream flows in a northeasterly direction along the North American continent and results from the confluence of the Florida Current and the Antilles Current, which originates from the North Equatorial Current. It is similar to the Kuroshio Current in the Pacific Ocean, becoming the western-boundary current of the Atlantic Ocean; its flow rate is 7–9 km/h in the offshore waters of the east coast of North America, where the flow is strongest, and it reaches a thickness of 1500–2000 m.

The movement of the Gulf Stream presents is serpentine, with such movement gradually developing eastward from Cape Hatteras, thereby forming a complex flow boundary accompanied by a vortex system. Some people refer to the movement of the Gulf Stream as multiple ocean currents. In the sea area in the vicinity of Nova Scotia in Canada, due to the influence of the surrounding topography, a very complex local vortex area forms, especially in the summer, and this oceanographic condition is deemed one of the major factors for the formation of many fishing grounds.

The flow range of the Gulf Stream expands when it reaches south of Grand Bank in southern Newfoundland, where it becomes the North Atlantic Current, which flows in a northeasterly direction. One branch of this current continues to flow in the northeasterly direction, becoming the Norway Current, which reaches the sea area off the west coast of Norway. This current is the main cause for the distribution of tuna fishing grounds in the vicinity of 70°N. The branch that flows northward moves south of Iceland and flows westward, becoming the Irminger Current, and most of the current forms a confluence with the East Greenland Current, which descends southward along the east coast of Greenland.

East Greenland Current

The East Greenland Current originates from the Arctic Ocean, and a current boundary is formed between the East Greenland Current and the Irminger Current. Part of the East Greenland Current combines with the Irminger Current and becomes the West Greenland Current. This current then merges with the southward-descending current from Baffin Bay and becomes the Labrador Current, which descends southward along the east coast of North America, forming a polar front at the intersection between the inshore waters of Newfoundland and the Gulf Stream, making this sea area rich in fishery resources; this area is traditionally one of the three major fishing grounds in the world.

North Atlantic Current

Due to the influence of the North Atlantic Current, a warm climate is present in the United Kingdom to places in northern Europe and along the coast of Norway. After the North Atlantic Current ascends northward along the west coast of Norway through the Faroe Islands, it divides into two branches: one branch ascends northward toward the western part of Spitzbergen, and the other branch flows into the Arctic Ocean along the north coast of Norway, warming the western and southern parts of the Barents Sea.

The North Atlantic Current, which ascends northward along the west coast of the United Kingdom, has a feeder current that descends southward along the east coast of the United Kingdom in the vicinity of the Shetland Islands in the north and another ocean current that flows in from the English Channel on the south coast of the United Kingdom. These are all major factors dominating the oceanographic conditions at the fishing grounds in the North Sea.

Canary Current

The southward-descending feeder current of the North Atlantic Current descends southward along the northwest coast of Europe and Portugal and the inshore waters of the northwest coast of Africa, forming the Canary Current. Changes in the flow direction and rate of the Canary Current are affected by wind. After the current arrives at the west coast of the African continent, it usually flows westward as a compensating current of the North Equatorial Current. Part of the Canary Current continues to descend southward along the west coast of Africa. Usually, this ocean current develops in the northern hemisphere in the summer and becomes the eastward-flowing Guinea Current. The Guinea Current also exists in the winter.

Brazil Current

The South Equatorial Current flows westward in the vicinity south of the equator and along the coast of South America and divides into two branches—a northward-ascending current and a southward-descending current. The southward-descending branch is the Brazil Current and has very high salinity. This ocean current merges with the Falkland Current, which ascends northward from the subantarctic waters at approximately 35 to 40°S, forming a subtropical convergence line; a surface temperature of 14.5 °C is an indicator in the summer. The Patagonian Seas are in the convergence zone between the Brazil Current and the Falkland Current; this sea area is rich in marine biological resources, and it is a major fishing ground.

Indian Ocean

The scope of the Indian Ocean is approximately 25°N to the north and approximately 40°S to the south, comprising sea areas of the subtropical convergence zone. The sea currents in the northern sea area of the Indian Ocean, particularly in the Arabian Seas, are greatly affected by monsoons. The main ocean currents in this sea area are the Southwest Monsoon Current in the summer and the Northeast Monsoon Current in the winter. In the southern sea area of the Indian Ocean, the main ocean currents are the Mozambique Current, the Agulhas Current, the West Australian Current, and the West Wind Drift (Fig. 2.6).

The circulation system in the northern sea area of the Indian Ocean is not quite the same as those in the Pacific Ocean and the Atlantic Ocean. From November to March of the following year, i.e., the northeast monsoon season, the following currents predominate: the westward-flowing North Equatorial Current from 8°N to the equator, the eastward Equatorial Countercurrent from the equator to 8°S, and the westward South Equatorial Current from 8°S to between 15 and 20°S. From May to September, i.e., the southwest monsoon season, the ocean currents north of the equator flow eastward, in reverse; together with the eastward-flowing Equatorial Countercurrent, this current is referred to as the (Southwest) Monsoon Current, which approximately occupies a scope from 15°N to 7°S. The South Equatorial Current flows west, as before, south of 7°S, but it is somewhat stronger than when the northeast monsoon is blowing. During the northeast monsoon season, there is an Equatorial Undercurrent at the depth of the thermocline east of 60°E, but it is weaker than those in the Pacific Ocean and the Atlantic Ocean; the undercurrent cannot be observed when the southwest monsoon is blowing.

In the coastal waters of Africa, from November to March of the following year, i.e., the northeast monsoon season, after the South

Equatorial Current flows near the coast of Africa, a branch turns north and enters the Equatorial Countercurrent, and the other branch heads south and merges with the Agulhas Current. This ocean current is deep and narrow, with a width of approximately 100 kilometers; it flows to the south along the coast of Africa and enters the Antarctic circulation by flowing eastward at the southern tip of Africa. When the southwest wind blows from May to September, part of the South Equatorial Current turns northward, becoming the Somalia Current, which extends to the east coast of Africa and continues ascending northward; most of current is within 200 meters of the surface layer. The South Equatorial Current, the Somalia Current, and the Monsoon Current constitute a considerably strong wind-blowing circulation in the North Indian Ocean.

During the southwest monsoon season from May to September, the Somalia Current is a low-temperature water area; it is similar to the Kuroshio Current and the Gulf Stream, and all are representative western-boundary currents. In the winter, the flow rate of the Northeast Monsoon Current in the offshore waters along the coast of Somalia is less than the flow rate of the Somalia Current. In other sea areas of the Indian Ocean, upwelling occurs when the southeast trade winds are strong; upwelling also occurs in the eastern Arafura Sea during the period in which the southeast trade winds are prevailing. During the upwelling development period, primary production is higher than that of the surrounding sea areas.

2.3   Sea Temperature Distribution in the Oceans

2.3.1   Basic Concept

Sea temperature is a physical quantity that represents the thermal state of sea water and, in oceanography, is generally expressed in degree Celsius (°C), with a required measurement precision of ± 0.02 °C. Solar radiation and the heat exchange between seas or oceans and the atmosphere are the two main factors that affect sea

temperature. Ocean currents also have an influence on temperature in the sea. In the open ocean, the distribution of isotherms for surface sea water is roughly parallel to the latitude. In nearshore areas, because of the effects of ocean currents, among other factors, the isotherms move in the north-south direction. The vertical distribution of sea temperature generally decreases with the increase in depth and presents seasonal changes.

Sea temperature is one of the most important factors in marine hydrological conditions and is often used as a basic indicator for studying the properties of water masses and describing the movement of water masses. Studying and mastering the spatial-temporal distribution and changing patterns of sea temperature is an important topic in fisheries science, and it has important significance for fisheries forecasting.

2.3.2   Sea Temperature Distribution Pattern

Horizontal Distribution Pattern of Sea Surface Temperature

The latitudinal distribution pattern for the average sea surface temperature is a progressive decrease from low latitude to high latitude because the amount of solar radiant heat received by the earth's surface is affected by the shape of the earth; the amount of solar heat received decreases progressively from the equator to the two poles.

The following are characteristics of changes in sea surface temperature: sea surface temperature is affected by season, restricted by latitude, and affected by ocean current properties.

Vertical Changes in Sea Temperature

The vertical distribution pattern for sea temperature is a progressive decrease as depth increases. From the sea surface to 1000 m of water layer, the decrease in sea temperature progresses quickly as depth increases; for water layer already below 1000 meters, the decrease in sea temperature slows. The main reason for this effect is that the surface of oceans is greatly affected by solar radiation, while deep water layers in oceans are less affected by solar radiation and heat conduction and convection currents that occur on the surface layer.

The sea temperature of the oceans generally varies between −2 and 30 °C, of which the area where the annual average sea temperature exceeds 20 °C accounts for more than half of the entire marine area. Observations show that the daily variation in sea temperature is very small and that the scope of variation occurs at water depths from 0 to 30 m; however, the annual variation can reach a water depth of approximately 350 m. At a water depth of approximately 350 m, there is a constant-temperature layer. However, as the depth increases, the sea temperature gradually drops (a drop of approximately 1°–2 °C per 1000 m in depth), and at a water depth of 3000–4000 meters, the temperature reaches 1 °C to 2°.

The following factors affect sea temperature.

(1) Latitude – the solar radiation at different latitudes is different, and thus, the temperature is different. The global distribution pattern of sea temperature is a progressive decrease from low-latitude sea areas to high-latitude sea areas.
(2) Ocean currents – in sea areas at the same latitude, the temperature of sea water where warm currents flow is higher, and the temperature of sea water where cold currents flow is lower.
(3) Seasons – sea temperature is high in the summer and low in the winter.
(4) Depth – the progressive decrease in temperature is significant as the depth of the surface sea water increases, with variations being more obvious within 1000 m. There is less variation from 1000 meters to 2000 meters, and low temperature is maintained all year at depths below 2000 m.

The annual average sea temperature at the surface of the three oceans is approximately 17.4 °C; that of the Pacific Ocean is the highest, reaching 19.1 °C, followed by that of the Indian Ocean, reaching 17.0 °C, and that of the Atlantic Ocean, at 16.9 °C. Sea temperature generally decreases as depth increases; the sea temperature is approximately 4–5 °C at a depth of 1000 m,

2–3 °C at 2000 m, and 1–2 °C at 3000 m. The sea temperature that accounts for 75% of the total volume of oceans is between 0 and 6 °C, and the average global sea temperature is approximately 3.5 °C. Sea temperature can also have daily, monthly, yearly, multiple-year, and other periodic variations and irregular variations.

2.3.3   Distribution of Global Sea Surface Temperature

The sea surface temperature is the sea temperature at the junction between the atmosphere and the oceans (sea surface). In fact, we cannot possibly measure the temperature of the sea surface itself. The measured surface temperature is in accordance with the observed depth difference and is generally a measurement of the temperature from the sea surface to a depth of 10 m.

Because the earth is spherical, the amount of heat (amount of solar radiation) the oceans receive from the sun varies with the latitude. Figure 2.7(a) shows the distribution of the mean sea surface temperature in the three oceans. Generally, the sea temperature distribution is higher in low-latitude areas and lower in high-latitude areas. Figure 2.7b, c is the sea surface temperature distributions for January and July, respectively. In mid-to-high-latitude areas, compared with the annual average sea surface temperature, the sea surface temperature in January is lower in the northern hemisphere and higher in the southern hemisphere. On the other hand, the sea surface temperature in July is higher in the northern hemisphere and lower in the southern hemisphere.

Sea surface temperature is also affected by atmospheric movement. For example, in the vicinity of the sea surface in the equatorial sea areas of the Pacific Ocean, an eastward trade wind exists. Under the action of this east wind, the warm water in the vicinity of the sea surface is blown to the western part of the Pacific Ocean; to compensate, in the eastern part of the South American sea area, ice-cold water surges to the vicinity of the sea surface. Figure 2.7a indicates that the sea surface temperature in the equatorial

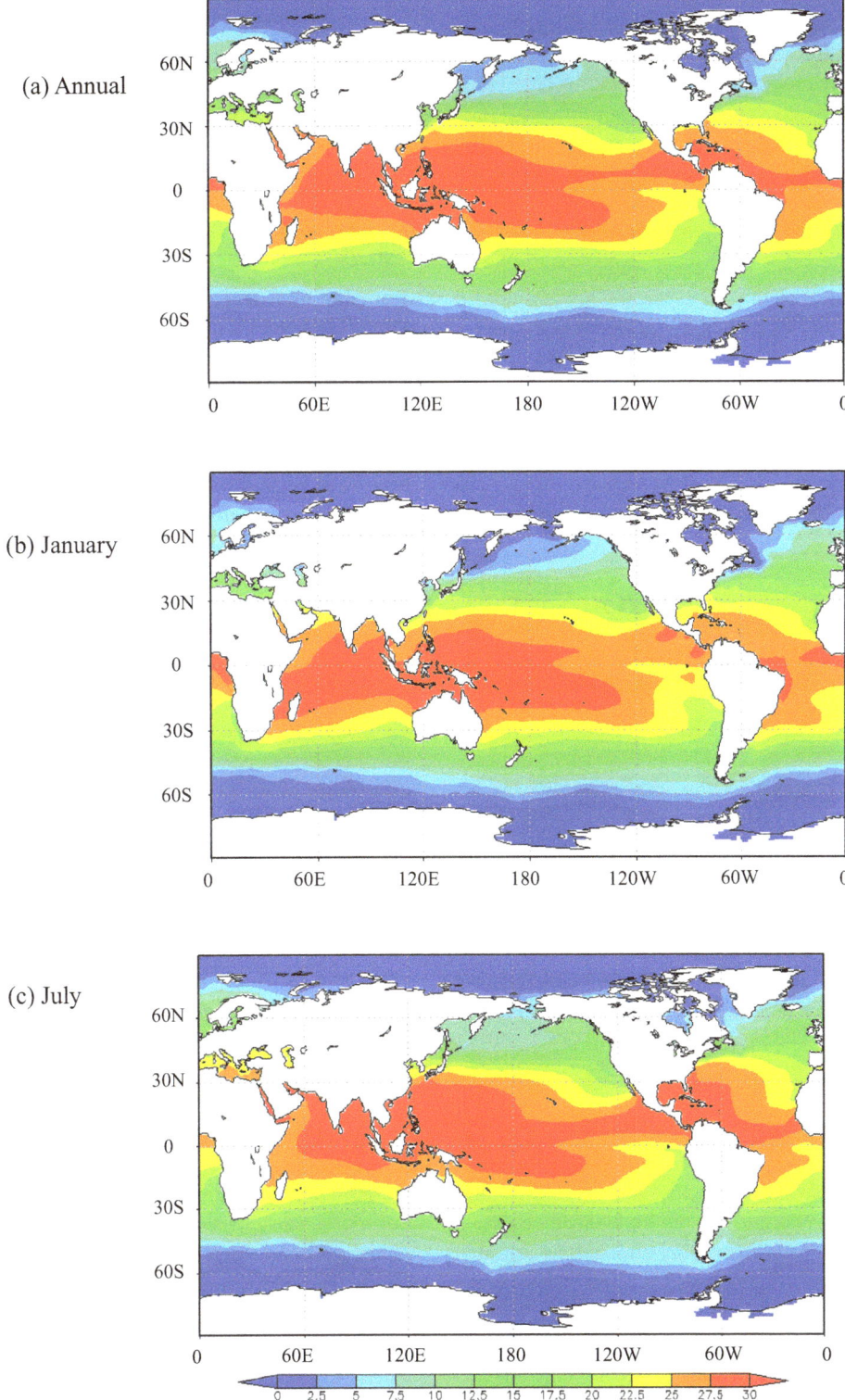

Fig. 2.7 Distribution of mean sea surface temperature in the three oceans (unit: °C). (**a**) Annual mean, (**b**) January, and (**c**) July

sea areas of the Pacific Ocean is higher in the western part and lower in the eastern part.

In addition, in the vicinity of the west coast of the continents in the northern hemisphere (southern hemisphere), when the south wind (north wind) blows along the coast, the warm sea water in the vicinity of the sea surface is acted on by the force of the wind direction and drawn into the sea due to the force generated by the earth's rotation. To compensate, cold deep water may surge into the vicinity of the sea surface. This is one of the reasons the sea surface temperature in the vicinity of the west coast of North America is lower than the surrounding areas in July (Fig. 2.7c).

As mentioned above, through various factors, such as solar radiation, atmospheric movement, sea water movement, and topography, sea surface temperature has a complicated distribution.

2.4   Distribution of Nutrient Salts and Primary Production in the Oceans

2.4.1   Nutrient Salts

Basic Concept
The nutrient salts in sea water refer to the elements dissolved in sea water that act as growth factors for marine plants. In addition to requiring carbon dioxide, oxygen, and other gases, marine plants also require phosphorus, nitrogen, silicon, sulfur, magnesium, calcium, and various other elements to constitute protein, cell nuclei, and other substances. Vertical convection currents and upwellings in oceans bring bottom sea water rich in nitrogen, phosphorus, silicon, and other nutrient salts to the upper layer; therefore, in estuaries, along fronts, at the confluence of warm and cold currents, and at upwellings, among other areas, the phytoplankton production volume is high, the fishery resources are abundant, and the fishing yield is high.

There are relatively small amounts of phosphate, nitrate, nitrite, ammonium salts, and silicate in sea water. There are many primary components and trace metals in sea water that are also nutritional components, but traditionally, in chemical oceanography, sea water nutrient salts only refer to salts such as nitrogen, phosphorus, and silicon because they are essential components for the growth and reproduction of marine phytoplankton and are also the basis of marine primary production and the food chain. The distribution of nutrient salt content in sea water is affected by marine biological activities, and such distribution is usually not greatly related to sea water salinity.

Sources of Nutrient Salts and Their Distribution
The sources of sea water nutrient salts are mainly weathered rock products, decomposed organic matter, and waste discharged into rivers and streams that have been brought by continental runoff. In addition, the decomposition of marine organisms, weathering, polar glaciers, volcanoes, submarine hot spring, and dust in the atmosphere all provide nutrients to the sea water and provide the basic conditions for the formation of fishing grounds.

The nutrient salt content in sea water has a vertical distribution and regional distribution. Generally, the water in oceans can be divided into four layers based on the vertical distribution of nutrient salts: (1) surface layer, low nutrient salt content, more even distribution; (2) sublayer, nutrient salt content increases rapidly with depth; (3) sub-deep layer, 500–1500 m, maximum nutrient salt content; and (4) deep layer, although this layer is thick, the phosphate and nitrate content variations are very small, and silicate content increases slightly with depth.

In terms of regional distribution, due to ocean currents, the activities of living organisms, and the characteristics of each sea area, nutrient salts in sea water are distributed differently in different sea areas. For example, the deep-water circulation between the Atlantic Ocean and the Pacific Ocean transports nutrient salts from deep water in the Atlantic Ocean toward deep water in the Pacific Ocean. A large amount of nutrient salts is consumed during phytoplankton growth and reproduction in the Antarctic sea areas, but because there are sufficient sources, the nutrient salts in the sea water remain considerably abundant. In inshore areas, because of the thriving reproduction and growth of phytoplankton during the summer, the nutrient salts in the water of the surface layer are consumed to exhaustion. The growth and reproduction of phytoplankton decline during the winter, and the vertical mixing of sea water intensifies, allowing the nutrient salts formed by the decomposition of organic matter and deposited on the seabed to replenish the surface layer via upwelling, increasing the nutrient salt content in the surface layer. Shallow seas near shores and estuarine zones are different from oceans; the distribution of the nutrient salt content in sea water is not only affected by the growth and dying of phytoplankton and seasonal variations but is also very largely related to variations in continental runoff, the growth and decline in thermoclines, and other hydrological conditions.

2.4.2   Primary Production

Related Concepts

Gross primary production refers to the total amount of organic carbon produced via photosynthesis. However, marine plants, like other living organisms, carry out continuous respiration, requiring the consumption of part of the organic carbon produced. Therefore, net primary production is the remaining yield after the respiratory consumption of the producers is deducted from gross primary production: net primary production = gross primary production – respiratory consumption of the autotrophic organism. Marine primary production [mgC/(m^2 day)] is often expressed as the amount of organic carbon produced per unit area (m^2) per unit time (day or year).

The factors that affect primary production are mainly light conditions and the content of nutrients required by plants, including other hydrological conditions related to the two. Under natural conditions, these factors are constantly changing. The main methods for determining primary production are (1) the ^{14}C tracer method, (2) chlorophyll fluorimetry, (3) the oxygen light and dark bottle method, and (4) water color remote sensing and scanning.

Water color remote sensing and scanning is one of the main methods for determining primary production. A coastal zone color scanner (CZCS) carried by satellites can record the color of sea water, determine the concentrations of chlorophyll and other pigments of algae and colored dissolved organic matter (CDOM) in a sea area, and identify particles suspended in the water. Therefore, the prominent contribution of the CZCS is overcoming the large-area sampling problem that is difficult to achieve in field surveys; the scope of its survey coverage can extend over the entire ocean. Furthermore, macro- and meso-scale physical processes that affect the spatial distribution and primary production of marine phytoplankton can also be analyzed through the CZCS, including the North Atlantic Oscillation (NAO) and the El Niño-Southern Oscillation (ENSO), among others. With the continuous improvement in remote sensing technology, it will be possible to more comprehensively understand the relationship between phytoplankton biomass and productivity and the features of marine hydrology. Products such as chlorophyll content determined by the remote sensing of marine water color have already gained wide application in fisheries science.

2.4.3   Distribution of Marine Productivity

The distribution of marine primary production is very uneven. Overall, the high-value areas of primary production are located in upwelling areas, continental shelves, and inshore sea areas with divergence, followed by the temperate subpolar region of the northern hemisphere and the Southern Ocean fronts. Low-value areas are located in the tropical and subtropical ocean regions of both the northern and southern hemispheres, with the lowest primary production in the sea areas of the Arctic Ocean (Fig. 2.8).

In the last several decades, satellite remote sensing technology has gained wide application in ocean monitoring, promoting the development of fisheries science and fisheries forecasting. The utilization of high-resolution satellite remote sensing enables continuous observation of the abundance of phytoplankton in different sea

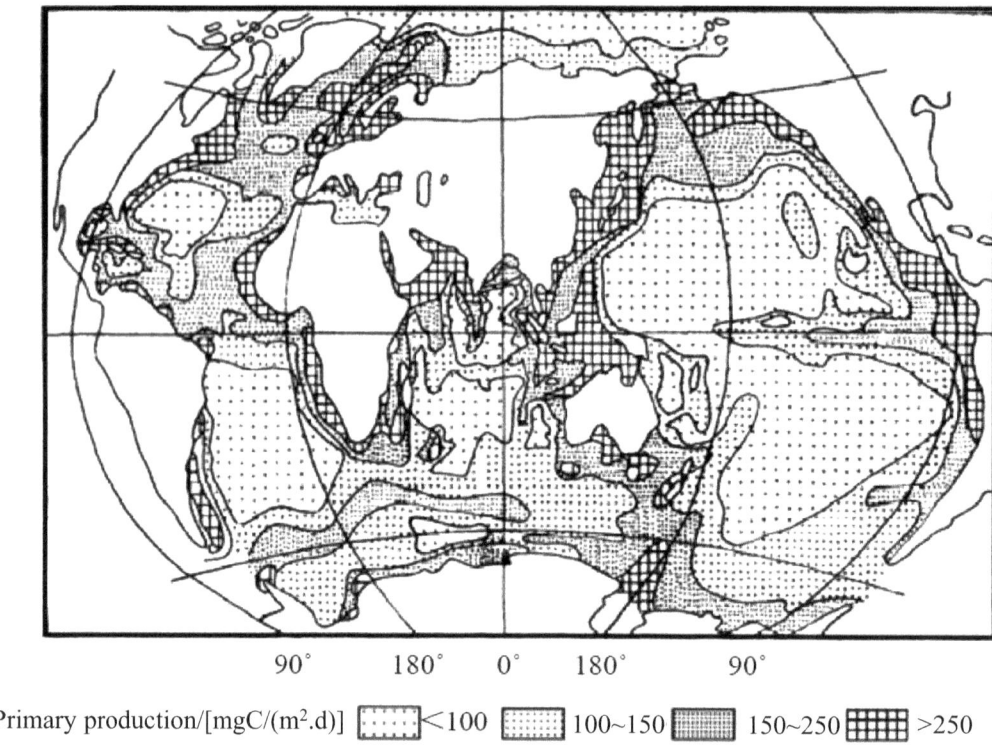

Primary production/[mgC/(m².d)] ▒▒▒<100 ▒▒▒100~150 ▤▤150~250 ▦▦>250

Fig. 2.8 Map of the distribution of primary production in the oceans (Koblentz-Mishke et al. 1970)

areas, the depth of the euphotic layer, and variations in primary production. Based on these satellite remote sensing data, combined with animal composition and physical oceanographic features such as wind and ocean currents, UK scholar Longhurst (1971) divided the seas and oceans into four basic biomes: the polar biome, the westerlies biome, the trades biome, and the coastal biome. Each biome is further divided into a number of ecological provinces. There are annual and seasonal variations at the boundaries of these subregions. Now, in combination with the characteristics of the aforementioned biomes, the geographic distribution of marine primary production is introduced.

Tropical and Subtropical Ocean Regions and the Equatorial Zone

Tropical and Subtropical Ocean Regions
The scope of the trades biome is between approximately 30°N and 30°S, and its boundary just passes through the central circulation area of the

subtropics, with the central axis located in the equatorial zone. The depth of the mixed layer in this sea area is mainly affected by macro-scale marine circulation.

The tropical and subtropical ocean regions receive sufficient solar irradiation and have high sea water transparency, and the depth of the euphotic layer exceeds 100 m, which is a favorable light condition for primary production. However, this sea area is located within the range of oceanic anticyclonic circulation (also referred to as the central circulation area); the surface sea water converges and sinks toward the center of the circulation, and the depth of the mixed layer exceeds the depth of the euphotic layer. The thermocline can reach 100–200 m in the summer and increases to approximately 400 m in the winter. Furthermore, because the high sea temperature strengthens the vertical stability of the sea water, it directly limits the deep-layer water from moving upward to replenish nutrient salts. Inorganic nutrient salts required by phytoplankton in the mixed layer are mainly (more than 90%)

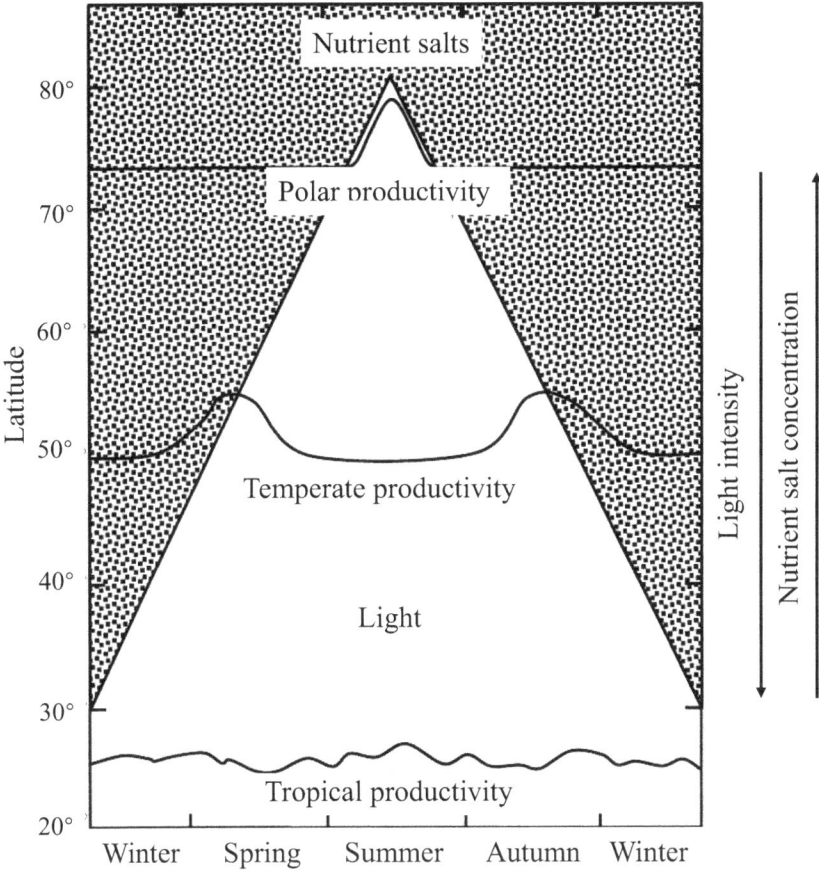

Fig. 2.9 Schematic diagram of the relationship between seasonal variations in the primary production of tropical and temperate sea areas and light and nutrient salts (Lalli and Parsons 1997)

sourced through recycling and regeneration within the system. The nitrate concentration in the euphotic layer is only 0–5 mg/m^3, the chlorophyll content is 5–25 mg/m^2, and the average annual primary production is only 50–100 mg C/m^2. Therefore, this sea area has the lowest primary production and is referred to as the "biological desert" of the ocean. Compared with that in temperate sea areas, the light intensity and vertical stability of the sea water in tropical sea areas do not have obvious seasonal variations, and primary production can occur all year; moreover, the depth of the production layer is deeper than that of the sea areas at mid to high latitudes, and the turnover rate is high; therefore, these ocean areas do not have obvious seasonal cycles

and maintain low production characteristics (Fig. 2.9).

Equatorial Zone

The North and South Equatorial Currents flow from east to west and are continuous within the anticyclonic-type circulation in the equatorial sea areas of the oceans in the two hemispheres; between the two currents is the west-to-east Equatorial Countercurrent (also referred to as the North Equatorial Countercurrent). Because the horizontal Coriolis force on the equator is zero, sea water in the vicinity of the Equatorial Countercurrent converges. Additionally, the northward and southward water transport components generated by the northeast trade-wind current

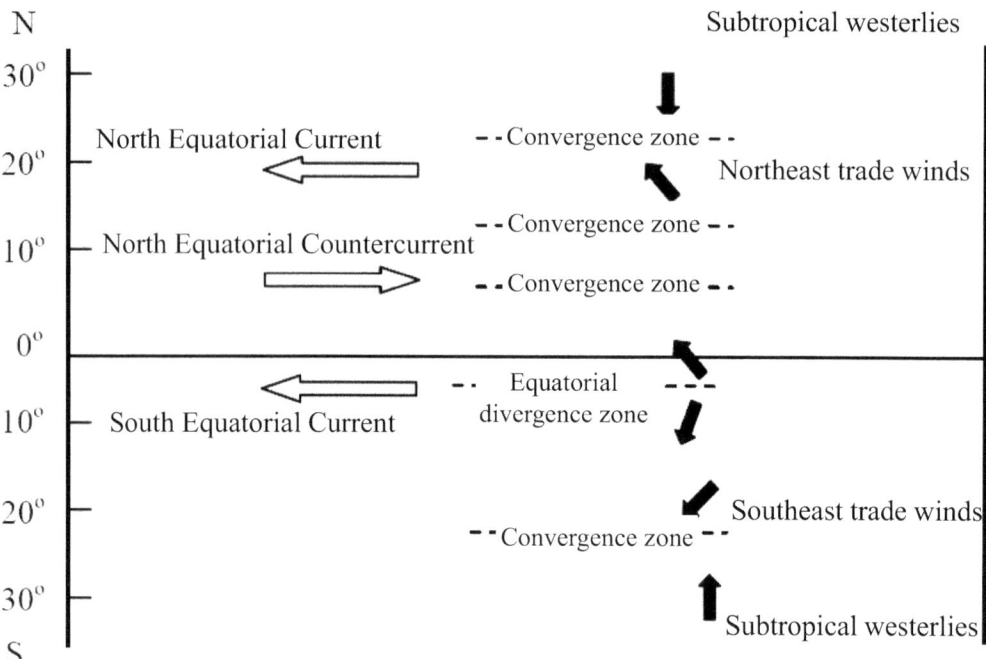

Fig. 2.10 Schematic diagram of the equatorial divergence area (Chen 2003)

and the southeast trade-wind current thereby form a divergence of sea water on both the north and south sides of the Equatorial Countercurrent (Fig. 2.10).

There is no lack of nutrient salts in the equatorial sea area; in particular, the nitrate concentration in the eastern mixed layer is generally higher than 2 μmol/L, that is, higher than the minimum nutrient salt concentration required to limit the growth of phytoplankton. The lack of iron (Fe) in the equatorial sea area is the main factor that limits the growth of phytoplankton. The main source of iron for this sea area is the small amount of iron provided by the near-surface mixed layer brought in by upwelling generated from the Equatorial Undercurrent and the sedimentation of atmospheric dust to the sea surface. Due to the lack of iron, the production volume of phytoplankton does not match the nutrient salt content; therefore, the equatorial sea area is a typical high-nutrient, low-chlorophyll (HNLC) sea area. The picophytoplankton in this sea area accounts for 90% of its gross biomass, and the consumers that

feed on the phytoplankton are mainly microflagellates, dinoflagellates, and ciliates.

There is a difference in productivity between the eastern part and western part of the equatorial zone. The trade-wind currents cause the water level at the western boundary to rise, generating a pressure gradient force, and the depth of the thermocline at the bottom of the mixed layer also rises from west to east. Furthermore, the Equatorial Undercurrent is blocked when it flows to the eastern boundary, and an upwelling that passes through the density-isolines occurs. In addition, the eutrophic water at the surface layer of the upwelling at the eastern boundary of the ocean extends westward with the trade winds. The comprehensive action of the above environmental features makes the productivity level on the eastern side of the equatorial zone higher than that on the western side, which can exceed 0.5 gC/(m^2·d) on average.

The western side of the equatorial zone does not have upwelling like the eastern side, and its primary production is similar to that of a subtropical ocean area. For example, the equatorial current in the Pacific Ocean pushes the flow of warmer sea water westward, forming a "warm pool" in the western part; the higher sea temperature promotes the evaporation of sea water and precipitation, creating a halocline above the thermocline. The double stratification increases the stability of the body of water in the "warm pool," and the constant quantity of nutrient salts is very quickly depleted; therefore, productivity in the western equatorial waters is very low.

Temperate (Subpolar) Zone in the Oceans
The temperate zone is situated between the westerlies and the polar oceans, and there are differences in the ecological characteristics in the temperate zone of the two hemispheres.

Northern Hemisphere
The temperate seas and oceans of the North Pacific Ocean and the North Atlantic Ocean are situated in a cyclonic-type divergent circulation area of the ocean, and the surface sea water spreads outward from the center of circulation, leading the deep-layer water moving toward the surface layer and replenishing the nutrient salts in the euphotic layer. The depth of the mixed layer here is mainly affected by wind force. The nitrate content is 5–25 mg/m^3, which is several times higher than that in the southern subtropical ocean area. The concentration of chlorophyll a is 15–150 mg/m^2, and the primary production of some sea areas can reach 300–500 gC/(m^2·a), much higher than that of the subtropical ocean area.

In the northeastern part of the Pacific Ocean, HNLC sea areas have emerged. For example, although the Gulf of Alaska contains abundant nutrient salts, it contains insufficient iron; in addition, the latitude where it is situated is comparatively high, the light conditions in winter are poor, and the level of primary production cannot be fully determined. The phytoplankton in many waters has a comparatively small particle size, and diatoms only occasionally dominate. Protozoans have become the main plant-eating animals; their growth and reproduction rates are

comparatively fast, and they often feed on phytoplankton in large quantities and promote the regeneration of nutrient salts in the early stages of water bloom. Medium-sized zooplankton are dominated by copepods, and many species undergo a dormant period after developing to copepod larvae stage V and sink below 400 meters to escape from predators. The situation in the North Atlantic Ocean is different from that in the North Pacific Ocean. Due to more iron replenishment sourced from the land, spring water blooms, dominated by diatoms, are significant. The phytophagous zooplankton here are calanoids, which are larger individuals, including *C. finmarchicus*, among others.

In the Northwest Pacific Ocean, a branch of the Kuroshio warm current ascends northward along the coast of Japan and generates an intense mixing action with the sea water at the confluence with the Oyashio cold current from the Kuril Islands in a southward direction. The level of productivity here is high, forming the famous fishing grounds in Hokkaido, Japan. Similarly, in the Northwest Atlantic Ocean, the famous Newfoundland fishing grounds are formed at the confluence of the northward Gulf Stream (warm current) and the Labrador Cold Current. The place of confluence for these warm and cold currents is referred to as the Northwest Convergence Zone of the northern hemisphere.

Southern Ocean
The difference between the southern hemisphere and the northern hemisphere is that the West Wind Drift is not blocked by the continents, thus forming the Antarctic Circumpolar Current, which surrounds the Antarctic continent. An important physical feature of the Southern Ocean is the mixing of strong winds and strong turbulence. Most of the sea area has a very high nitrate content, and an adequate phosphate content, but the primary production is not high. It is a major HNLC sea area among the oceans. The main reason that has caused the mismatch between primary production and the content of nutrient salts such as nitrogen and phosphorus is a low iron content (range, 0.2–0.5 nmol/L) and a very low biogenic silicon content. However, at

the Antarctic front, that is, the sea area of the Antarctic Convergence Zone (located in the vicinity of 60°S in the Pacific Ocean and the vicinity of 50°S in the Atlantic Ocean and the Indian Ocean, on average), due to the rise of sea water generated by wind-driven mixing and vertical turbulence, comparatively sufficient iron is provided to the euphotic layer in the frontal zone. The iron content in the surface layer can be ten times higher than that in most of the Southern Ocean, the biogenic silicon is replenished, and the near-surface mixed layer is comparatively shallow (50 to 100 meters); therefore, this frontal zone is conducive to the emergence of water blooms of phytoplankton. The Antarctic front is a high-productivity area in the Southern Ocean. The phytoplankton is dominated by diatoms, but in other sea areas, nondiatoms or species with lower silicon content are dominant.

The zooplankton most abundant in numbers in the HNLC sea area of the Southern Ocean is a new calanoid, *Neocalanus tonsus*. This species is similar to the aforementioned dominant species in the North Pacific Ocean, but it also stores lipids and goes dormant.

In many temperate sea areas, the light conditions, temperature, and vertical stability of the sea water have obvious seasonal cycles, and the comprehensive action of these physical factors causes primary production to also have seasonal variation features, which usually present as a peak in spring and a secondary peak in autumn (Fig. 2.9).

Polar Zone

The main environmental features of the polar zone are that most of the sea areas are covered with ice, the average sea temperature is very low (<5 °C), the light conditions are poor, and the production season is short, critical factors that affect primary production.

The Arctic Ocean is basically surrounded by continents, most of the zones are situated in high-latitude areas north of 75°N, and sea ice is present all year round or seasonally. The primary production of the Arctic Ocean is not limited by a lack of nutrient salts but, rather, by poor light conditions. There are very long periods of time during the

year when light is very weak (the angle of the sun is low, and the daylight hours are short) or when it is dark for several consecutive months (polar nights). This sea area only has a net yield of phytoplankton during the period with light, with only one peak period of production; it is a single-cycle production area. For example, in areas close to the Atlantic Ocean, from Greenland to the Barents Sea, abundant nutrient salts are present from the mixing of the bodies of water before freezing, and water blooms appear in the spring when there is light. The water blooms start from epiphytic communities under ice floes, and as the sea ice melts, there is a net yield in primary production. Closely following the retreat of the sea ice, the scope of the water blooms also moves quickly northward. It is generally held that the annual average primary production of the Arctic Ocean is lower than that of the oligotrophic subtropical ocean areas, but actual surveys have deemed that the traditional view of the Arctic Ocean as a desert of biological productivity is worth debating.

Due to the short production season of phytoplankton, the generation cycle of Arctic Ocean herbivores is comparatively long. For example, the life history of the Arctic calanoid *Calanus glacialis* is as long as 2–3 years, and they overwinter in deep-water areas at a depth of 1000 meters, whereas Arctic cod (*Boreogadus saida*) is the key species in the process of energy and material transfer from phytoplankton to birds and mammals.

The polar seas and oceans of Antarctica refer to the sea areas from south of the Antarctic front to the Antarctic continent. Between the easterly circulation on the Antarctic continental margin and the Antarctic Circumpolar Current on the north side, an Antarctic divergent front forms due to the dynamic action, and the rise of the deep-layer water brings abundant nutrient salts (especially inorganic salts such as nitrogen and phosphorus). In addition, the convergence and sinking of sea water along the shelf form between the easterly circulation and the continent, that is, the Antarctic continental convergence zone. Diatom blooms appear in polar seas and oceans with the seasonal melting and southward retraction of sea ice and the rapid growth of epiphytic algae. The primary production of the polar seas and oceans in the southern hemisphere is higher than that of the Arctic Ocean. Surveys have deemed that the primary production in Prydz Bay and its adjacent sea areas presents the following pattern: high in the bay and low outside the bay, with the lowest at the continental slopes and the deep sea. The survey results also showed that the photosynthesis rate of phytoplankton is highest in the subsurface layer, then decreasing progressively as depth increases. It is generally believed that the supply of trace elements, iron in particular, determines the scale of water blooms. Among the zooplankton, the Antarctic krill (*Euphausia superba*) is the most important. In addition, the Antarctic silverfish (*Pleurogramma antarcticum*) also feeds on phytoplankton such as diatoms. There are also very abundant seal, whale, and bird (penguins) resources.

Coastal Zones

The zones along the coastlines are high-productivity areas within the seas and oceans because of the existence of various marine fronts, which provide abundant nutrient salts for photosynthesis by phytoplankton.

Upwelling Fronts

The largest coastal upwellings in oceans occur in the eastern upwelling areas of the oceans. The West Wind Drift in the eastern main current of the ocean flows toward low latitudes along the west coast of the continent and finally merges into the trade-wind currents. When surface sea water gradually enters an area affected by trade winds, it is subjected to the actions of the southeast trade winds and the northeast trade winds and flows away from the coast, causing sea water from the lower layer to surge toward the surface layer, resulting in upwelling and the transportation of nutrient salts toward the euphotic layer, which promotes phytoplankton growth (Fig. 2.11). This type of upwelling mainly includes the Peru upwelling and the California upwelling in the Pacific Ocean and the Canada upwelling and the Benguela upwelling in the Atlantic Ocean. For most of the year, these upwellings bring eutrophic water from the deep layer to replenish the surface layer. They are all situated between latitudes 10° and 40° and have sufficient solar irradiation to thereby support the high productivity of phytoplankton and maintain large quantities of fish and bird species. For example, the Peruvian anchovy yield in the Peru upwelling area once accounted for approximately 20% of the world's catch.

Coastal upwellings also appear in other shallow sea areas of the shelf, for example, wind-driven upwelling, bottom-water upwelling caused by topography (such as islands and shoals), and upwelling caused by a combination of the two. These sea areas are also eutrophic, high-

Fig. 2.11 Schematic diagram of coastal upwelling (Nybakken 1982)

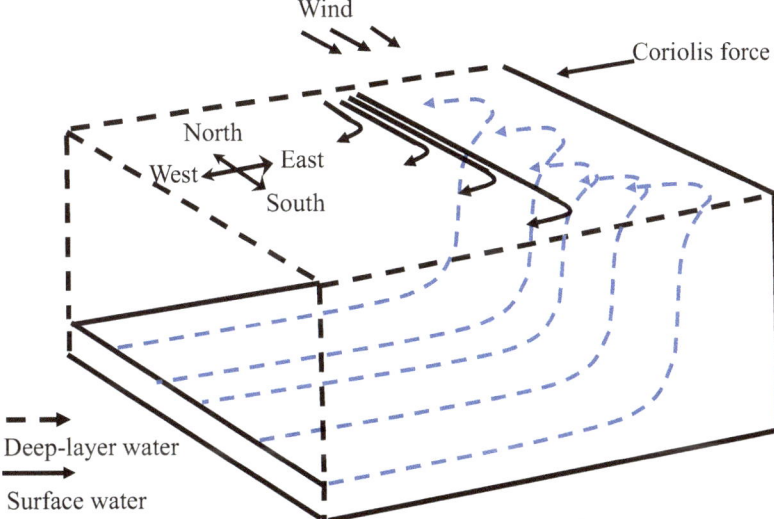

productivity areas, but their localization and sea-sonality are more apparent.

Shelf-Slope Break Frontal Zone

Many shelf-slope break frontal zones appear at marine continental shelf-slope breaks. For example, the Kuroshio Current front at the shelf-slope break in China's seas in the West Pacific Ocean and the Gulf Stream front at the boundary between the Gulf Stream and the shelf water in the North Atlantic Ocean are both typical shelf-slope break fronts. The Kuroshio Current and the Gulf Stream both flow from the high-temperature, high-salinity water in the vicinity of the equator, with very low nutrient salt content at the surface layer. They converge with low-temperature, low-salinity coastal water at isobaths that span the continental shelf-slope breaks of the western boundary, generating a narrow zone of sudden change in temperature or salinity, that is, the shelf-slope break front of the Kuroshio Current or the Gulf Stream. In addition, when these two sections of warm currents flow northward in a meandering manner, they often generate cold and warm frontal vortices on both sides of the continental shelf water boundary. When the frontal vortex of a cyclonic warm center emerges at the shelf-slope break, the water near the surface layer rotates in a counterclockwise manner, pulling away from the Kuroshio Current or the Gulf Stream near the shore; the main body enters the open sea, generating a divergent upwelling at the shelf-slope break. Such a vortex front brings deep-layer water that is colder but rich in nutrient salts to the surface layer, replenishing the inorganic salts required for the growth of phytoplankton in the euphotic layer. Under sufficient light conditions, a water bloom of phytoplankton will occur here. The primary production of phytoplankton within the continental shelf sea area of the southeastern United States depends on the nutrient salt transport process of upwelling at the shelf-slope break. The dynamic mechanisms generated by upwelling are the frontal vortex of the Gulf Stream and the prevailing southwesterlies in spring and summer. Research revealed that at the frontal vortex of the Gulf Stream at the shelf-slope break along the southeastern shore of the United States, when upwelling brings nutrient salts from the deep layer to the surface layer, the average primary production is usually about four times higher than that in the mixed layer at the shelf-slope break. After the nutrient salts replenished by the upwelling are consumed, the primary production also decreases. This type of frontal vortex is also commonly seen in the Kuroshio Current system at the shelf-slope breaks in China's seas.

Not all shelf-slope breaks are boundary transition zones resulting from the meeting of ocean water and shelf water. Broadly, any front structure that appears at the shelf-slope break can be classified as a shelf-slope break front.

Low-Salinity Fronts and Tidal Mixing Fronts

Continental river runoff feeds a steady stream of freshwater into the shallow sea areas of the continental shelf, thereby generating a rapid transition zone between low-salinity water and high-salinity sea water in sea areas; these areas are referred to as low-salinity fronts. Because the input volume from rivers varies seasonally (rainy season and dry season), the intensity and the scope of influence of low-salinity fronts also have seasonal variation features.

River runoff brings large amounts of suspended organic particles, dissolved organic matter, and inorganic nutrient salts to estuaries, and these nutrients are mainly sourced from farmland fertilization and sewage discharge from coastal cities. Among them, inorganic nutrient salts such as nitrogen and phosphorus can be directly absorbed by phytoplankton in estuaries, and organic matter is utilized by phytoplankton through the internal recycling and regeneration of nutrient salts and the release of nutrient salts from suspended sediments. Furthermore, when the freshwater provided by rivers in flowing in shallow seas near shores, vertical circulation will be generated locally in estuaries, bringing the nutrient salts released by the seabed to the upper layer. In estuaries within low-salinity fronts, the high degree of turbulent mixing, the turbid sea water, the very shallow euphotic layer (some are even only approximately 1 m), and the shading effect of phytoplankton themselves will affect the full

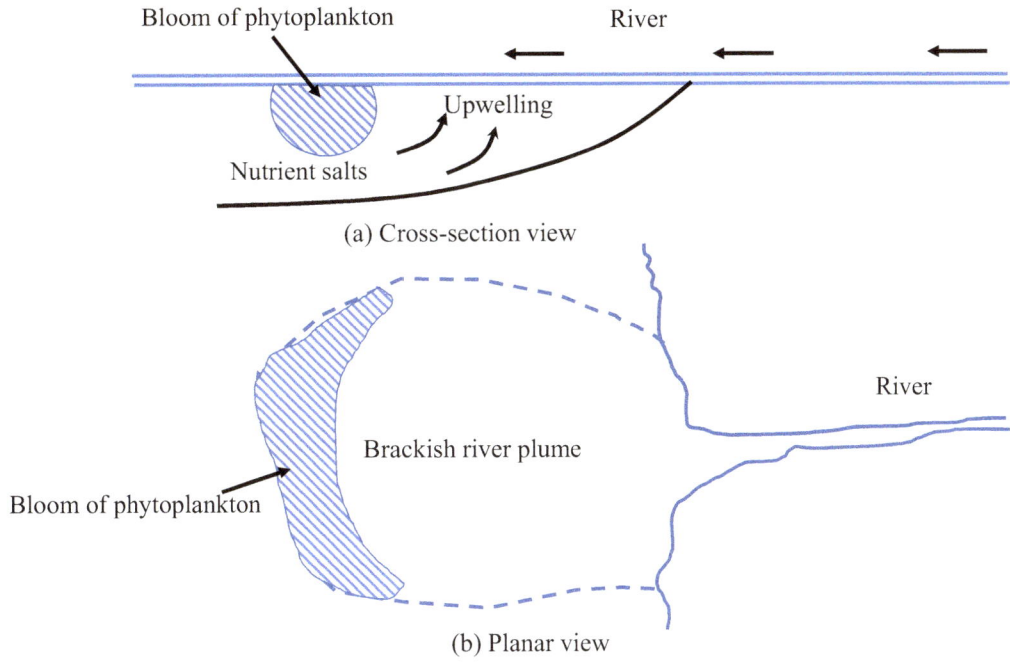

Fig. 2.12 Entrainment of nutrient salts in an estuary (Lalli and Parsons 1997). (**a**) Cross-section view; (**b**) planar view

utilization of abundant nutrient salts by phytoplankton. However, at low-salinity fronts, the euphotic layer deepens, there are sufficient nutrient salts, and water blooms of phytoplankton are often prone to occur (Fig. 2.12).

Tidal mixing fronts refer to the temperature or salinity (or density) transition zone generated by tidal mixing. Tidal fronts appear in the shallow sea areas of the shelf where tidal currents and seabed friction are comparatively high. Notably, in many estuaries, low-salinity fronts and tidal mixing fronts exist simultaneously, and the fronts observed are really a comprehensive portrayal of the mixed interactions of diluted river water, tides, and the wind.

It can be determined through the above discussion that there are great differences in the primary production of sea areas with different hydrological features (Table 2.1); these differences are mainly related to whether surface nutrient salts can be replenished by the divergence (rising) or convergence (sinking) of surface sea water.

Table 2.1 Scope of annual primary production in different sea areas (Lalli and Parsons 1997)

Sea area type	Average annual primary production/[g/C/m^2.a]
Upwelling areas of continental shelves (such as the Peru Current and the Benguela Current)	500–600
Continental shelf-slope breaks (such as the European Shelf, the Grand Shoal, and the Patagonia Shelf)	300–500
Subarctic areas (such as the North Atlantic Ocean and the North Pacific Ocean)	150–300
Anticyclonic vortex areas (such as the Sargasso Sea and the Pacific subtropical sea area)	50–150
North Pole (ice-covered)	<50

References

Cenedese C, Gordon AL (2021) Ocean current. Encyclopedia Britannica. Invalid Date, https://www.britannica.com/science/ocean-current. Accessed 10 November 2021

Chen CS (2003) Marine ecosystem dynamics and modeling. Higher Education Press. (In Chinese)

Chen XJ (2014) Fisheries resources and fisheries oceanography. Ocean Press. (In Chinese)

Chen XJ (2016) Theory and method of fisheries forecasting. Ocean Press. (In Chinese)

Koblentz-Mishke OJ, Volkovinsky VV, Kabanova JG (1970) Plankton primary production of the world ocean. In: Wooster S (ed) Scientific exploration of the South Pacific. National Academy of Sciences, Washington, D.C., pp 183–193

Lalli CM, Parsons TR (1997) Biological oceanography: an introduction, 2nd edn. Butterworth –Heinemann, Oxford

Longhurst AR (1971) The Clupepod researches of tropical seas. Oceanogr Mar Biol Ann Rev 9:349–385

Nybakken JW (1982) Marine biology-an ecological approach. Harper & Row Publishers, New York

United nations (2005). Convention on the continental shelf 1958

Shoaling and Migration of Fish and Their Relationships with Environment

Xinjun Chen

Abstract

Fish, cephalopod, shrimp, and other marine economic species generally have the habits of shoaling and migration in the ocean, which is the result of their adaptation to the changes of living and non-living environment in the long-term living process. Generally, in the process of fish and other marine economic species gathering and migration, the fishing ground is often formed in the ocean. Therefore, it is very important to understand and master the law of fish shoaling and migration and the relationships between fish behavior and marine environment. In the research of fisheries forecasting, people are interested in the following questions: when, where, and in what areas the fish appear and aggregate, how long the fish shoaling time is and how large the fish stock is, what is the marine environmental condition of a fish shoaling, what is the movement route behind the fish shoaling, and so on. Therefore, it is one of the important contents of fisheries forecasting for us to study the aim of fish shoaling and migration and master the law and mechanism of fish shoaling and migration. In this chapter, we will briefly introduce the concept of fish shoaling and type of fish shoaling and the behavior mechanism of fish shoaling. We also summarize the concept and type of fish migration, the mechanism and biological significance of fish migration, and methods of studying fish migration. Finally, the general rule of the marine environment on the distribution of fish and the relationship between fish behaviors and water temperature, current, salinity, dissolved oxygen, and other environmental factors were analyzed. Among them, we focus on the relationship between fish behavior and water temperature and fish behavior and ocean current. The above basic theory and knowledge about fish shoaling and fish migration will provide a theoretical basis for the scientific research of fisheries forecasting.

Keywords

Shoaling · Migration of fish · Marine environment · Tagging

3.1 Significance of Fish Shoaling and Migration

Fish and some other economic animals that inhabit oceans (hereinafter referred to as fish) often shoal and migrate. These behaviors, which are conditioned reflexes related to the physiological and ecological characteristics of fish, result

X. Chen (✉)
College of Marine Sciences, Shanghai Ocean University, Lingang Newcity, Shanghai, China
e-mail: xjchen@shou.edu.cn

from the long-term adaptation of fish to both biotic and abiotic environmental changes.

Fish shoal and migrate for various reasons. They may spawn and reproduce through shoaling and spawning migration for physiological needs and for the continuation of the species. Because water temperature changes with season, fish, as poikilotherms, shoal and migrate to avoid low-temperature areas unsuitable for living and search for areas suitable to live in winter. To replenish the tremendous amount of nutrients and energy consumed during reproduction and overwintering migration (i.e., migration for a suitable temperature), fish shoal and migrate to sea areas with nutrient-rich organisms for survival. In addition, fish often encounter sudden attacks from predators and sudden climate changes; hence, they gather in groups and form temporary shoals to escape from predators or when they perceive environmental stimulation from sound, light, electricity, etc. In summary, fish shoaling and migration occur for various reasons, differ in duration and shoal size, and can involve either only a single species or mixed species. Shoaling can be either regular or irregular. Fish shoaling and migration are complex behavioral phenomena, resulting from the adaptation and selection of fish to the natural marine environment.

For marine fishers, fishery resource researchers, and fisheries forecasting, the following are the most pertinent questions. (1) When, where, and in what sea area fish shoals will appear? (2) How long will fish shoals last? (3) How large are fish shoals? (4) What marine environmental conditions drive the shoaling of fish? (5) What are the movement routes after fish form shoals? (Chen 2016). Therefore, the purpose of studying fish shoaling and migration is to understand the pattern and mechanism of fish shoaling and migration, thus providing a basis for fisheries forecasting. Behavioral research related to fish shoaling is of direct practical significance because fish shoals are the major target of marine industrial fishing. In addition, such research may help find ways to improve fishing efficiency by artificially increasing fish shoal density or

controlling the shoaling behavior of fish, for example, purse seine and log fishing to target tuna shoals and using light to lure fish shoals.

3.1.1 Concepts and Categories of Shoaling Fish

The shoaling of fish is a phenomenon during which individual fish with similar physiological status and common living needs gather in groups due to their physiological requirements and survival needs. The species in and size, time, and form of fish shoals vary with life stage and marine environments. Generally, fish shoals can be divided into four categories based on the purpose of shoaling: breeding shoal, feeding shoal, overwintering shoal, and temporary shoal.

Breeding Shoals
Breeding shoals, also known as spawning shoals, consist of sexually mature fish. Generally, the members of a breeding shoal are similar in regard to gonad development and body length. Breeding shoals are very dense, relatively concentrated, and stable.

Feeding Shoals
Fish with similar feeding habits form feeding shoals to prey on the same types of bait organisms. The members of a feeding shoal have the same feeding habits. Generally, fish of the same species with the same feeding habits have similar body lengths, but fish of different species also often gather to feed on the same types of baits. The density of a feeding shoal depends on the abundance and distribution range of bait organisms. In addition, the shoal may reorganize as the fatness of fish increases and the marine environmental condition changes. For example, feeding shoals distributed in tropical and subtropical zones form breeding shoals when the gonads of fish members become mature or re-mature; feeding shoals distributed in temperate and frigid zones form overwintering shoals when water temperature decreases.

Overwintering Shoal

Prompted by a water temperature change, fish may gather and form overwintering shoals to seek a new environment suitable for living. Fish of the same species with similar fatness, regardless of their age and body length, may shoal for overwintering. On the way to their overwintering ground, the primary shoal may split into smaller shoals based on the fatness of fish members; after arriving at the overwintering ground, most small shoals may merge into larger shoals, which are often tremendous in size. Varying by feeding habits and fatness, fish may stop feeding or reduce feeding at overwintering grounds. Generally, overwintering shoals are relatively dense and vary with environmental conditions.

Temporary Shoals

When encountering a sudden environmental change or ferocious predators, individual fish may temporarily gather to form a temporary shoal. In general, a sudden change in environmental conditions (e.g., a dramatic change in temperature and salinity gradients or the emergence of ferocious predators) and stimuli in habitats (e.g., sound, light, and electrical signals) can cause a temporary concentration of scattered feeding shoals or moving shoals, regardless of the life stage of the fish. The fish in temporary fish shoals separate, returning to non-shoal life once the environmental conditions return to normal.

3.1.2 General Patterns of Shoaling Fish

Fish have different shoaling patterns at different stages of life. At the juvenile stage, fish of the same species born in the same sea area during the same period often shoal. All members in the shoal are in the exact same biological state and have the same biological developmental pace. The shoal during this stage is considered a basic species group. As the fish members grow at different speeds, with varied gonad development, the basic fish group begins to differentiate and reorganize. The growth rate of juvenile fish often

varies; individuals that grow faster and are more sexually mature due to sufficient food intake and a strong ability to absorb nutrients often leave the original group and preferentially join a group whose members are older and more sexually mature. In contrast, some basic group members that grow slower and are less sexually mature may join in a group whose members are born later and have similar gonadal maturity. The remaining basic group members that have an average growth rate and similar gonadal maturity maintain the original basic group. The fish groups formed from reorganization and recombination after the differentiation of basic species groups are called the shoal or stock of fish. In a fish shoal, the members may not be the same age but have similar biological status and unified movements, and they will stay together for a long time. Occasionally, fish in the same shoal may disperse into several small shoals temporarily to chase bait or escape from a predator; these small temporary shoals will automatically merge once the environmental conditions become suitable.

3.1.3 Functions of Shoaling Fish

Although researchers have conducted numerous studies and have proposed many hypotheses, our understanding and research regarding the role, biological significance, and mechanism of shoaling fish remain insufficient. The following are the known roles of shoaling.

Defensing Against Predators

The most certain role and biological significance of the shoaling behavior of fish is defense against predators. Shoaling can reduce the probability of a fish being targeted by a predator and the probability of the targeted fish being killed by the predator. A shoal of thousands or even millions of fish may seem exposed, but actually, the chance of the shoal being detected is not higher than a single fish being found by predators in the ocean. Due to light absorption and scattering by particles suspended in seawater, the visible distance of an object in water is extremely limited and only approximately 200 m in extremely clear water. In fact, the visible distance can be even much smaller than this value in a real environment. In addition, the visible distance does not change with the size of the object. Fish shoaling, a social form developed during long-term evolution, can reduce the probability of prey fish being found, successfully captured, and killed by predators; undoubtedly, the defensive process of shoaling fish also involves other forms of defensive mechanisms. In an experiment using an aquarium, cod could eat pollock juveniles moving alone at an average rate of one every 26 seconds but could eat shoaling pollock juveniles at an average rate of only one every 135 seconds (Chen 2014).

In addition, shoaling helps fish escape from a moving fishing net. The entire shoal is often able to escape when only part of the fish shoal is in the net. Experienced fishers know that fishing can be most effective only if an entire fish shoal is trapped in a purse seine because the members of a fish shoal are extremely sensitive and can react extremely fast; as a result, when one fish member is frightened and changes direction, the entire fish shoal can coordinately turn direction almost at the same time. Therefore, shoaling can reduce the risk of fish being preyed on or purse-seined and increase the chance of detecting predators as early as possible, thereby playing an important role in defense.

Enhanced Foraging Success

Forage-related relationships are one of the most basic forms of interspecific and intraspecific relationships. Shoaling helps fish discover and find food more easily. Existing studies have found that similar to prey fish, certain predatory fish also gather in shoals. Therefore, it is believed that shoaling behavior also plays a role in the life of predatory fish species. However, up to now, little in-depth research has been conducted to explore this topic.

A small number of existing studies have suggested that the shoaling of predatory fish can increase the total number of their sensory organs and expand the range they are able to search. When one shoal member discovers bait, all members can prey upon the target. The search range maximizes when the members maintain the furthest distance apart that allows them to see each other. As a result, shoaling fish can find food more easily and quickly compared to when they act alone. In fact, for fish species in the middle of the food chain that are both prey and predators, shoaling helps with both defending and foraging.

Reproductive Advantages

To meet the needs of spawning and breeding, fish with mature gonads gather to form breeding shoals to improve reproductive effectiveness and reproduce offspring. Previous studies have shown that breeding shoals have extremely strict requirements for water temperature, which is usually limited to a small range because fish with mature gonads will not spawn in an inappropriate water temperature. Breeding shoals are usually highly dense, which can improve the fecundity of fish. For most fish species, shoaling provides a necessary condition for spawning and promotes gene spread to a certain extent when many individuals gather to spawn and mate. Undoubtedly, shoaling is critical for fish to breed and continue their species.

Other Aspects

A large number of studies have revealed that in addition to the roles and biological significance in defense, foraging, and breeding, shoaling is important in other aspects in the life of fish. For example, compared with fish acting alone, shoaling fish are more resistant to adverse environmental changes. Shoaling enhances the resistance of fish against toxins and reduces their oxygen consumption. From the perspective of hydrodynamics, swimming in shoals decreases the energy consumption of each individual because the eddy energy generated by fish swimming in the front can be utilized by the fish behind and reduce their effort required to swim forward.

In summary, shoaling behavior has broad biological significance, and it is inappropriate to only consider one aspect. The biological benefits of fish shoaling jointly contribute, in multiple aspects, to the survival of shoaling fish. For example, mutual benefits to the group, the chaos effect, mimicry (e.g., pretending to be a large number and a large animal), and energy-saving effects can be all integrated to benefit the survival of fish.

3.1.4 The Behavioral Mechanism and Structure of Shoaling Fish

The Behavioral Mechanism of Shoaling Fish

What mechanism do fish use to form and maintain shoals? Existing studies have suggested that sound, posture, water flow, chemicals, light scintillation, and electric fields are the major channels through which fish signal each other. Therefore, vision, lateral sense, auditory perception, olfaction, and electrical sense all play significant roles in the formation and maintenance of fish shoals. However, no agreement has been reached regarding the behavioral mechanism of fish shoaling among all existing studies.

The Role of Vision in Shoaling Fish

Many researchers propose that vision is the most important sense involved in fish shoaling, and some even believe that vision is the only sense involved in the shoaling behavior of fish. However, these views are one-sided. In addition to vision, hearing, lateral sense, and olfactory organs of fish are closely related to their shoaling behavior and have importance comparable to that of vision.

As revealed by existing studies, vision provides mutual attraction signals between shoal members and enables the members to attract and approach each other, thereby playing a significant role in the shoaling behavior of fish. Recent studies have further revealed that the visual system seems to be an important sensory organ of a fish to maintain distance and orientation with the nearest fish.

The Role of Lateral Sense in Shoaling Fish

Most fish have lateral line systems on both sides of the body. Although its role has been confirmed in past studies, the lateral line system is generally believed to be less important than vision in the formation of fish shoals. However, some studies have claimed that the lateral sense might be as important as vision in the shoaling behavior of fish, and it is the most important sensory organ for a fish to determine the speed and direction of adjacent fish. There is sufficient evidence to prove that fish use both vision and lateral sense during swimming.

The Role of Olfaction in Shoaling Fish

Olfaction also plays an important role in the shoaling behavior of fish. For example, the skin exudates of living and dead loaches (*Misgurnus anguillicaudatus*) have the same lure effect on other loaches, but that of dead loaches does not trigger the fear-related reaction of others. Studies on tench (*Tinca tinca*) found that olfaction is critical for shoaling when vision is unavailable. Similarly, olfaction plays an important role in the shoaling of eel catfish (*Plotosus lineatus*) (Chen 2014).

In conclusion, fish shoaling behavior is realized by using information from multiple sensory sources. The reason for this may only be revealed from the perspective of evolution. As a result of natural selection, fish and other aquatic animals are able to use a variety of information, and it is reasonable to believe that there may be other sensory systems involved in fish shoaling in addition to vision, lateral sense, and olfaction. For example, some researchers have proposed that hearing and electrical sensation might be involved in the shoaling behavior of fish; however, these hypotheses have been systematically studied only rarely.

Fish Shoal Structure

Understanding the structure of fish shoals is of great significance for further elucidating fish shoaling behavior, the reconnaissance of fish shoals, and fisheries forecasts. The structure of a fish shoal can be investigated from two aspects: external and internal structures. External structure refers to the shape and size of a fish shoal, etc. Internal structure refers to the species composition, body length composition, density, swimming pattern, spacing, and moving speed of shoal members. Shoals formed by different types of fish have different external structures because of variations in fish shape and size. For shoals formed by fish of the same species, the external structure may also change with time, location, the physiological state of shoal members, and environmental conditions.

The external traits of fish shoals are the major characteristics of concern in fish reconnaissance and fishery production. The external traits of fish shoals vary with species and the life stages of the fish members, as well as environmental conditions. They also vary between pelagic and demersal fish. The differences mainly manifest in the shape, size, population color, etc., which are most distinct among pelagic fish species. For example, the mackerel (*Pneumatophorus japonicus*) shoals distributed in the northern seas of China and the shoals of *Decapterus maruadsi* and round sardinella (*Sardinella aurita*) in the northern continental shelf of the South China Sea have nine different shapes: triangle, transverse-straight-line-like, crescent-like, trinacriform, "crew cut"-shaped, oval, square, round, and dumbbell-like (Figs. 3.1 and 3.2).

Generally, the number of fish in a shoal of pelagic fish species can be estimated based on the shape, size, color, and moving speed of the shoal. Fast-moving shoals often have a relatively small size, and slow-moving shoals likely have a relatively large number of members. Fish shoals with a darker color are often larger than those with a light color. Taking the *P. japonicus* shoals in Fig. 3.1 as an example, the first three types of fish shoals generally have a small or relatively small size and do not show any color. These shoals move quickly and often generate noticeable ripples on the water surface when the weather is clear, wind is calm, and ocean waves are low. The first and second types of shoals consist of hundreds of fish, no more than 1000 or 2000, and the third type has a larger size and moves more slowly. The fourth, fifth, and sixth types of shoals have a slightly larger size. Compared with these three types, the seventh type, a square shape, has a larger number of members and moves more steadily; the number of members can be estimated according to color and is in the range from thousands to more than 10000. The eighth type has a round shape with dark red or purple-black color. The shoals appear small at the sea surface but become larger with depth and generally consist of more than 20,000 or 30,000 members, occasionally reaching 60,000

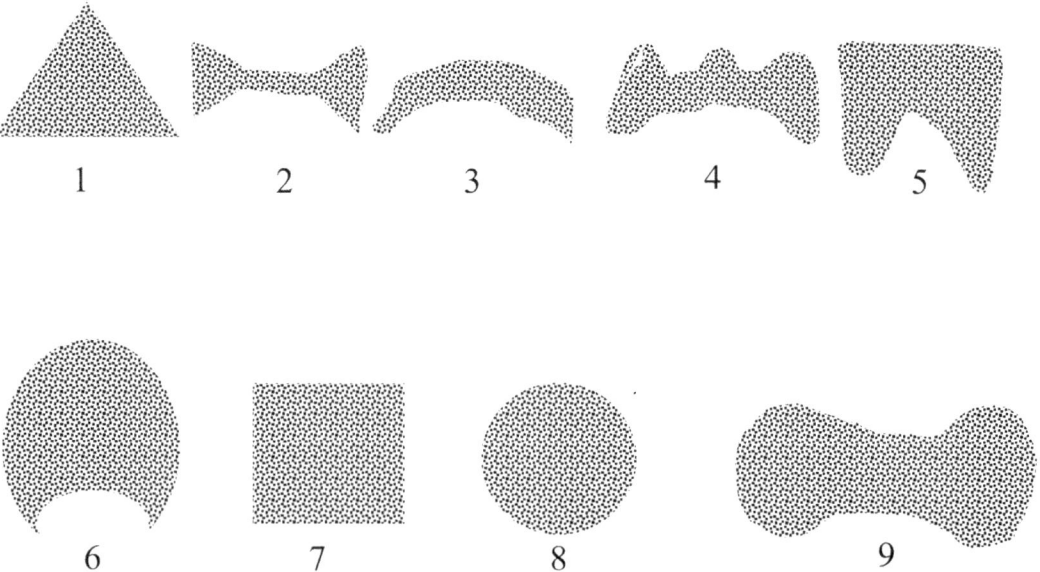

Fig. 3.1 The shapes of *P. japonicus* shoals distributed in the northern seas of China (Chen 2014, 2016)

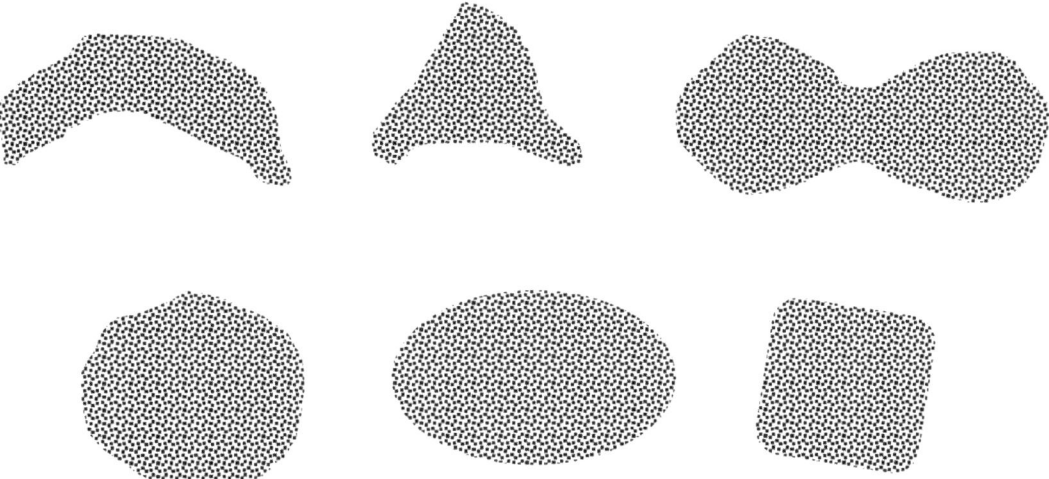

Fig. 3.2 The shoaling shapes of *Decapterus maruadsi* and *Sardinella aurita* shoals in the northern waters of South China Sea (Chen 2014, 2016)

or 70,000 members. This type of shoal moves very slowly, making purse seine fishing easier. The ninth type of fish shoal, with a dumbbell shape, has the largest size. They generally do not appear on the water surface, and their movement is the slowest and cannot be easily disturbed. For example, the shoal can immediately split when a ship passes through and merge together after the ship has passed. Each shoal usually consists of more than 30,000 or 40,000 fish, but those with a dark purple or dark black color may have more than 100,000 members. However, this type of shoal is difficult to detect.

3.2 Fish Migration and Distribution

3.2.1 Definition and Types of Fish Migration

Definition of Fish Migration

Due to genetic factors, physiology, and environmental impact, fish shoal and move periodically and directionally on a regular basis, known as fish migration. Migration is an adaptive attribute for fish to expand their distribution scope and living space to ensure survival and increase the number of species. Migration is a large-scale, periodical, and directional movement, and generally, it

occurs annually. As a social behavior involving the movement from one environment to another, migration is a necessary and adaptive habit of species that results from evolution and natural selection. Fish migration follows a certain route, i.e., a migration route.

During migration, a large number of fish regularly appear in a certain sea area and form dense shoals that can be fishing targets; that is, the sea area where fish migrate may form a fishing ground. Therefore, studying and understanding patterns of fish migration is critical for detecting changes in fishery resources and shoal composition to ensure more accurate fishing season and fishing ground forecasting. Furthermore, fish migration information is of great significance for formulating effective measures for fish reproduction protection and management, improving fishery production and resource protection and management, and promoting fishery stock enhancement. Other aquatic animals, such as prawns, have the same migratory habits.

Fish migration is an innate instinctive behavior and has a certain biological significance. After gradually forming and stabilizing during the long evolutionary process, migration has gained high heredity stability. Due to differences in inherited migration habits, different fish species and different populations of the same fish species have their own migration routes and their own places for breeding, feeding, and overwintering. The inherited migration behavior results from natural selection and is highly stable; therefore, changes do not occur easily. Under certain internal and external factors, fish migrate following their inherited habits. However, existing studies have shown that the increase in water temperature caused by climate change is affecting the migration and distribution of some marine fish species, either expanding or reducing the distribution of these fish.

Not all fish species migrate as a result of seasonal changes or their own needs. Based on whether they migrate, there are two types of fish: migratory and resident/straggling. For most fish, migration is an indispensable part of their life cycle. Only a small number of fish often settle in places and do not move over a long distance on a regular basis, such as some species of the family Gobiidae. Some fish, such as salmon, migrate only after they become sexually mature. Their juveniles swim from the spawning ground to the feeding ground and live there, and they do not move long distances until they are sexually mature. Generally, juvenile migratory fish complete a migration cycle of spawning ground → fattening ground → feeding ground → spawning ground, while adult migratory fish migrate directly between the spawning ground and feeding ground. Fish living in temperate and frigid zones also undertake overwintering migration because their survival requires a certain temperature range.

Types of Fish Migration

The migration of fish is a residency-changing movement in a certain direction, over a certain distance, and at a certain time. This periodic movement is usually performed on a large scale on a regular basis and is an inherited trait of fish. Using different standards, fish migration can be divided into the following categories.

Based on Driving Force of Migratory Movements

Fish migration can be divided into active and passive migration. Active migration refers to migratory activities actively conducted by fish using their own ability to move, such as migrations to spawning grounds when fish near sexual maturity, migrations to overwintering grounds when fish reached a certain degree of fatness, and migrations returning to feeding grounds after fish have reproduced or passed the winter. Conversely, the floating eggs, larvae, and juveniles have a low ability to move and are often carried far away by water currents, i.e., passive migration. A typical example of passive migration is the long distance that larval eels are carried by ocean currents.

Based on Purpose of Migration

The purpose for migrations can range from breeding (or spawning migration) to foraging (or feeding migration) and overwintering (temperature-adaptive migration).

Breeding Migration or Spawning Migration

Breeding migration refers to the movement of fish from a feeding ground or overwintering ground to a spawning ground. In fish with mature gonads, sex hormones released into the blood stimulate the nervous system to prompt ovulation and reproduction; as a result, the fish often gather in shoals and migrate to search for water areas suitable for spawning as well as for the growth, development, and residency of their offspring. Spawning migration can be divided into three types based on migration path and the ecological environment for spawning: anadromous, catadromous, and landward migration.

Landward migration, which is the most common type, refers to breeding migration from deep ocean to coastal shallow-water areas. Anadromous migration refers to the breeding migration of fish that grow in the ocean and migrate from the sea up rivers for spawning after sexual maturity, for example, salmon and shad (*Tenualosa reevesii*). Catadromous migration refers to the breeding migration of fish that grow in rivers and migrate from rivers to the sea to spawn after sexual maturity. A representative species is eels, which migrate in a direction opposite to that of salmon.

The following are characteristics of breeding migration:

(a) Fish shoals move fast, travel a long distance, and are not susceptible to environmental impacts. Knowledge regarding the moving speed and direction of breeding migrations will allow accurate forecasting on the next fishing ground and fishing season based on the current fishing status.

(b) During breeding migrations, fish show the strongest shoaling behavior and gather in different shoals based on age or body length during breeding migrations.

(c) During breeding migrations, fish experience dramatic gonadal changes, showing significant differences in both development degree and size/weight.

(d) The destination of breeding migrations is the spawning ground in a relatively fixed sea area every year, which might change to a certain extent under the influence of hydrological conditions, including changes in temperature and salinity.

Feeding Migration (Also Called Foraging or Fattening Migration)

Feeding migration refers to the movement of fish from a spawning or overwintering ground to a feeding ground. After overwintering, both sexually immature and adult fish that have consumed a large amount of energy due to migrating and breeding activities swim to food-rich sea areas where they can feed and fatten, replenish physical strength, and accumulate nutrients to prepare for overwintering and reproduction in the following year.

The following are characteristics of feeding migrations:

(a) The route, direction, and time of feeding migrations may change constantly because the purpose of feeding migrations is to forage. Therefore, the range of feeding migrations is relatively unstable compared to that of reproductive migrations.

(b) Nutrient condition is the main factor determining feeding migrations, and hydrological conditions (temperature, salinity, etc.) are a secondary factor.

The distribution and movement of bait organisms decide the dynamic migrating activities of predatory fish species. The density of bait organisms decreases to a certain extent due to consumption by predatory fish. This can cause the energy for feeding to exceed the energy obtained from food; thus, the fish will continue to migrate and look for new bait organisms. The time and area of fish migration often change with the number and distribution of bait organisms. Therefore, understanding the distribution and movement patterns of bait organisms can help fishers generally predict changes in fishing grounds and fishing seasons. For example, in the northern coastal areas of China, largehead hairtail (*Trichiurus lepturus*) like to eat prey fish such as *Ammodytes personatus* and *Setipinna tenuifilis*; hence, the largehead hairtail fishing season begins about 10 days after their prey fish arrive at the largehead hairtail fishing ground every year, during which a large number of largehead hairtail can be captured.

(c) Generally, feeding migrations have relatively shorter routes and involve relatively scattered fish shoals. For example, many fish species in China that spawn in spring and summer usually forage in nearby seas after spawning.

Overwintering Migration (Also Known as Seasonal Migration or Temperature-Adaptive Migration)

This refers to the migration of fish from feeding grounds to overwintering grounds. Fish are poikilothermal animals that are extremely sensitive to changes in water temperature. Water temperatures suitable to life vary among fish species. In pursuit of suitable areas for survival, fish will move in shoals, namely, migrate for overwintering, when the environmental temperature changes.

The following are characteristics of overwintering migrations:

(a) Fish usually migrate from low-temperature areas to higher-temperature areas for overwintering; therefore, marine fish in China generally migrate from north to south and from shallow seas to deep seas.

(b) During overwintering migrations, fish usually reduce or stop feeding and mainly consume the nutrients that have accumulated in the body during the feeding period; therefore, to a certain extent, the distribution of and changes in bait organisms do not affect the movement of fish during this period.

(c) Only the fish that reach certain degrees of fatness and fat content can migrate for overwintering. That is, a relatively good biological status is the basis for the overwintering migration of fish. Fish begin overwintering migrations after they reach a certain biological state and are stimulated by external environmental conditions (e.g., decline in water temperature). Therefore, environmental change is a major factor triggering the overwintering migration of fish. Fish that have not reached a certain body fullness and fat content will continue to forage and fatten instead of participating in overwintering migrations.

The purpose of an overwintering migration is to move to an area with a water temperature that is suitable for enduring the winter. Therefore, the overwintering migration process is easily affected by water temperature, especially the distribution of isotherms.

Breeding migrations, feeding migrations, and overwintering migrations are interrelated, and one stage prepares for the next stage in the life cycle of fish (Fig. 3.3). Entering the migratory stage is related to the biological state of fish, including body fullness, fat content, gonad development, and blood osmotic pressure. The initiation of migration mainly depends on the biological state of fish as well as changes in environmental conditions.

However, not all migratory fish species in the ocean undertake all three types of migrations. Some fish species undertake only breeding migrations and feeding migrations but no overwintering migrations. For some fish, the three types of migration overlap to a certain extent and cannot be clearly distinguished. For example, fish species that spawn in batches often begin small-scale feeding migrations in their spawning grounds, and their feeding migrations might also be mixed with overwintering migrations due to changes in bait biomass or seasons.

According to Ecological Environment (Nature of Waters)

Fish migration can be divided into four types: oceanodromous, anadromous, catadromous, and potamodromous.

Migration of Oceanodromous Fish

Oceanodromous fish account for the largest proportion of migratory fish species and include approximately 500 species in 104 families/genera that migrate in international waters. The most important families/genera, known as "highly migratory species," include the families Scombridae and Bramidae, *Makaira* spp., the family Istiophoridae, *Xiphias gladius*, the families Scomberesocidae and Coryphaenidae, and 17 genera of the Elasmobranchii family.

Oceanodromous migratory fish live and migrate completely in the ocean. Fish of the same species are often divided into different groups, each of which has its own migration route and does not mix with other groups. The species group in each sea area has its unique characteristics, and each sea area has a migratory group of this species. For example, *Larimichthys polyactis* (*L. polyactis*, known as "small yellow croaker") in the East Sea and the Yellow Sea of China can be divided into four species groups, and each has its own migration routes for overwintering, spawning, and feeding. The simplest migrations of oceanodromous fish are seasonal migrations between offshore (overwintering grounds) and nearshore waters (spawning and feeding grounds).

Migration of Anadromous Fish

Anadromous fish live in the ocean but migrate to the middle and upper reaches of rivers to breed.

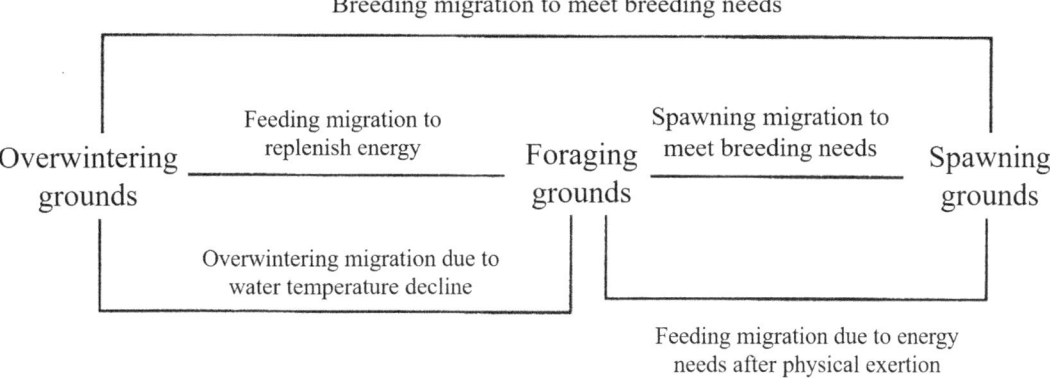

Fig. 3.3 The relationships among different migratory types (Chen 2014, 2016)

These types of fish are highly adaptive to the ecological conditions of habitats, especially water salinity. The representative species include Pacific salmon, shad, *Coilia nasus*, *Coilia mystus*, and Chinese sturgeon (*Acipenser sinensis*). For example, chum salmons (*Oncorhynchus keta*) in the North Pacific Ocean stop feeding once they begin their upstream migration into the river. They migrate against the river current, which can have a speed of tens of kilometers per hour, for more than 10 kilometers every day, thus utilizing a large amount of energy. After arriving at the spawning ground, the parent fish perish after reproducing, and juvenile fish enter the sea in the same year or the following year. Some shallow-water fish living near river estuaries only migrate short distances to estuaries to breed, for example, *Coilia mystus* living in the Yangtze River estuary.

Migration of Catadromous Fish
Catadromous fish, such as eels, live in freshwater most of the time and migrate to the sea to breed. European eels and American eels stop feeding after migrating downstream into the sea and travel thousands of kilometers within sea areas to spawn. All parent eels perish after breeding, and the juveniles return to freshwater at different times, i.e., 3 years for European eels and 1 year for American eels. The representative catadromous fish in China are Chinese eels and Songjiang perches.

Migration of Potamodromous Fish
Potamodromous fish migrate entirely in only freshwater and seasonally return to their spawning grounds, which are usually located upstream. Migration in rivers is referred to as river migration; fish that have baiting or spawning grounds in lakes are called limnodromus fish, such as some perch species. Freshwater fish live and migrate completely in inland waters, with short migration distances and diverse patterns. Some fish species live in running water and spawn in still water; others live in still water and spawn in running water. Four notable fish species in China, i.e., grass carp, black carp, silver carp, and bighead carp, are all semi-migratory fish. These fish species feed and fatten in lakes connected to the main stream of rivers and migrate upstream in shoals to spawning grounds in the main river stream during their breeding season. Parent fish return to the food-rich lakes for feeding after breeding.

Due to differences in salinity and osmotic pressure between seawater and freshwater, both anadromous and catadromous fish usually need to stay in salty freshwater areas for a period when crossing the estuary so that they can adapt to physiological changes.

3.2.2   Mechanism and Biological Significance of Fish Migration

Mechanism of Fish Migration

Factors Affecting Fish Migration

The factors that affect fish migration processes are complicated and include both internal and external factors. The migration process results from the joint action of internal and external factors. That is, fish initiate migrations only when they reach a certain physiological status and are stimulated by corresponding environmental factors.

Internal Factors

Internal factors dominantly affect the migration of fish; such factors include changes in biological status, including gonad development and hormonal effects, as well as changes in body fat, fat content, and chemical composition of the blood. When gonads develop, sex hormones secreted from the gonads act on the nervous system and trigger breeding migration. Certain levels of body fullness and fat content are necessary conditions for overwintering migration, and the need for feeding after reproduction or overwintering prompts feeding migration.

Changes in chemical composition and regulatory mechanisms of blood osmotic pressure are also migration-affecting internal factors. Taking freshwater eel (*Anguilla japonica*) as an example, prior to entering the sea, the level of carbon dioxide in the blood gradually increases, leading to elevated blood osmotic pressure, which makes moving to the sea become an urgent physiological need. When fish of the family Salmonidae enter freshwater, their blood osmotic pressure gradually decreases, their digestive tract shrinks, and they stop feeding, enabling more active breeding migration.

Regardless of whether they have reached reproductive age, fish with poorly developed gonads will not initiate breeding migrations. Similarly, fish with insufficient body fullness and fat content will not start overwintering migrations even if winter has come. Therefore, internal factors are the dominant factors affecting the migration process of fish, while changes in external environmental conditions play a role in stimulating or inducing the beginning of migrations.

External Factors

Fish that are ready to migrate do not necessarily begin to migrate immediately; instead, migrations are generally triggered by certain external factors. An ongoing migration process is also affected by external factors. For example, adverse external factors may cause fish to pause their migration activities or deviate from their current migration route. Therefore, external factors not only serve as signals or stimuli to initiate migration but also affect the entire migration process.

Many external factors affect fish migration but vary largely in the degree of effect; therefore, such factors can be divided into primary factors and secondary factors. Notably, primary and secondary factors are not fixed because they may vary among species and switch among different developmental stages and life periods for fish of the same species. The main purpose of breeding migrations is to find a suitable place for breeding; therefore, external factors such as water temperature, salinity, transparency, and flow velocity often significantly affect the behavior of fish as they migrate to the spawning ground. For feeding migrations, because the main purpose is to seek bait, bait becomes the primary external factor that determines the behavior of fish. During overwintering migrations, water temperature and topography are the main influencing factors because the main purpose of these migrations is to find a suitable place for overwintering, and thus, fish gradually swim to the sea areas with higher water temperature or deep water with a slower water current.

Taking the overwintering migration of largehead hairtail in China's inshore waters as an example, previous research has revealed that water temperature, salinity, water mass, wind, current, water transparency, and water color are the main external factors affecting the overwintering migration of largehead hairtail to the Shengshan fishing ground. (1) Water temperature has the strongest impact. Largehead hairtail enter the Shengshan fishing ground when the water temperature (surface temperature) is approximately 20 °C and leave the fishing ground when the water temperature decreases to 13 °C. The peak fishing season occurs when the water temperature is between 15.3 °C and 18.6 °C. Therefore, a fishing season can be predicted based on water temperature forecasts. (2) Largehead hairtail generally migrate south along the 30 m to 40 m isobaths, and the water color of the fishing ground is rated No. 11–14, often called "white rice"-like color by fishers. (3) Largehead hairtail shoals are often concentrated in the mixing zone of the low-temperature low-salinity water system along the coast and the high-temperature high-salinity water system offshore. When the current along the coastal is strong, fish migrate closer to the offshore side; otherwise, they migrate closer to the costal side. (4) Wind also affects fishing. Eastward wind strengthens the influence of the high-temperature high-salinity system offshore, causing the migration route of largehead hairtail to be closer to the coast; conversely, westward wind causes the migration route to be closer to the offshore side. When there is a lack of strong wind for a long time, water temperature shows a clear vertical stratification, affecting the shoaling of largehead hairtail and thus reducing fishing production. Conversely, fish shoals become more concentrated due to distinct flow separation after storms, a process that is conducive to fishing production. Continuous storms prompt fish to move southward and shorten the fishing season, decreasing fishing production. As the above discussion shows, the key to understanding the effect of external factors on the overwintering migration of largehead hairtail is to have knowledge of the primary factors, which can be used as a scientific basis for forecasting largehead hairtail fishing and guiding fishing operations.

The influencing factors of each type of fish migration are briefly summarized and presented in Table 3.1. Generally, internal factors play a dominant role, and external factors are conditional.

Mechanisms of Orientation During Migration

Almost all fish species migrate in shoals. Generally, migratory fish shoals are composed of fish with similar body length and biological status. There is no fixed leader in a migratory fish shoal as the fish originally in the front will fall behind, being replaced with other lead fish after a period of time. Migratory fish shoals usually have a certain shape to ensure the most favorable dynamic conditions. In addition to the role of hydrodynamics in movement, the adaptive behavior of fish shoals during migrations helps fish identify their direction. The shoal size of migratory fish varies among fish species and is undoubtedly related to its role in providing the most favorable migration conditions.

Fish can use their sensory organs for navigation so that they can successfully complete migrations that sometimes can be thousands of kilometers. At present, there is no satisfactory answer to why fish can migrate in a certain direction and route and spawn in the same place. The research on this aspect is insufficient and has

Table 3.1 Internal and external factors of fish migrations (Chen 2014, 2016)

Migration type	Major internal factors	Major external factors
Breeding migration	The physiological condition of the fish must reach a certain level, and the gonads have begun to mature. For example, salmon change body shape and color and enter the river when their gonads are stimulated	The external environmental stimulus is mainly temperature. When the temperature does not meet requirements for breeding, fish cannot spawn even if their gonads are fully mature
Overwintering migration	Body fullness and fat content also meet certain requirements	Water temperature decreases
Feeding migration	Physical exhaustion due to spawning or starvation after overwintering	Distribution of bait organisms

proposed mostly only hypotheses. In general, the following are believed to be the main reasons.

Hydrochemical Factors

Chemical properties of water, especially salinity, are an important factor affecting fish migration. The change in salinity in water can cause the change in osmotic pressure of fish, which stimulates their nervous system to respond. In addition, water quality greatly impacts fish migration.

A large number of studies have demonstrated that salmon can rely on their olfaction to easily find the river where they were born and where they spawn; water smell plays a guiding role. In addition, it has been suggested that fish also determine their migration direction based on salinity gradients and dissolved oxygen content. Based on gradient changes, fish may perceive whether they are leaving or approaching the shore and thus make correct decisions to ensure migration success. Similarly, fish may also use water temperature gradients for navigation. However, some studies have stated that gradient changes in salinity, dissolved oxygen content, and water temperature are too small to be sensed by fish sensory organs, and therefore, they might have little significance in migration navigation.

Suspended sediments in water can increase water turbidity, which is an important characteristic of water. Salmon, white bass (*Morone chrysops*), and sculpin (Cottoidea) usually avoid turbid water; carp (*Cyprinus carpio*) and catfish (Siluriformes) are the opposite and often migrate when water is the most turbid. Turbidity, together with other characteristics of water, can be sensed by fish and is conditionally associated with spawning migration. Suspended sediments, which are sensed by fish in freshwater rivers and the ocean, can move along certain routes and become a stable factor guiding fish to migrate from oceans to rivers.

Water Current

The main organs of fish that can sense water currents are eyes and some skin sensory organs, e.g., the lateral lines can sense water currents and guide the migration direction of fish. In general,

water current is a directional indicator of long-distance migration; in particular, larval fish depend entirely on water currents for passive migration. Existing studies have found that anadromous, oceanodromous, and potamodromous fish all rely on a sense of water currents for migration navigation. Examples include cod and Atlantic herring; chum salmon, which rely on the Amur current for navigation after arriving in the Sea of Okhotsk on their way to the Heilongjiang River for spawning; and flathead asp and roach, which rely on water currents for navigation.

Tropism of Fish

Under certain conditions, fish generally have positive electrotropism. Some studies have suggested that natural electric currents in the sea, which are generated by the Earth's magnetic field, may play a role in migration navigation. However, some researchers argue that it seems unlikely that fish can be directed by the natural electric current in the ocean because the electric current that causes positive electrotropism of fish must be four to nine times higher.

Temperature

Latitude plays a substantial role in the direction and route of fish migrations. Fish are poikilotherm animals that require an extremely strict temperature range for spawning; as a result, fish usually migrate along isotherms.

Topography

Many fish species may also rely on coastline and seafloor topography for navigation during migration. This may be related to water pressure perception to a certain extent.

Genetic Factors

Fish migration is heritable, and every species/population has its own characteristic migratory behavior. Migration heritability involves the nervous system and is a trait that began to be exhibited in ancestors during the initial formation stage of species and is shaped by constant selection in a long historical process. The variations generated during long-term historical evolution contribute to the formation of heritability. The stimulation of internal and external conditions can trigger a specific behavior, namely, instinct reactions, which serve as one of the main drivers of annual spawning, feeding, and overwintering migrations.

Cosmic Factors

Environmental hydrological factors play an important role in directing the migration of fish; in particular, periodic changes in ocean currents lead to periodic fish migrations. The periodic changes in ocean currents are related to periodic changes in geophysics and the universe, among which the change in heat obtained from the sun is the most prominent. Solar radiation is related to sunspot activity, which has an 11-year cycle. When sunspot activity increases, thermal radiation increases. As a result, the ocean absorbs a very large amount of heat, and the water temperature increases, leading to changes in temperature and in the strength of annual warm currents; consequently, this has a direct impact on the development and migration of marine fish.

The Biological Significance of Fish Migration

The migration behavior of fish gradually developed throughout long evolutionary processes and is the result of the long-term adaptation of fish to the external environment; therefore, it is bound to have certain biological significance. The existing studies generally believe that migration ensures favorable living and breeding conditions. Breeding migration is an adaptive behavior to ensure optimal conditions for the development of fish eggs and larvae, especially to protect them against predators in the early developmental stage. Feeding migration helps fish obtain abundant bait organisms to ensure high growth and development rates and maintain large population sizes. Overwintering migration is a unique behavior to ensure the most favorable abiotic conditions and adequate defense against predators when fish are in a low-activity low-metabolic-intensity state. Overwintering is characterized by decreased activity, complete cessation or a significant decline in feeding activity, and decreased metabolic levels; during overwintering, fish mainly

depend on the energy accumulated in the body to maintain metabolic functions. Overwintering is an adaptive behavior to ensure the survival of the population in the season unsuitable for activity.

Epipelagic fish can be used as an example. This type of fish usually migrates from offshore to coastal areas to feed and breed. Coastal areas have high water temperatures and strong currents as well as rich nutrients and organic substances, offering sufficient food for fish. In addition, compared to the vast open ocean, narrow coastal areas are much more conducive to interactions between male and female fish. A sharp temperature increase and sufficient food can shorten the developmental period of fish eggs, enabling them to better survive dangerous periods and hatch into larvae earlier. The water current from the mainland to ocean also affects the migration of these fish. Nevertheless, coastal areas are not always good habitats for fish. In winter, the sharp decrease in water temperature and food sources prompts fish to go to deeper seas, necessitating the process of overwintering migrations.

Anadromous migration is biologically significant for salmonids. The lack of bait organisms in rivers inevitably limits population growth if salmonids born in a river stay in the river without entering the sea to forage and fatten. This is clearly unfavorable to the survival and continuation of the population. Instead, long-distance migrations to seas rich in bait organisms can provide good nutritional conditions to ensure a stable population size. However, the ocean environment is unfavorable for salmonids spawning and breeding because salmonids bury eggs in gravelly riverbeds and the development of their eggs is slow; the oxygen level is relatively low in deep seas, and the gravelly area near the shore is subject to impact from strong waves. Therefore, remaining in rivers to breed is an adaptation for salmonids to ensure a higher survival rate of juveniles and maintain a large population size. However, this leads to another question: why do eels spawn in the sea rather than in rivers, where the conditions seem better for the development of fish eggs and larvae? According to existing studies, the spawning ground of European eels is in areas of the Atlantic Ocean with high salinity and the least number of predators that eat fish eggs and larvae, that is, the most suitable area for the development of eel eggs.

3.2.3 Vertical Migration of Fish

The rhythmic behavior of fish diurnally moving between the upper and lower water layers to feed or breed at different times of day is known as diurnal vertical migration or diel vertical migration, which is an adaptive behavior of fish to a living environment. The vertical migration of fish is directly related to fishing production. For example, trawl fishing is not effective for catching demersal fish when they move vertically from the sea floor to the middle and upper layers of water; in this case, purse seine or midwater trawl fishing should be applied. Similarly, when pelagic fish sink to deep water, purse seines are not useful for catching fish unless they are high enough to reach the depth of the fish shoals. Therefore, knowledge regarding the circadian rhythm of the vertical movement of fish and its changes is important for fisheries and is an important research subject in fish behavioral and fishery oceanography.

Characteristics of Vertical Migration

The range, speed, ascending time, and descending time of diurnal vertical migration vary among fish species. This variation is due to different biological traits and environmental conditions, especially differences in gonadal maturity, fullness, age of fish and light, temperature distribution, and food-related factors in habitats.

Most fish stay in deep water during the day, ascend around dusk, remain in the middle and upper layer of the water at night, and migrate to deep water at dawn, i.e., deep water during the day and middle and upper water layers at night. However, there are some species that exhibit the opposite pattern, such as *Navodon septentrionalis*. The diurnal vertical migration of fish does remain consistent during different life phases; for example, such migration is not observed during the breeding period of *L. polyactis*, the feeding period of *Navodon septentrionalis*, and the overwintering and feeding periods of Pacific herrings. Migrations also change with seasonal changes in the external environment. For example, in the Arctic, diurnal vertical migration only occurs in autumn when the day/night cycle is obvious; fish do not undertake diurnal vertical migrations and live only in a certain water layer during periods with polar days and polar nights. However, diurnal vertical migration does not change with the seasons in tropical waters.

The range of vertical migration varies with species and the hydrological conditions of habitats. For many fish species, the descending depth is related to the light distribution in the water. The migration depth is generally less than 500 m and might be only a dozen meters in shallow waters; however, it can be greater than 500 m in some cases. In addition to seasonal changes, migration depth is also related to moon phases, weather, and the physiological state of the fish. For example, *Navodon septentrionalis* descends less deep when there is a full moon (the 15th day of a lunar month) and deeper when there is a new moon (the first day of a lunar month) or dark night. Largehead hairtail descend deeper on sunny days and less deep on cloudy, rainy, or foggy days. The diurnal vertical migration of Pacific herring becomes more prominent, with a larger depth range, with the development of their gonads.

Vertical Migration Speed

In natural conditions, the speed of vertical migration is difficult to measure and generally calculated indirectly based on the rate of movement of the dense center of the fish shoal. Due to the impacts of gravity and light that gradually weakens from top to bottom, the diurnal vertical migration of aquatic animals generally exhibits a slow ascending and faster descending pattern, slower ascending before dusk than after dusk pattern, and faster movement at the beginning compared to the end of ascending or descending.

In addition, speed varies in different habitats and with age among fish species. Generally, tropical fish ascend faster than frigid- and temperate-zone fish. In natural waters, the actual migration speed is faster than the measured speed because the route is tortuous rather than straight up and down. The vertical movement may be limited by the distribution of water temperature and salinity, especially by thermoclines, which block many fish from crossing.

Influencing Factors

Existing studies have indicated that the main factors affecting the vertical migration of fish are as follows.

(1) Light conditions – Aquatic animals generally inhabit the water layer where light conditions are most suitable for their survival. Their vertical distribution changes correspondingly and is consistent with the diurnal rhythm of light. Some fish species, such as anchovy and Pacific herring, ascend to the surface during daytime because they need good light conditions to promote gonadal maturation during spawning.

(2) Food factors – Zooplankton have a remarkable diurnal vertical migration, ascending to the surface at night and descending during the day; correspondingly, fish move vertically in pursuit of bait. For many fish species, feeding activities are most intense in the evening and predawn hours. Some fish descend to deep water during the day to avoid predators that use vision for prey detection.

(3) Temperature factors – The behavioral habit of fish ascending to the upper layer of water in the night is associated with suitable conditions for food digestion. The higher temperature of the upper water layer accelerates the digestion process. Studies have found that a slight temperature change between 0.03 and 0.07 °C can stimulate fish to migrate. The internal tidal waves that cause periodic changes in water temperature also directly promote the diurnal or semidiurnal vertical migration of some pelagic fish species.

3.2.4 Research Methods for Fish Migration

Fish migration distribution is the main content of fisheries forecasting and fishery oceanography research, the purpose of which is to understand the pattern of fish migration, the relationship between fish migration and the marine environment, and the underlying mechanisms. The main methods used to study the distribution of fish migrations include exploratory fishing surveys, tagging, statistical catch analyses, direct reconnaissance with instruments, distribution model predictions, and analyses and prediction of trace elements and stable isotopes. Each of these methods has its own advantages and disadvantages. Statistical catch analyses are the most practical if detailed, accurate, and continuous long-term production data can be collected from a large number of fishing vessels. However, due to various factors, statistical data often have substantial errors and do not meet the requirements for the intended purpose. Survey vessels for specific missions can provide accurate and targeted data, but such operations are costly and time-consuming and can cover only a limited range. Tagging is a traditional method that provides the most intuitive and effective data. With the application of satellite remote sensing technologies and the development of electronic technology, such as the emergence of data-storage tags and pop-up satellite archival tags, tagging technology has boomed. Here, each of the research methods for studying fish migration is described below.

Statistical Catch Analyses

First, based on massive long-term data collected from fishing vessels, catch statistics are generated for fishing areas, fish species, and the early, middle, and late periods of each month. Subsequently, the catch distribution of each fish species can be mapped for each fishing area. Finally, the migration routes and distribution ranges of fish can be analyzed based on the map. The continuous development of spatial analysis methods (e.g., geographic information system) and marine monitoring technology has bolstered statistical catch analyses. Through long-term statistical catch analyses, fishing maps for all economic fish species can be obtained, which is of important reference value for analyzing fishing grounds and fishing seasons. Combined with satellite remote sensing data, the relationship between fish migrations and the marine environment can be preliminarily clarified. These methods have advantages of low cost and high effectiveness, but they require long-term fishing log data, extremely precise position information for the vessels, and records regarding the yield and biological traits of each species that has been caught.

The Tagging Method

The Definition of Tagging

Tagging refers to attaching a tag or electronic device to or marking a captured fish and then releasing it. Migrations are analyzed using the tagging and recapturing records. Tagging is an important tool in the study of fishery resources and fishery oceanography. This method has a long history and was used as early as the sixteenth century. With the technical development of sensors and satellite remote sensing, tagging techniques have improved, the number of tagging targets has increased, and the application scope of tagging has expanded. At present, in addition to economic fish, tagging has been used for other aquatic animals, including crab, shrimp, shellfish, and cetacean species.

In terms of the approach used to label fish, there are two types of methods: marking and tagging. Marking, one of the earliest methods, involves marking the body of a fish, e.g., removing all or part of its fins. Tagging involves attaching a special tag, which indicates the organization, date, and place of tagging, to an aquatic animal. Tagging is the most common method in modern operations. There are several types of tags, including external tags, internal tags, bio-remote sensing tags, data-storage tags, and pop-up satellite archival tags (PSAT) (Chen 2014, 2016).

The Significance of Tagging

When a tagged fish is released and then recaptured after a certain period of time, the time and location of tagging and recapture can be analyzed to trace the movement and growth of the tagged fish in water. Therefore, tagging is a common method used to investigate fishing grounds and study the migration distribution of fish shoals and the growth of fish. This method can also be used to estimate fish resources, which is of great significance to fishery production and management. The significance of tagging mainly manifests in the following aspects:

1. The data obtained from tagging provide information regarding the direction, route, speed, and distribution range of fish migrations. After a tagged fish migrates with a fish shoal and is recaptured in a certain sea area after a certain time, a comparison of the time and place between tagging and recapture can reveal the direction, route, distribution range, and speed of its migration. This method is the most effective way to directly estimate fish migrations. However, the migration speed calculated based on the distance from the tagging site to the recapture site can only be used as a conceptual reference and cannot be considered the exact migration speed.

2. Tagging can be used to estimate the growth rates for the body length and weight of fish via a comparison between the time of tagging and the time of recapture.

3. Tagging can be used to estimate the approximate catch ratio and decline rate as well as the resource quantity. When a large number of fish are tagged and released, there is a high chance that the tagged fish will return to the original fish shoal and may be recaptured later. Assuming the tagged fish are properly mixed into the original fish shoal, the recapture rate for the tagged fish should be similar to the catch ratio of the fish resource, namely, the ratio of the catch amount to the resource quantity. Therefore, the approximate catch ratio in a given fishing ground during a fishing season can be estimated using the total number of tagged fish and the number of recaptured fish, in

combination with corrections to the results. Subsequently, based on the total catch amount, the resource quantity can be estimated, which is a valuable reference for fishery production and management.

If the number of the tagged fish is X, the total number of fish captured during the fishing season is Z, and the number of the recaptured tagged fish is Y, the resource quantity N of the tagged fish species is calculated as follows:

$$N = XZ/Y$$

4. This method can be used to determine environmental indicators that are necessary for the formation of a fishing ground. By combining the measured environmental conditions and the environmental factors obtained by satellite remote sensing, the relationship between fish migrations and the marine environment can be analyzed to explore the environmental indicators necessary for the formation of a fishing ground.

Types of Tagging

External Tags

These tags are most commonly used to study growth, migration, variations in resources, and the effect of stock enhancement by release in natural sea areas. External tags with a clear color are pierced through or tied to an appropriate outer body part of the fish. This traditional method is simple and relatively low cost, but it has various shortcomings, and only a limited amount of effective data can be obtained.

To ensure the success of tagging, the design of external tags must consider the resistance force in water and tag material corrosion. At present, small metal tags with a plate- or pin-like shape are commonly used. Most of these tags are made of silver, aluminum, or plastic, with others made of nickel, stainless steel, etc. (Fig. 3.4). All tags are engraved with the code of the institution in charge of the project and the tag number; the information regarding the release site and time is

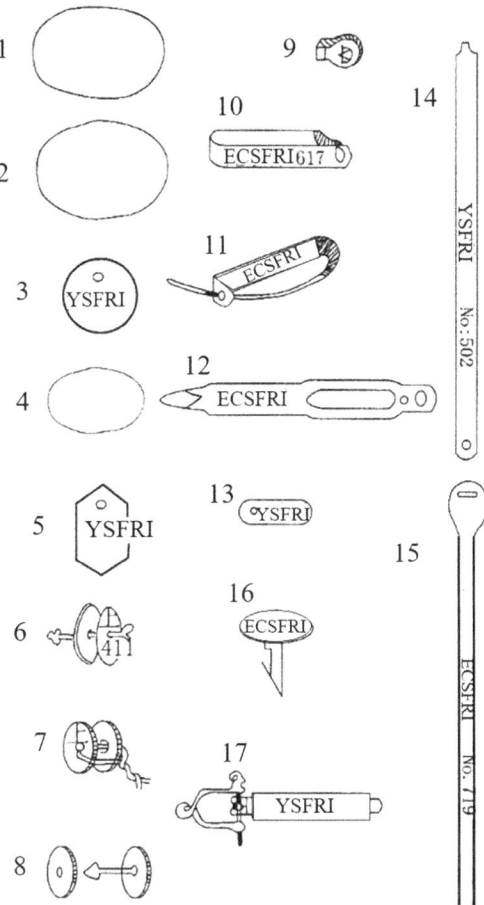

Fig. 3.4 Types of tags (Chen 2014, 2016). Yellow Sea Fisheries Research Institute (YSFRI); East China Sea Fisheries Research Institute 1–5: Hanging. 6–8: Button. 9–12: Toggle clamp. 13: Internal. 14–15: Belt. 16: Flip button. 17: Hydrostatic

logged in a record sheet, serving as the baseline data of the tagged fish after recapture. Tag position varies according to the shape of the fish (Fig. 3.5).

Isotopic Tags

Radioisotopes with a long radiation period (generally 1–2 years) and harmless to fish are introduced into fish as tags, and recaptured tagged fish are detected using an isotope detector. Currently, the most used isotopes are P^{32} and Ca^{43}. There are two ways to introduce radioisotopes

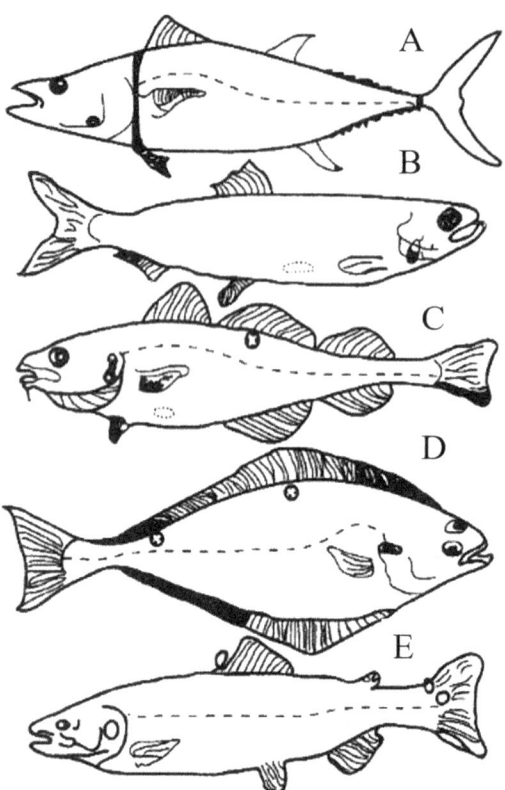

Fig. 3.5 Tag positions in fish of different sizes (Chen 2014, 2016). (**a**). Tuna; (**b**) herring; (**c**) cod; (**d**) righteye flounder; (**e**) trout

into fish: one is to feed fish with bait containing radioisotopes, and the other is to put fish into water containing radioisotopes for direct infusion. These methods are easy for tagging and releasing fish, but the tagged fish are difficult to recapture because they are difficult to detect.

Bio-remote Sensing Tags

These tags are remote sensors. An ultrasonic or electric wave generator is installed on the fish body, and the tagged fish can be tracked and recorded by a survey ship equipped with a receiver. The tagged fish are under continuous observation, tracking migration routes, migration speed, and diurnal activity patterns. This method is simple and can be used to record the activity patterns of fish in detail; however, this type of tag generally does not remain attached for a long period of time.

Data-Storage Tags

These devices are controlled by a microcomputer and record fish activities. A data-storage tag installed in the body cavity of a captured fish is activated every 128 s after the fish is released. Consequently, 675 records from multiple sensors are obtained daily, reporting water pressure, light intensity, and internal and external temperature data. Based on the recorded data, the geographical location of the fish can be calculated and stored in the tag. Tags retrieved from recaptured fish can provide detailed information regarding migrations and vertical movements of fish. The challenge of this method is recapturing fish and retrieving the tags.

Installation of data-storage tags: Small tags can be installed in the body cavity, and large tags can be inserted into the dorsal muscle close to the first dorsal fin. Real-life experience has confirmed that muscle insertion is faster and less dangerous than is the placement of a tag in the body cavity.

PSAT

A pop-up satellite archival tag is composed of streamlined pressure-resistant housing made of epoxy hydroxy resin and equipped with an antenna, a corrosion-resistant detachment device, and a floating ring that forces the antenna to stand vertically once the tag has detached from the tagged fish (Fig. 3.6). A microprocessor installed in the tag collects temperature and depth data as soon as the fish is released and can continuously record data for more than 1 year. After detaching from the fish body, the tag transmits archived data to the ground receiving station via one of the Argos satellites. The operation principle is shown in Fig. 3.7.

The main components of a PSAT include a clock, sensors, a control and archival system, a floating control system, a power supply device, and housing. The clock provides time control. The sensors, which commonly included temperature sensors, pressure sensors, and light-level sensors, can obtain various environmental parameters. The floating control system controls the tag after releasing from the fish body, and its

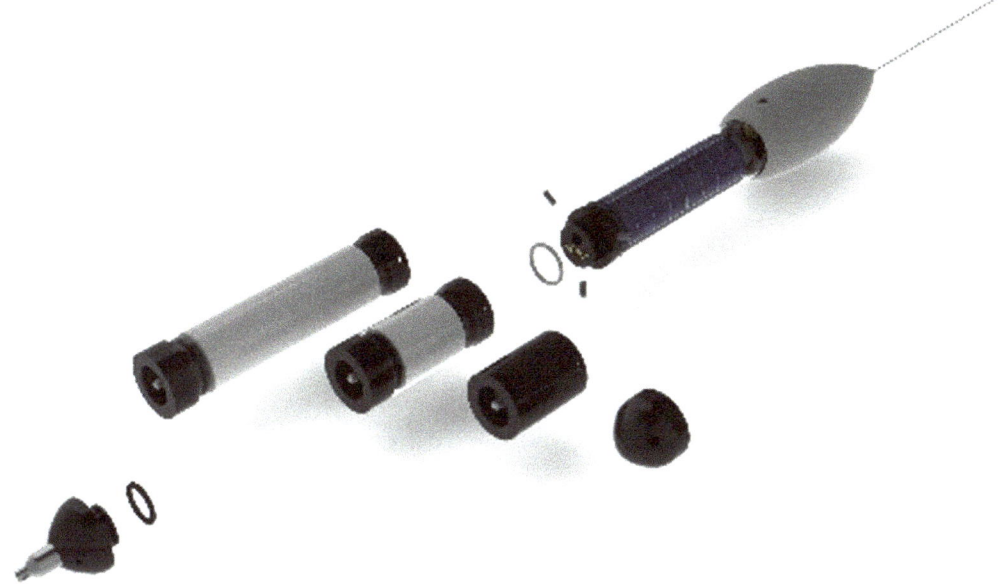

Fig. 3.6 Schematic diagram of a PSAT (PAT model)

operation indicates the completion of the release process. The floating control system is mainly composed of a floating ring and an antenna to ensure that the tag can communicate with an Argos satellite. The power supply device provides power to the tag during the entire operation, from data acquisition to communication with the satellite after the tag floats to the surface;

Fig. 3.7 Illustration showing the principle of using pop-up satellite archival tags (PSATs) to track fish

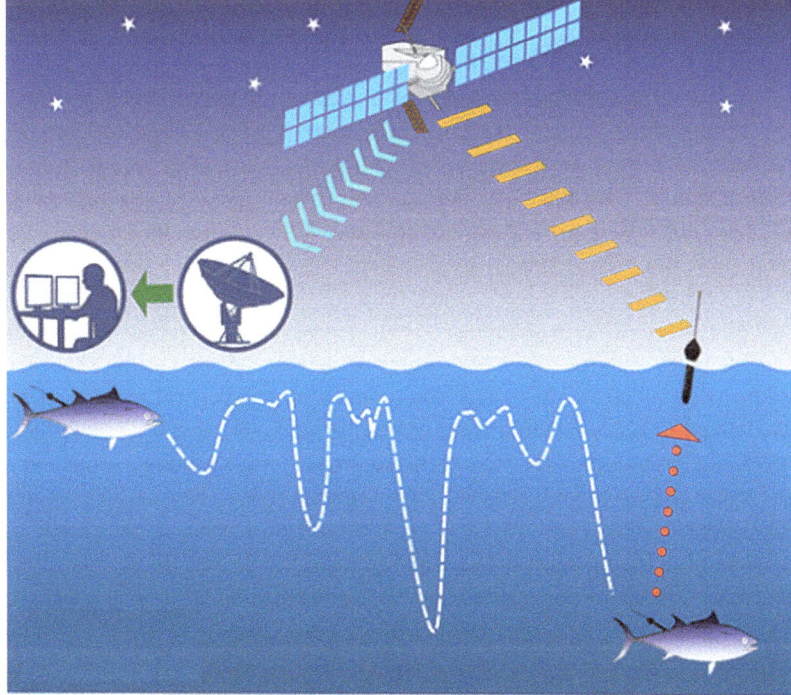

Table 3.2 Comparison of the PTT-100 archival pop-up tag and the pop-up archival transmitting tag (Chen 2021)

Company	Microwave Telemetry, Inc.	Wildlife Computers
Product name	PTT-100 archival pop-up tag	Pop-up archival transmitting tag
Maximum water depth	2000 m	1000 m, 1750 m
Water temperature	0–35 °C	−40 to 60 °C
Longest deployment duration	500 days, 12 days, 24 days	1.5 years
Reading interval	1 h, 2 min, 4 min	Preset by user, 1 s to 9 min
Pop-off date	Predefined at purchase	Predefined by user
Abnormal release	Depth is unchanged for 4 days (predefined)	The depth is greater than 1000 m, or the depth is unchanged for 24 h to 96 h (predefined by user)
Data acquisition	The data are processed by the manufacturer to a readable format and then sent to the user	The data are processed by user
Data	Water depth, water temperature, and light-based geolocation estimates	Graphical water depth, water temperature, and light level

therefore, the power system is long-lasting and has a high capacity. The control and archival device is the central system that controls the normal operation of all the abovementioned parts. The housing is usually made of epoxy hydroxyl resin that is resistant to corrosion and high pressure, and it is made in a streamlined shape to reduce drag during movement. At present, the companies producing PSATs mainly include Microwave Telemetry, Inc. and Wildlife Computers of the United States, both of which use Argos satellites for data transmission (Table 3.2).

PSATs have been widely used in studies on the large-scale movement (migration) of marine animals, such as marine mammals, seabirds, sea turtles, sharks, and tunas, and the physical traits of their habitats (e.g., water temperature). For example, from September to October 1997, PSAT technology was used to track tuna for the first time in the North Atlantic Ocean. Twenty bluefin tunas were released after being tethered with a PTT-100 PSAT, and the data were predefined to be transmitted from March to July 1998. Seventeen tags were recovered, from which the collected data were successfully extracted. The recovery rate was 85%, and an average of 61 days of recorded data was obtained from each tag. This tagging project provided valuable data, including vertical and horizontal distributions, migration direction and route, and the water

temperature of tuna habitats during different time periods.

The use of PSATs can provide a large amount of data on the tagged organism, including migration distribution and movement speed, diurnal vertical movement patterns, patterns of habitation in different water layers and the optimal water layer, the relationship between habitat distribution and temperature, suitable water temperature, optimal water temperature, etc. In addition, it can provide a scientific basis for more accurate assessments of fish resources.

Distribution Model Prediction

Data-based species distribution models (SDMs) are effective tools for analyzing the relationship between species and habitats and predicting the migration range of fish. Based on distribution (presence/absence) and environmental data, SDMs can estimate the preference of the study subjects for each environmental factor using a specific algorithm and express the preference level in the form of probability. The results can be used to explain the probability of the presence of the study subject in a habitat and the suitability of a habitat, as well as other related aspects. SDM studies on species distribution can be traced back to the 1920s. Initially, it was used to investigate the relationship between plant communities and environmental gradients. Most researchers believe that the application of SDMs to the distribution and spreading of aquatic organisms started from research on the seasonal and vertical distribution of planktonic foraminifera in the 1980s, noted as the first case of application in the field of marine biological research. In the 1990s, the rapid development of computer technology and the continuous in-depth study of statistical science further expanded research on distribution models. Many programming languages and computing software programs provided strong support for the internal computing of distribution models. In the twenty-first century, the rapid development of computer science has enabled the free sharing of SDM computer programs (packages), further promoting SDM-related studies.

There is no unified standard for SDM classification. The models used are divided into statistics-based models (e.g., generalized additive models and generalized linear model), artificial intelligence models (e.g., artificial neural network), and machine learning models (e.g., random forest, support vector machines, and classification and regression tree).

Prediction of Migration Based on Trace Element and Stable Isotope Analysis

Trace Element Analysis

Fish organs, such as otoliths, can record habitat "fingerprints" and are often used to elucidate the life history of individual fish. The migration distribution of fish or other marine animals can be estimated by analyzing fish tissues (e.g., otoliths) using the following procedures. First, the distribution of a trace element in a cross section of ground hard tissue is determined by using laser ablation inductively coupled plasma mass spectrometry (LA-ICP-MS). Subsequently, the ratio of the trace element to calcium (e.g., Sr/Ca) is used to establish a distribution model of the trace element for the whole time series from the core area (formed at the time of birth) to the marginal area (formed around the time of death) of the hard tissue section. Therefore, the period from birth to death (fishing- or sampling-related death) of the fish can be deduced based on the sampling time and age. Finally, combining environmental data,

e.g., water temperature, salinity, and element concentration, during the same time period as well as the relationship between the deposition of the measured element and environmental factors, such as water temperature and salinity determined in the laboratory or in the field, the probability that fish (shoals) appear in water areas with certain environmental conditions (e.g., high/low temperature, high/low salinity, and high/low element concentration) during its main life stages from birth to death can be determined. For example, Zumholz (2005) used LA-ICP-MS to analyze the time series of nine trace elements in the otoliths of *Gonatus fabricii*. The Ba/Ca changes confirmed that juvenile *Gonatus fabricii* lived in surface water and adults lived in deep water. Furthermore, gradual increases of U/Ca and Sr/Ca from the center to edge of otoliths suggested that adult *Gonatus fabricii* migrated toward cold-water areas.

Among all elements, Sr and Ca have been used most widely. Most existing studies report that otolith Sr/Ca is positively correlated to water salinity and that seawater has a largely higher Sr concentration than does freshwater. Therefore, Sr/Ca has been widely used to infer the migration of marine animals (e.g., fish) between freshwater and seawater. In China, studies have mainly focused on fish species, including the genus *Coilia*, the genus *Anguilla*, largehead hairtail, and tuna, as well as other aquatic animals such as cephalopods.

Stable Isotope Analysis

Stable isotope analysis has gradually been applied to the prediction of the migration trajectory of fish and other marine animals. Stable isotopic imprints in tissue reflect the food web in a water habitat because the stable isotope content in the food web exhibits spatial heterogeneity between habitats due to variations in biochemical processes. Fish and other marine animals that migrate between food webs with different isotopes retain stable isotope information related to their previous feeding grounds. The creation of isotopic imprints is related to the efficiency of the conversion of chemical elements by tissues. The stable isotopic content in a tissue sample or in a tissue cross section can be measured by a thermal ionization mass spectrometer or a laser ablation plasma mass spectrometer, respectively, and compared with the content in water areas to determine the possible geographical locations where the fish (shoal) has lived. The stable carbon isotope ratio (δ ^{13}C) reflects the spatial difference of the environment, and it is mostly used to indicate the migration experience of individual fish (shoals) in open-sea/nearshore waters, seafloor/surface waters, and high-/low-latitude waters. In addition to δ ^{13}C, the stable oxygen isotope ratio (δ^{18}O) is an effective supplementary tool to indicate the temperature and salinity changes that a fish (shoal) has experienced. This method can be used to extract information from various information carriers, including otoliths, fin rays, bone, and muscle, and is mostly used for studying the migration trajectories of fish and other marine animals between coastal and offshore waters and between oceans and rivers.

3.3　Characteristics of the Marine Environment and Its Impact on Fish Distribution

3.3.1　Characteristics of the Marine Environment

Fish and other animals are widely distributed in the oceans, and their spatial distribution is closely related to the marine environment. There are three environmental gradients in oceans, i.e., the latitudinal gradient from the equator to the poles, the depth gradient from the sea surface to the bottom, and the horizontal gradient from the coast to the open ocean. These gradients have significant impacts on the life of marine fish species and the spatiotemporal distribution of fish productivity.

In addition to the latitudinal gradient from the equator to the poles, the solar radiation intensity gradually decreases and exhibits a more remarkable seasonal difference. The difference in daily light duration directly affects the seasonal difference in photosynthesis and the thermocline between sea areas at different latitudes. The depth gradient is mainly caused by the light that penetrates the surface layer of the water, with deeper layers having only faint light or no light. Temperature also varies vertically as it is higher in the surface layer due to solar radiation and lower and more stable in the bottom layer. In addition, pressure increases with depth, and organic food is scarce in deep water. In the horizontal direction, the gradient from the coast to open ocean mainly involves changes in water depth, nutrient content, and seawater mixture, as well as the weakening fluctuations of other environmental factors (e.g., temperature and salinity). These gradients have a substantial impact on the shoaling, horizontal distribution, and vertical distribution of fish. Dense fish shoals often emerge in sea areas with drastic environmental gradients, thereby forming fishing grounds.

Compared with land, the marine environment is relatively stable. Because they are large bodies of water and have high specific heat and seawater mixing effects, oceans have relatively even heat distribution, resulting in only small temperature differences and slow temperature changes. In addition, the composition and pH value of seawater are relatively stable. These environmental conditions are similar within a considerably large range, allowing marine fish to distribute throughout a broad range and form dense shoals in suitable areas. Furthermore, due to the contact between the ocean surface and the atmosphere and the O_2 produced by photosynthesis, the O_2 content is basically saturated in the surface layer but varies largely in deep water, potentially affecting the distribution of fish. After cooling and descending, the surface water at high-latitude areas flows to the sea floor at low-latitude areas, through which surface water with high O_2 content can reach the bottom, enabling the presence of marine life at all depths of the ocean.

3.3.2　The Impact of the Marine Environment on Fish Distribution

Marine Environmental Factors
Environment is defined as the sum of natural conditions surrounding a given subject, but not the organisms contained therein. It includes the

space where organisms inhabit and various environmental factors that directly or indirectly affect organisms. These factors interact with each other and form a system.

Environments include major environments and small regional environments. A major environment refers to the combination of various natural factors, including atmospheric circulation, geographical latitude, sea and land distribution, and large-scale topography, in a large area. A collection of all organisms living in a major environment is called biome. A small regional environment refers to a small-scale regional ecological environment where a specific fish species or biological population or community reside, also called a habitat.

Environmental elements that directly or indirectly influence the growth, development, reproduction, behavior, clustering, and distribution of organisms are called ecological factors. Ecological factors exert varying degrees of effects on fish and other organisms. Some of them, such as food and physical and chemical factors that affect metabolism, intra- and interspecies relationships, and living space, play decisive roles in the survival and reproduction of fish and other marine organisms. Some of them exert only indirect rather than direct impacts on a specific species of fish or other marine organisms. For any fish species, there are no completely irrelevant factors in the surrounding environment because all factors are interrelated and affect each other. Ecological factors comprehensively affect the spatial distribution of fish as well as the formation of fish shoals and fishing grounds.

The external environment of fish involves various abiotic and biotic environmental factors. Abiotic environmental factors refer to water bodies with different properties, physical and chemical factors of water, and abiotic environmental conditions caused by human activities, including temperature, salinity, light, ocean currents, seafloor topography, sea floor sediments, meteorology, etc. Biotic environmental factors of fish refer to bait organisms and the interspecific relationships between fish and other animals/plants that share the same environment and

among fish themselves. Most organisms living in the same environment with fish are fish food, while some may prey on fish. Understanding the influence of external environmental factors on the behavior of fish can provide a basis for analyzing fishing conditions, identifying fishing grounds, and improving fishing gear and fishing methods.

The following are the main effects of marine environmental factors on fish distribution and their characteristics.

Comprehensive Effects

Environmental factors in the ocean are interrelated and interact with each other; therefore, it is necessary to analyze their comprehensive influence on the distribution of fish and other marine organisms. For example, various environmental factors including ocean currents, water temperature, and salinity comprehensively affect the distribution and shoaling of fish.

Effects of Dominant Factors

There are many environmental factors in the marine environment, but their effects on fish and other marine organisms are different. The factors that play decisive roles in the living, growth, and development of fish and other marine organisms are called dominant factors, which change at different life stages of fish. For example, temperature is the dominant factor affecting spawning and individual death in the spawning stage of many marine fish, and food supply after hatching becomes the dominant factor for the survival of larvae.

Direct and Indirect Effects

Marine environmental factors may directly and indirectly affect fish distribution. For example, the free CO_2 content in seawater itself generally does not directly affect the growth and distribution of fish, but it alters the pH value of the seawater through changing the equilibrium process of the CO_2 system; as a result, the increase in CO_2 content in seawater indirectly affects the spatial distribution of fish through a decrease in the pH value.

The Limiting Mechanism of Marine Environmental Factors

For any fish species, a continuous supply of necessary substances from the environment for growth and reproduction and appropriate physical and chemical conditions are essential for survival and reproduction in an environment. The marine environmental factors that approach or exceed the tolerance limits for a fish species and hinder its survival, growth, reproduction, or spread are called limiting factors.

Marine environmental factors may exert different impacts on fish due to property differences (e.g., temperature, salinity, and light). In addition, the survival, reproduction, and distribution of fish are affected if the level of an environmental factor is too high or too low. In other words, a fish species can only adapt to a marine environmental factor within a certain range. The range between the lowest and highest levels that fish can tolerate is called the limit of tolerance. The marine environment where an environmental factor is close to or even beyond the limits can affect the growth and development of fish or even cause death. As a

result, fish can only survive in the marine ecological environment within the tolerance limits. The principle regarding the limiting effect exerted by the maximum and minimum levels is Shelford's law of tolerance. Fish can avoid tolerance limits through changing the range of activities, such as overwintering migration. The relationships between the law of tolerance and the distribution of fish shoals and population levels are shown in Fig. 3.8. Fish have an optimum range for each marine environmental factor within the zone of tolerance, e.g., an optimal temperature, which allows the fish to consume the least energy while maintaining normal metabolism. In addition, the limits of tolerance of fish for marine environmental factors are often not fixed; instead, they change with age and other conditions in the life history of fish. Most fish have a lower tolerance during breeding periods as well as the egg, embryo, and larval stages.

Limits of tolerance reflect the adaptability of fish to changes in the marine environment. Some fish species can adapt to environmental changes within a large range, while others can only adapt

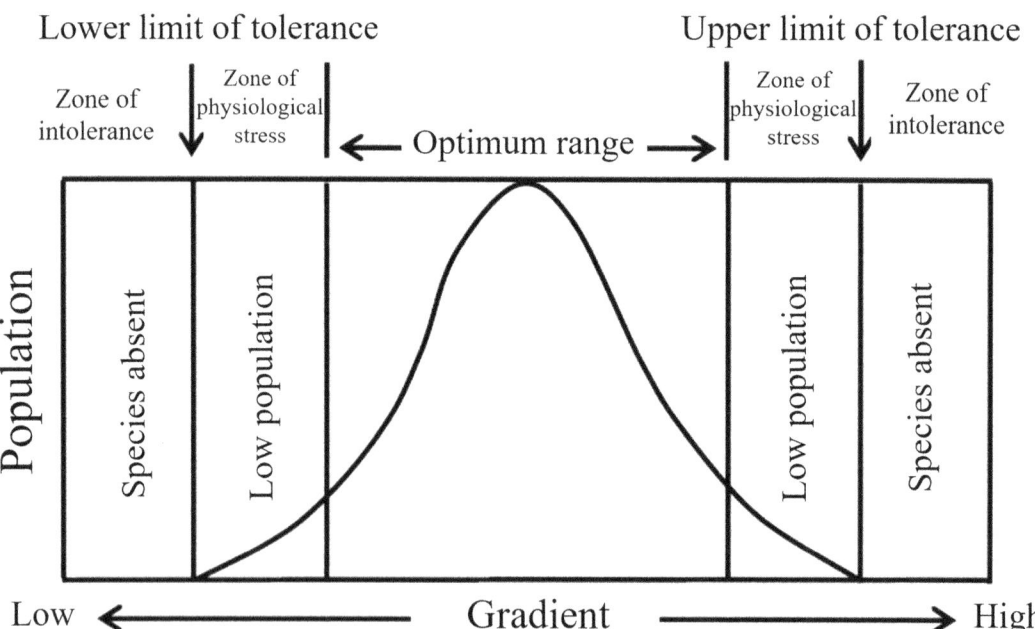

Fig. 3.8 The relationships between the law of tolerance and the distribution of fish shoals and population (Shelford 1911)

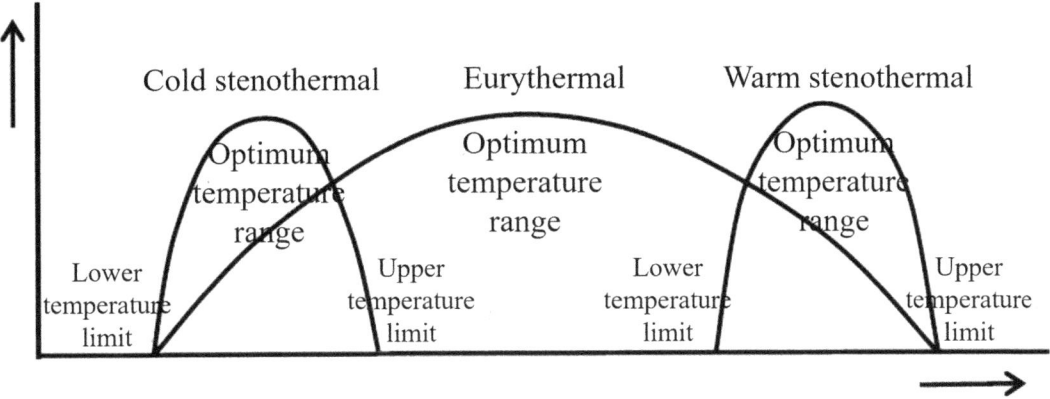

Fig. 3.9 Comparison of the marine environment ranges in which eurythermal and stenothermal fish species can adapt (Sun 1992)

to environmental changes within a small range. The species that are widely distributed in a variety of environmental conditions and adaptable to a wide range of marine environmental changes are called eurytopic species; conversely, species adaptable to a narrower range are called stenotopic species. In terms of temperature, there are eurythermal and stenothermal species. For salinity, there are euryhaline and stenohaline species. Fish that live in an environment with dramatic changes can tolerate a wider range of marine environmental changes. For example, species that live along the coast are more adaptable to changes in temperature and salinity than species that live in the open ocean (Fig. 3.9).

General Effects of Marine Environmental Conditions on Fish Behavior

The migration, distribution, and movement of fish are determined by their adaptability and tolerance limits to marine environmental factors, which therefore should be the focus of relevant relationship studies. When the conditions of the external environment necessary for the survival and activities of fish change, the activities of fish will change accordingly to adapt to the new environmental conditions. Although environmental changes inevitably affect the behaviors of fish such as feeding, reproduction, migration, movement, and shoaling, the impact primarily depends on the status of the fish itself, including size, life

stage, and physiological conditions. Concurrently, fish activities can affect the changes in environmental conditions. In addition to the mutual impact between fish and various environmental factors, the factors are interrelated and interact with each other. Therefore, the relationship between fish and the environment is a unity of opposites where the two influence each other and are dynamically balanced.

In general, the relationship between abiotic environmental factors and fish behavior is as follows: fish are often scattered and unlikely to form dense shoals in areas where the sea conditions are stable or uniform for a long time. Fish shoals and good fishing grounds can be formed only when environmental conditions (such as water temperature) form a sharp spatiotemporal distribution gradient or change dramatically. Therefore, the relationship between abiotic environmental factors and fish behavior can be summarized as follows: "changes cause fish to move, and the fish move to form shoals." This relationship is expressed by the following equation:

$$D = f_n(S, \Delta S, S')$$

where N represents the shoaling of fish, S represents a group of environmental factors, ΔS is the spatial gradient distribution of environmental factors, and S' is the rate of change of environmental factors.

3.4 The Relationship Between Fish Behavior and Temperature

Temperature is the most important physical factor of the marine environment. Seawater temperature fluctuates in a 37 °C range, between 35 and −2 °C, while land temperature fluctuates in a 130.5 °C range, between 65 and −65.5 °C. Even a temperature fluctuation of only a few degrees is significant in the ocean; in particular, in the sea areas within the frigid zone and near the poles, a water temperature change of 0.1 °C would be considered remarkable. Water temperature changes drastically affect fish shoaling and migration and the formation of fishing grounds; it is probably true to say that all habits of fish are directly or indirectly affected by changes in water temperature in every season of the year. Therefore, water temperature can provide critical information for reconnoitering fish shoals and searching for central fishing grounds.

3.4.1 Response of Fish to Temperature

Fish are poikilotherms, commonly known as "cold-blooded animals." Almost all the heat generated in their bodies is released into the environment. Due to a lack of regulatory mechanism, the body temperature of fish changes with the environmental temperature and often remains roughly the same as the ambient temperature. Nevertheless, the temperature of a fish is not exactly equal to its ambient temperature. Generally, the body temperature of fish is slightly higher than the external water environment, generally not exceeding 0.5–1.0 °C.

In general, the body temperature of fish changes with the ambient temperature. A large number of studies have confirmed that the body temperature of highly active pelagic fish is generally higher; for example, the body temperature of tuna is usually 3–9 °C higher than its external water temperature. A general explanation is that these fish have a structure similar to a heat exchanger. However, thermoregulation abilities vary significantly among different species. For example, tuna can be classified into two types based on significant differences in their thermoregulation ability: warm-water species, including skipjack tuna (*Katsuwonus pelamis*), yellowfin tuna (*Thunnus albacares*), blackfin tuna (*Thunnus atlanticus*), etc., which mainly inhabit the water layer above the thermocline in the tropical ocean and cold-water species, such as bigeye tuna (*Thunnus obesus*), albacore (*Thunnus alalunga*), and southern bluefin tuna (*Thunnus maccoyii*), which inhabit the higher-latitude sea areas or below the thermocline in tropical waters. Researchers have concluded that the body temperature of fish can indirectly reflect the water temperature, thereby providing scientific evidence for the search of fishing grounds and the reconnaissance of fish shoals.

3.4.2 Adaptation of Fish to Water Temperature Changes

With the changes in water temperature in oceans, the body temperature of fish changes, and fish develop adaptations to the change. However, the adaptability of fish to water temperature change is limited. Based on adaptability, fish can be classified as eurythermal and stenothermal, with most fish species being stenothermal. Generally, coastal or anadromous fish have the widest temperature range suitable for living, followed by inshore fish and oceanic or demersal fish, which live in the narrowest temperature range. Tropical and subtropical fish are more stenothermal than temperate- and cold-water fish. Stenothermal fish can be further divided into cold and warm stenothermal species. Warm-water fish species mainly reside in tropical waters, with some residing in temperate waters, while cold-water fish are common in cold and temperate waters.

Fish have an upper limit (highest temperature), lower limit (lowest temperature), and optimal water temperature range in which they can live. The limits of tolerance and the optimal range of temperature for living vary among fish species and among life stages of fish of the same species.

In general, the optimal temperature is closer to the upper limit and further from the lower limit of tolerance. Temperature changes usually prompt fish to actively migrate to an environment with an optimal temperature and avoid detrimental water temperatures so that they can maintain their body temperature within a certain range. This is known as behavioral regulation of body temperature. The overwintering migration of fish is mainly caused by a decrease in water temperature.

Water temperature in the sea has a significant impact on the distribution of fish, leading to a close relationship between fish distribution and water isotherms. Based on the adaptability of fish to water temperature, fish species living in the upper water layer of ocean can be classified into three types.

1. Warm-water species: These fish generally grow and reproduce within a temperature range higher than 20 °C, and the monthly average water temperature in their natural distribution area is higher than 15 °C. These fish are predominantly tropical and subtropical species. The temperature range suitable for living is higher than 25 °C for tropical fish and 20–25 °C for subtropical fish. The southern part of the South China Sea, the eastern part of the East China Sea, and the eastern coast of Taiwan are tropical sea areas. The western and northeastern coasts of the East China Sea and the northern part of the South China Sea are subtropical sea areas. In the sea area south of Hainan Island, warm-water species are dominant; these fish originate from tropical sea areas near the equator and spread to mid-latitude sea areas mainly due to warm currents (e.g., the influence of the Kuroshio Current and its branch currents).

2. Temperate-water species: These fish generally grow and reproduce in a wide temperature range, from 4 to 20 °C. The monthly average water temperature in their natural distribution area varies greatly from 0 to 25 °C. These fish include cold-temperate and warm-temperate species. The temperature range suitable for

living is 4–12 °C for cold-temperate species and 12–20 °C for warm-temperate species. The Bohai Sea and Yellow Sea in northern China are warm-temperate sea areas and contain many warm-water species that originate from mid-latitude temperate sea areas and spread both north and south.

3. Cold-water species: These fish generally grow and reproduce in temperatures lower than 4 °C, and the monthly average water temperature in their natural distribution area is 10 °C or below. These fish include cold-zone species and subcold-zone species. The temperature range suitable for living is approximately 0 °C for cold-zone species and 0–4 °C for subcold-zone species. Although there are no cold, subcold, or cold-temperate zones, the coastal areas of the Bohai Sea and the Yellow Sea have very low water temperature due to the influence of continental climate and coastal currents in winter and, therefore, are resided with some cold-water species. These cold-water species originate in the polar oceans and adjacent cold sea areas and spread to mid-latitude sea areas mainly due to cold currents.

3.4.3 The Optimal Temperature Range for Fish Species

Fish thrive in their optimal temperature range and exhibit limited activities or even die when the water temperature is too high or too low. Therefore, fish always actively migrate to an environment within their optimal temperature range to avoid detrimental water temperatures. As a result, large numbers of fish are distributed in environments within their optimal temperature range. This pattern is extremely important to the marine fishing industry. Previous research has indicated that fish prefer a select water temperature range, which varies with their ability to adapt to temperature changes. Furthermore, it can be inferred that the selected water temperature range of fish also changes with other environmental factors and the biological status of fish. In other words, selected water temperature ranges are not fixed even for fish of the same species.

In a natural environment, fish select a certain water temperature to live. Due to the complex influence of various environmental factors and changes in the biological status of fish themselves, the selected water temperature is often a range rather than a fixed temperature. This range is considered the optimal temperature range for fish species from the behavioral perspective. From the perspective of fishery production, the water temperature at which the yield of a certain fish species is high is considered the optimal temperature for this fish species; this temperature occurs within a range rather than at a fixed level and corresponds to the temperature range most suitable for living per fish species.

The experience gained from fishing indicates that both the temperature range suitable for living and the width of the range vary among fish species. Existing studies have revealed that the suitable temperature range is generally 9–26 °C for *Pseudosciaena crocea* (known as large yellow croaker), 6–20 °C for *L. polyactis*, and 10–24 °C for largehead hairtail in the inshore waters of China.

For fish of the same species, the suitable temperature range also varies among different habitats and populations. For example, the suitable temperature ranges are 14–22, 18–24, and 18–26 °C, and the optimal ranges are 16–19.5 °C, 19.5–22.5 °C, and approximately 22 °C for *P. crocea* of the Daiqu stock in the East China Sea, the Min-yue stock, and the Nao-zhou stock, respectively, during their spawning period.

The suitable temperature range varies among different life stages of adult fish (Table 3.3). For example, for *P. crocea* in the coastal waters of Zhejiang, the suitable range is 14–22 °C for

spawning and 9–12 °C for overwintering offshore of Zhoushan. For *P. japonicus*, the optimal water temperature is 14–12 °C for spawning in the Yanwei fishing ground, 17–19 °C for postspawning foraging in the sea area of Haiyangdao, and 8–9 °C or above for overwintering. In summary, the temperature range suitable for fish varies among different species, populations, habitats, developmental stages, and life stages. Therefore, investigating and identifying the suitable and optimal ranges of water temperature for fish species is critical for exploring fish behavior, determining central fishing grounds, and predicting fishing seasons.

However, although water temperature can indicate the possibility of the existence/appearance of a fish species, a water area within a suitable temperature range for a given fish species does not necessarily contain that fish. For example, common small sardines can be found in sea areas at a temperature of 6–22 °C; however, in reality, they inhabit less than 1/10 of the ocean area within this temperature range, which accounts for approximately 3/8 of Earth's oceans. Therefore, water temperature should be used in combination with other relevant environmental factors and the biological characteristics of fish to predict the existence/appearance of a fish species. Nevertheless, the use of water temperature information can certainly improve the predication accuracy regarding the distribution of fishing grounds and the scope of fish shoals.

3.4.4   Impact of Water Temperature on Fish Shoaling and Migration

Seasonal changes in ocean temperature lead to the periodic migration of fish, i.e., they migrate south

Table 3.3 The suitable water temperature ranges for different fish species during different life stages (Unit: °C) (Chen 2021)

Fish species	Overwintering period	Spawning period	Foraging period
L. polyactis	8–12	12–14	16–23
Largehead hairtail	14–21	14–19	8–25
P. japonicus	12–15	12–18	19–23
Scomberomorus niphonius (S. niphonius)	8–14	10–12	15–18
Black carp	5–8	2–3	8–12

or north and offshore or to deep sea areas in the Northern Hemisphere. Therefore, water temperature is an important factor affecting fish migration. For example, the gradually increasing water temperature in the spring prompts *Fenneropenaeus chinensis* (known as Chinese white shrimp) and *L. polyactis* that inhabit overwintering grounds in the central and southern Yellow Sea to migrate north in shoals, most of which pass the Shandong Peninsula and enter spawning grounds in the Bohai Sea.

The beginning of fishing seasons, the size of fish shoals, and the length of fishing periods are often closely related to the water temperature in fishing grounds. Therefore, before the fishing season, water temperature can be used as an indicator to predict the area and time for fishing. Taking *P. crocea*, which migrate to the inshore of Zhejiang to spawn, as an example, when the water temperature (the top 5-m water layer) increases to 13–16 °C, the first batch of fish shoals begin to appear and spawn, indicating the beginning of the fishing season; when the water temperature increases to 17–19.5 °C, fish shoals are dense and large, indicating the peak of the fishing season; and when the water temperature increases to 22–23 °C, fish complete spawning, and the fishing season nearly ends. In this case, water temperature can be used as an indicator to effectively predict the time and area of fishing. In China's inshore fishing grounds, the fishing season generally comes earlier when the water temperature rises quickly and ends earlier when the water temperature drops quickly.

The movement and aggregation of fish shoals are closely related to the horizontal water temperature gradient. The best fishing grounds are usually located where two different water systems converge or areas with steep horizontal water temperature gradients; in particular, fish shoals are highly dense in water areas where the isotherm curves and presents a pouch (front)-shaped distribution. Generally, in a fishing ground, fish shoals are more concentrated in areas with a steep horizontal temperature gradient and relatively scattered in areas with mild horizontal temperature gradients. For example, Pacific saury, tuna, and sardines distributed in the seas near Japan often shoal where the Kuroshio (warm current) and Oyashio (cold current) currents converge, forming a central fishing ground mostly in area where the surface water temperature gradient is the steepest.

The effect of water temperature on fish is particularly obvious during the pre-spawning and spawning periods. Maturation and spawning require a certain temperature range. Generally, the temperature range for spawning (reproduction) is narrower than that for survival. Within the temperature range suitable for maturation and spawning, a higher water temperature accelerates gonad maturation. Water temperatures exceeding the suitable range hinder gonad development or lead to an inability to spawn. Therefore, the time when fish appear and spawn in a spawning ground is often determined by water temperature. Abnormal water temperature in a spawning ground would force fish to leave and go to nearby areas where the water temperature is suitable. Furthermore, a long-term change in water temperature would lead to a northward or southward shift of spawning grounds (fishing grounds). Water temperature also has a significant effect on the shoaling behavior of fish during the process of spawning migrations. For example, the temperature range suitable for *P. crocea* spawning is 14–22 °C in the inshore area of Zhejiang; when the temperature of the fishing ground reaches 16–19.5 °C, fish shoals become concentrated, i.e., the peak of the fishing season. Therefore, the water temperature during the peak season when fish shoals are dense is within the optimal temperature range for fish shoaling.

In addition to bait, the intensity of fish foraging is directly associated with water temperature. Water temperature below the optimal level lowers the ability of fish to forage. Similarly, water temperature that is too high also decreases the feeding activity of fish. The foraging intensity of many fish species, especially those in temperate zones, shows seasonal variations and is closely related to water temperature changes. The foraging activity of fish in temperate zones is intensive in spring and summer and stops or significantly declines in winter. In contrast, the foraging intensity of cold-water fish decreases at higher temperatures. For example, only 3.2–4.9% of herrings still feed during winter when the monthly average water temperature is 4.7–9.1 °C, and most herrings intensively forage during late spring and early summer when the water temperature increases to 22.7–30.6 °C.

3.4.5 Vertical Structure of Water Temperature and the Distribution of Fish

In addition to the horizontal structure of water temperature, the vertical structure of water temperature can also affect fish distribution and fishery yield. The vertical and horizontal structures of water temperature are closely related to the movement and shoaling of fish. A thermocline with a sharp vertical water temperature gradient often appears in the water layer where the water temperature decreases sharply. In the Northern Hemisphere, the vertical distribution of the thermocline shows the following pattern: near the sea surface in high-latitude areas, in the deepest water layer in the subtropical area near 25°N–30°N, gradually rising to the shallow water near the equator and to

the shallowest water layer near 10°N, and then in deeper water in further south. In the sea area north of the subtropical zone, the seasonal thermocline generally emerges in spring and summer and disappears during the vertical convection period in autumn and winter; as a result, nutrient salts in the bottom layer move via a seawater convection cycle to supplement the surface water. Therefore, the presence of a thermocline is closely related to the vertical distribution of plankton and fish species as well as fishery production; in particular, the distribution of pelagic fish in different water layers is more closely related to the formation and change in the thermocline.

A thermocline refers to a water layer where the water temperature changes sharply in the vertical direction. Thermoclines can be divided into two types based on the cause of formation: external environmental conditions, such as warming and wind power, and the superimposition of water systems with different natures. Figure 3.10 provides a schematic diagram of a thermocline. Points A and B, with the largest curvature on the vertical distribution curve for water temperature, are the upper and lower boundaries of the thermocline, respectively. The depth Z_a at point A is the

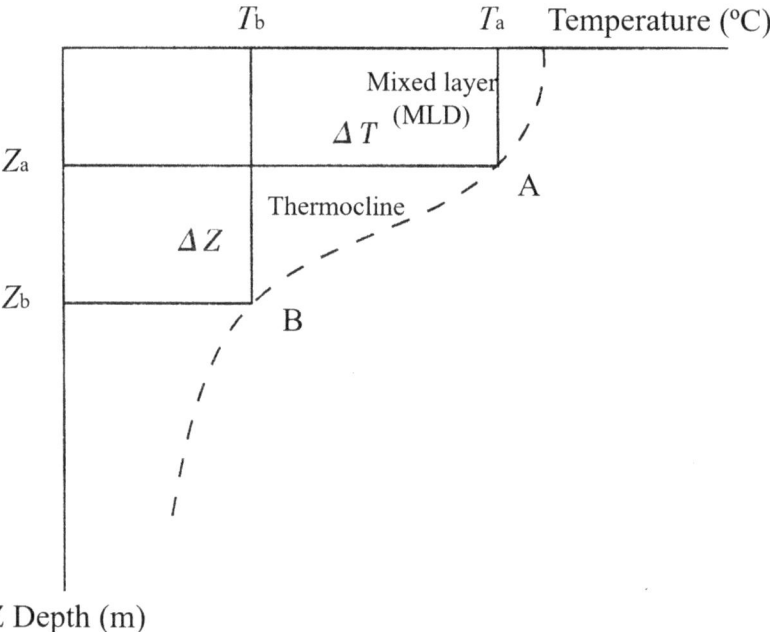

Fig. 3.10 A schematic diagram of a thermocline (Chen 2021)

upper boundary depth of the thermocline, i.e., the mixed layer depth (MLD). The depth Z_b at point B is the lower boundary depth of the thermocline. $\triangle Z$ is the thickness of the thermocline. When the water temperature difference between points A and B is $\triangle T$, $\triangle T/\triangle Z$ is the intensity of the thermocline, which is a positive value when the water temperature decreases from top to bottom; otherwise, the value is negative. The minimum standard value of the intensity of the thermocline $\triangle T/\triangle Z$ depends on the needs and specific conditions of the sea area; generally, it is 0.2 °C/m for shallow seas and 0.05 °C/m for deep seas.

The water layer where pelagic fish reside depends on, to a great extent, the vertical structure of water temperature. For example, *P. japonicus* in the Yellow Sea of China can tolerate a water temperature of 8 °C, with a minimum temperature of 12 °C for shoaling. From May to June, *P. japonicus* usually migrate to the Yantai, Weihai, and Haiyangdao fishing grounds with the shift of the 8–10 °C isotherm of surface water. They often form shoals due to the impact of a thermocline within the water layer in which they reside. To avoid the cold-water mass in the lower water layer, *P. japonicus* concentrate in the surface layer; consequently, *P. japonicus* shoals are larger, and the catch yield is higher when the thermocline is closer to the surface.

Some fish live above thermoclines, some live in thermoclines, and some mainly live in the deep water below thermoclines. In addition, many fish species undertake diurnal vertical migration.

However, because the water temperature differs remarkably between the water layers above and below a thermocline, the thermocline itself acts as a natural environmental barrier limiting the vertical migration of fish, especially pelagic fish. Therefore, the vertical distribution of water temperature is critical for the formation of fishing grounds, which is of particular significance for tuna purse seine fisheries. Figure 3.11 shows the relationships between the vertical distribution of several tuna species and the thermocline in tropical waters.

3.5 Relationship Between Fish Movement and Ocean Currents

Investigations have revealed that the horizontal movement of ocean currents is the main factor that causes regional changes in a marine environment and exerts a great impact on the distribution, migration, and shoaling of fish.

3.5.1 Types of Ocean Currents

Ocean circulation, including surface ocean circulation and deep ocean circulation, is the most important current. Surface ocean circulation is caused by the surface wind field and thus is referred to as wind-driven circulation. Deep ocean circulation originates in the polar or sub-polar sea areas. The cooling and freezing of

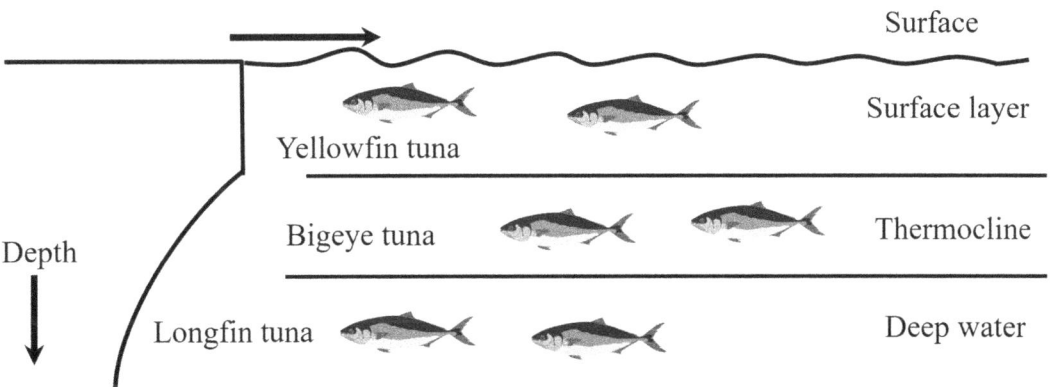

Fig. 3.11 Vertical distribution of several tuna species in tropical waters (Chen 2021)

seawater cause cold water with a high salt content to descend and flow along the deep ocean floor, leading to deep ocean circulation. Ocean circulations have a nearly consistent flow direction, but their speed and volume may vary with the seasons.

Ocean currents can be divided into cold and warm currents based on their temperature characteristics. A cold current refers to an ocean current with a water temperature lower than that of the sea area through which it passes. Cold currents usually flow from high latitudes to low latitudes, e.g., the cold Kuril Current, and they generally have low temperatures, low salinity, and low water transparency. A warm current refers to an ocean current with a water temperature higher than that of the sea area through which it passes. Warm currents usually flow from low latitudes to high latitudes, e.g., the warm Kuroshio Current, and they generally have high temperatures, high salinity, and high water transparency.

The abovementioned ocean currents or circulations involve the constant movement of seawater. Another form of periodic seawater movement is tidal currents. Under the tide-generating force of celestial bodies (mainly the moon), the periodic horizontal and vertical movements of seawater are called tidal currents and tides, respectively. Tidal currents and tides are the most important environmental characteristics of intertidal zones.

In addition, ocean currents where deep water ascends due to the action of wind or topographical factors are called upwelling, and currents where surface seawater converges and descends toward the sublayer are called downwelling.

3.5.2 The Impact of Ocean Currents on the Survival Rate of Fish Larvae

The size of each generation in a fish population depends on the number of larvae and juveniles that survive. The environment has a great impact on size changes in a species population, especially on the mortality of eggs, larvae, and juveniles. Fish have a limited ability to adapt to the environment and high mortality in the early developmental stages. During these stages, appropriate environmental conditions and sufficient food are necessary for the development and hatching of fish eggs as well as for the survival and growth of larvae. A species population can be strengthened only if larval fish have a high survival rate; otherwise, the population will weaken. Therefore, the size of a fish population does not necessarily depend on the number of eggs produced; instead, it depends heavily on the conditions for the early development of fish and survival of larvae. A large number of studies have demonstrated that poor conditions for the early development and survival of larvae are primarily responsible for most fish resource fluctuations, especially for squids, which have a short life cycle.

In the early developmental stages of fish, the normal ocean currents carry floating eggs, larvae, and juveniles from spawning grounds to nursery grounds. The larvae and juveniles that develop and grow in nursery grounds with suitable environmental conditions and sufficient food migrate to the feeding grounds along ocean currents to forage after they grow to a certain level. If the ocean current carrying the larvae and juveniles changes and transports them to sea areas unsuitable for development and growth, it is possible that a large number of larvae of the same generation will die, substantially affecting the multiplication and growth of offspring and leading to a fluctuation in the volume of fish resources. For example, the El Niño phenomenon kills a large number of anchovies in the sea area of Peru. Because fluctuations in the quantity of fish are closely related to ocean currents and ocean currents are affected by wind fields each year, some researchers have investigated wind field changes to predict the survival rate for larvae each year.

3.5.3 Relationship Between Ocean Currents and the Range of Fish Migration

Different marine organisms reside in ocean currents with different temperature and salinity levels and various chemical properties, indicating that fish have a certain adaptability to different water systems, water masses, and ocean currents. Generally, warm-water fish live in sea areas affected by warm currents, and their migration changes with changes in warm currents. This relationship is similar to that between cold-water fish and cold currents and between coastal fish and coastal water systems. In the inshore region of China, fluctuations in the open-sea water system (the Kuroshio Current) and the coastal water system have a substantial impact on inshore fisheries. The fishing season comes earlier, and the fishing grounds are closer to the shore when the open-sea water system is strong; otherwise, the fishing season comes later, and the fishing grounds are closer offshore.

Mixed water areas where different ocean current systems meet as well as the fronts where different water masses make contact often form a boundary with significantly different water color and are usually referred to as "current barriers." Eddy currents and upwellings that are often generated at current barriers bring nutrients from the lower water layer to the surface water layer, a process that is conducive to the growth and reproduction of plankton. Therefore, fish tend to cluster near current barriers to feed. There are many types of current barriers, including those caused by topography obstacles near islands, reefs, headlands, etc. and those formed by the convergence of water systems with different water quality and temperature, in addition to barriers between cold and warm currents and between coastal water and open-ocean water. For example, the current barrier formed by the convergence of the Oyashio Current (cold current) and Kuroshio Current (warm current) in the northwest Pacific Ocean is a good fishing ground for Pacific saury, Ommastrephidae, tuna, and cetaceans; the current barrier between the North Atlantic warm current and the Arctic cold current is a good fishing ground for cod and herring in the northeast Atlantic Ocean.

During the winter fishing season for largehead hairtail in the inshore area of Zhejiang, shoals usually gather near the mixed water area where the coastal water and the open-sea water converge, making it a central fishing ground. In the Shengshan fishing ground of Zhejiang, the horizontal and vertical gradients of the isotherm are very sparse at the beginning of the fishing season around mid-November. At this time, the central fishing ground is mainly distributed near the edge of the high-salinity tongue (salinity 33‰) offshore. During the last 10 days of November, the isotherms become dense, and a current barrier forms. In the area 25 nautical miles northeast of Huaniao, there are two northwest-southeast mixed frontal areas where horizontal temperature and salinity gradients gradually increase, near which are relatively stable fishing grounds for largehead hairtail. According to fishers, largehead hairtail often gather in water barriers with a "white rice"-like color, which refers to the mixed water areas where water systems converge. Therefore, the location of the frontal area where two water systems converge and its pattern of change are effective indicators for determining the movement of central fishing grounds and fish shoals.

As a large amount of fishing experience has demonstrated, changes in water systems, water masses, and ocean currents have a very close relationship with the concentration, dispersal, and distribution of fish shoals and affect changes in fishing grounds, the time and duration of fishing seasons, and catch yields. Taking the fishing grounds in the northeastern Atlantic Ocean as an example, the warm current of the Atlantic Ocean was particularly strong in 1938 and caused the migration range of herring to shift 100 nautical miles east from the previous years. In other years, due to a weak Atlantic warm current, the cold water in the Arctic flow to the coast of Norway results in a concentration of cold-current fish near the coast of Norway. Sudden abnormalities in ocean currents often bring unexpected losses to fisheries. For example, the annual catch of Peruvian anchovies living in the coastal sea area controlled by cold currents was more than 12 million tons in 1971 but sharply decreased, by half, in 1972 because the invasion of the equatorial current from the north caused a sharp temperature elevation in the area.

3.5.4 Relationship Between Ocean Currents and Tuna Distribution

To some extent, the distribution of tuna is closely related to that of ocean currents. Because ocean currents transport many substances, including fish eggs, larvae, and juvenile fish, and transfer heat and salt (i.e., transfer seawater characteristics such as water temperature and salinity), similar ocean characteristics can be found in the same ocean current system. From the biological point of view, the circle of life of organisms forms based on ocean current systems.

Existing studies have found that in the tropical waters of the western central Pacific Ocean, the distribution area of bigeye tunas is located in the sea area centered on the equatorial countercurrent; it is near the current boundary on the north side of the equatorial undercurrent in the west and near the south side of the equatorial undercurrent in the east. However, with the expansion of the longline fishing ground for bigeye tuna to the east, no fishing ground for bigeye tuna forms in the equatorial countercurrent area of the eastern Pacific Ocean; accordingly, longline fishing operations have moved toward the south. In the tropical waters of the western and central Pacific Ocean centered on the equatorial countercurrent, the depth of the water layer with a temperature (10–15 °C) suitable for bigeye tuna is the same as the setting depth of longline fishing hooks. There is no fishing ground for bigeye tuna in the equatorial countercurrent area of the eastern Pacific Ocean because the dissolved oxygen in this sea area is less than 1 mL/L in the water layer within a depth of 100 m and in deep water, making it impossible for bigeye tuna to survive in the area.

Tuna, marlin, and sailfish are all migratory fish species in the Pacific Ocean, and their distribution and migration are closely related to ocean currents. Examples can be found for almost every fish species. For example, more longfin tuna and less yellowfin tuna are distributed south of the 8°S–10°S line in the Pacific Ocean, while more yellowfin tunas are distributed north of the line. The distribution is divided by the southern equatorial current and the equatorial countercurrent between October and March of the following year and by the southern equatorial current and nondirectional ocean currents from April to September. Another example is the distribution difference between yellowfin tuna and longfin tuna. Longfin tunas are distributed in the southern equatorial current system, while yellowfin tunas are distributed in the sea areas of the equatorial countercurrent and nondirectional currents.

3.5.5 Relationship Between Tidal Currents and Fisheries

The tides and the resultant tidal currents change most significantly in shallow coastal seas, especially the areas between islands and adjacent to headlands, harbors, and estuaries. The changing tides and tidal currents adjust the differences between water bodies by changing the distributions of the temperature and salinity gradients and altering the substance content difference between adjacent water bodies. In addition, they cause regular periodic horizontal and vertical changes in the water level, water depth, flow direction, flow velocity, etc. As a result, they affect fish habitats to a certain extent and lead to changes in the density of fish shoals, the water layer inhabited by fish, and the movement direction and speed of fish. Therefore, the impact of tides and tidal currents must be considered when studying the relationships among the marine environment, fish behavior, and fishing ground changes.

Study findings have shown that fish behavior is closely related to tides, especially during spring and neap tides. Taking herring as an example, research has revealed a negative correlation between tides and herring catch; the smallest shoals occur when there is a new moon and full moon, and the largest shoals occur at the first and last quarter of the moon. The interaction between the circadian behavior of fish and strong tidal currents affects the movement of fish species.

Some fish require a certain water flow rate to stimulate ovulation during the spawning period.

For example, *P. crocea* in the inshore area of Zhejiang require a certain temperature and a certain flow rate to spawn. In the Daiqu fishing ground, an ocean current flow rate of 2–4 kn is necessary for fish to shoal and spawn in large quantities. Therefore, a large catch of *P. crocea* is usually obtained during spring tides. Because changes in tidal levels are positively correlated with the flow rates of tidal currents, they are often used as indicators for fishing season prediction. It has been observed that the gonads of *P. crocea* in the coastal area of Jiangsu are basically maintained at a certain maturity stage during neap tides, which have a low flow rate. During spring tides, the gonads of *P. crocea* develop rapidly from stage IV to stage V within 3–5 days. During this period, *P. crocea* often form large shoals and swim, making loud noises, toward the fast-moving tide current to reproduce.

The size and direction of tidal currents directly affect the intensity of fish shoaling, but the effect varies among fishing grounds, terrains, and fish species. In nearshore fishing grounds, especially areas with strong runoff inflow, the flow direction and speed of tidal currents are often inconsistent between the upper and lower layers as tides rise or ebb. This phenomenon is called "tide separation" ("Chao Ge Luan" in Chinese) by fishers and "stratified currents" in oceanography. This phenomenon often results in the absence of large fish shoals and makes fish operations inconvenient, thus greatly reducing catch amounts.

3.6 Relationship Between Fish Movement and Water Salinity

3.6.1 Distribution of Salinity in the Ocean

Seawater can be regarded as a solution containing various substances, including inorganic and organic substances as well as gases, dissolved in pure water. Salinity, a measurement unit of the total salt content in seawater, refers to the total amount (grams) of inorganic salts dissolved in 1 kg of seawater.

Although seawater salinity varies due to the difference in the balance of evaporation and precipitation, the ratios between major ionic components are nearly consistent among different sea areas. The ocean surface water salinity ranges approximately from 34 to 36‰, primarily resulting from the ratio of precipitation to evaporation in areas at different latitudes.

The water salinity is low (approximately 34.5‰) in the equatorial sea areas due to high precipitation and low evaporation (low wind speed). Subtropical sea areas (between the latitudes of 20° and 30° in both hemispheres) have the highest salinity (approximately 36‰), gradually decreasing toward the temperate sea areas to a level comparable to that in the equatorial sea areas. The polar seas have the lowest salinity (approximately 34‰), primarily because of the melting of polar ice.

The salinity below the sea surface results from the convergence, descent, and spread of surface water at different latitudes. The salinity of subsurface seawater is related to temperature (and density) in addition to the salinity of the surface water above it. Therefore, the salinity distribution in the subsurface seawater layer is stratified based on the abovementioned factors.

Subsurface seawater can be roughly divided into three layers based on the vertical stratification of salinity.

(1) Ocean subsurface water (high salinity): High-salinity surface water in the subtropical regions in both the Northern and Southern Hemispheres descends and then expands toward the equator.

(2) Ocean mid-layer water (low salinity): This layer of seawater in middle- and high-latitude areas in the Northern and Southern Hemispheres descends from the surface and extends to low-latitude areas. In the middle- and low-latitude areas, the isodensity lines are particularly dense between the high-salinity subsurface water and the low-salinity middle water, forming a significant halocline in the vertical direction that corresponds to the thermocline.

(3) Deep ocean water and bottom seawater: In these layers, the water descends from the low-salinity low-temperature upper water in high-latitude areas and polar sea areas and then spreads to the ocean floor. Deep ocean water has a salinity of approximately 35 and temperature of approximately 3 °C. Bottom layer seawater has lower salinity and temperature than those of deep ocean water (approximately 34.6‰ and −1.9 °C, respectively).

The above discussion describes only the gross vertical stratification status of seawater salinity in the ocean; however, the depth and distribution scope of salinity layers vary among oceans. Affected by continental freshwater, the water salinity in shallow seas is often lower and fluctuates in a wider range (27–30‰) than in the oceans, and semiclosed sea areas (e.g., the Baltic Sea) have a salinity below 25‰. In estuarial areas, the impact of freshwater is more prominent and results in greater salinity fluctuations (0–30‰). In the above cases, the mixture of seawater and freshwater forms seawater with a lower salinity, referred to as brackish water. In addition, the salinity of seawater in some sea areas (such as the Red Sea and tropical coastal lagoons) can exceed 40‰, referred to as hypersaline water.

3.6.2 Responses of Fish to Salinity Changes

Fish can sense salinity changes with their lateral lineal nerves and respond to changes as small as 0.2‰. Fish, particularly anadromous and catadromous fish such as salmon, trout, and eel, have the ability to distinguish small differences in water salinity.

Significant salinity changes are important factors controlling fish behavior. They affect the osmotic pressure of fish and the ability of eggs to float. Generally, the water salinity changes very little in oceans but varies substantially in coastal areas due to the influence of continental runoff. Therefore, fish species that often inhabit oceans have a strong adaptability to high-salinity water,

and their adaptability varies significantly once they arrive in nearshore areas or along the coast. Some fish have limited migration and distribution ranges due to significant salinity decreases beyond the regulatable osmotic pressure range or can die from sudden salinity changes. Only a small number of intermediate-type fish species can inhabit waters with low salinity (0.02–15‰). These brackish fish species are mainly present near the coast. The number of brackish fish species is small because stable salinity levels are not maintained in these water areas.

Marine fish species differ in their adaptability to salinity. Marine fish species can be classified into two types based on tolerance and sensitivity to salinity changes: stenohaline fish and euryhaline fish. (1) Stenohaline fish are very sensitive to salinity changes and can only live in environments with stable salinity levels. For example, fish in deep seas and oceans are typical stenohaline species. These fish will die quickly if they are brought by wind or current to a coastal sea area and estuarial area where the salinity changes substantially. (2) Euryhaline fish have a strong adaptability to seawater salinity changes and can tolerate drastic salinity changes. The fish and migratory animals in coastal and estuarial areas are euryhaline species. For example, mudskippers can live in both freshwater and seawater because their resistance to salinity changes has been greatly enhanced after living in an environment with a constantly changing salinity. Mugil (*Mugil cephalus*) and barracuda (*Sphyraena*) also have strong adaptability to salinity.

The adaptability to salinity also varies among different populations of the same fish species and among fish in different life stages of the same species population. For example, for *P. crocea* distributed in the inshore areas of China, during the spawning period, the population living in the Daiqu fishing ground can adapt to 17–23.5‰ salinity, and the population in the Maotou fishing ground can adapt to 26–31‰ salinity; during the overwintering period, the population in the Zhoushan offshore fishing ground can adapt to

32–33.5‰ salinity, the population in the northern inshore area of Fujian can adapt to 27.5–28.7‰ salinity, and the population in the Naozhou sea area of Guangdong can adapt to 30.5–32.5‰ salinity.

3.6.3 Fish Respond to Salinity Changes by Regulating Osmotic Pressure

Fish can respond to seawater salinity changes primarily because their osmotic pressure is affected by seawater salinity. Generally, the osmotic pressure of a solution increases with the increase in solute concentration. The greater the seawater salinity is, the higher the osmotic pressure is. The osmotic pressure of a solution, $\triangle$, is usually indicated by the decline in the freezing point in degrees centigrade. For example, the freezing point of seawater with a salinity of 35‰ is -1.91 °C; therefore, its osmotic pressure is $\triangle = 1.91$. Among fish species, as well as other aquatic animals, the difference in body fluid concentration leads to the difference in the freezing point and different $\triangle$ values.

Aquatic animals can be classified by comparing the $\triangle$ values of internal and external media:

1. Aquatic species isosmotic to the water they live in: In these species, the $\triangle$ value for the internal medium is equal to that for the external medium; mainly includes invertebrates.
2. Aquatic species hyperosmotic to the water they live in: These species have a greater $\triangle$ value for the internal medium than that for the external medium; mainly includes freshwater fish species.
3. Aquatic species hypoosmotic to the water they live in: These species have a lower $\triangle$ value for the internal medium than that for the external medium; mainly includes seawater teleost species.
4. Aquatic species slightly hyperosmotic to the water they live in: These species have a slightly higher $\triangle$ value for the internal medium than that for the external medium

due to the presence of urea in the blood; they can be classified as either hyperosmotic or isosmotic and mainly include seawater cartilaginous fish species.

Euryhaline fish (e.g., eel) have a strong ability to regulate osmotic pressure. When moving from freshwater to the ocean, they lose water and weight in the first few days; when entering freshwater from the ocean, they absorb water and gain weight, and their body weight returns to normal due to mechanisms that regulate osmotic pressure.

3.6.4 The Relationship Between Salinity and Fish Movement

The impact of salinity on fish movement is mainly indirect and mediated by water masses, ocean currents, etc. (Fig. 3.12). For example, warm-water fish species migrate with warm currents (high temperature and high salinity); cold-water species migrate with cold current (low temperature and low salinity). Research has confirmed that salinity rarely exerts any direct impact on most fish species.

In sea areas with a sharp horizontal salinity gradient, salinity affects the distribution of fish shoals or the location of fishing grounds to a certain extent, sometimes becoming a limiting factor. Generally, the measured isohaline distribution can be used to predict whether a fishing ground is closer to the offshore side or coastal side. However, for fish species that can adapt to wide salinity ranges, salinity does not affect the formation of central fishing grounds offshore but becomes a dominant factor affecting fishing ground formation only in estuarial areas with large runoff or areas where different water systems converge. For example, in the Daiqu fishing ground outside the Qiantang River estuary during the spring flood of 1954, an excessive amount of rain caused massive continental runoff to flush into the fishing ground, reducing the salinity sharply to 10–11‰, exceeding the lower limit of salinity (17‰) required by *P. crocea*. As

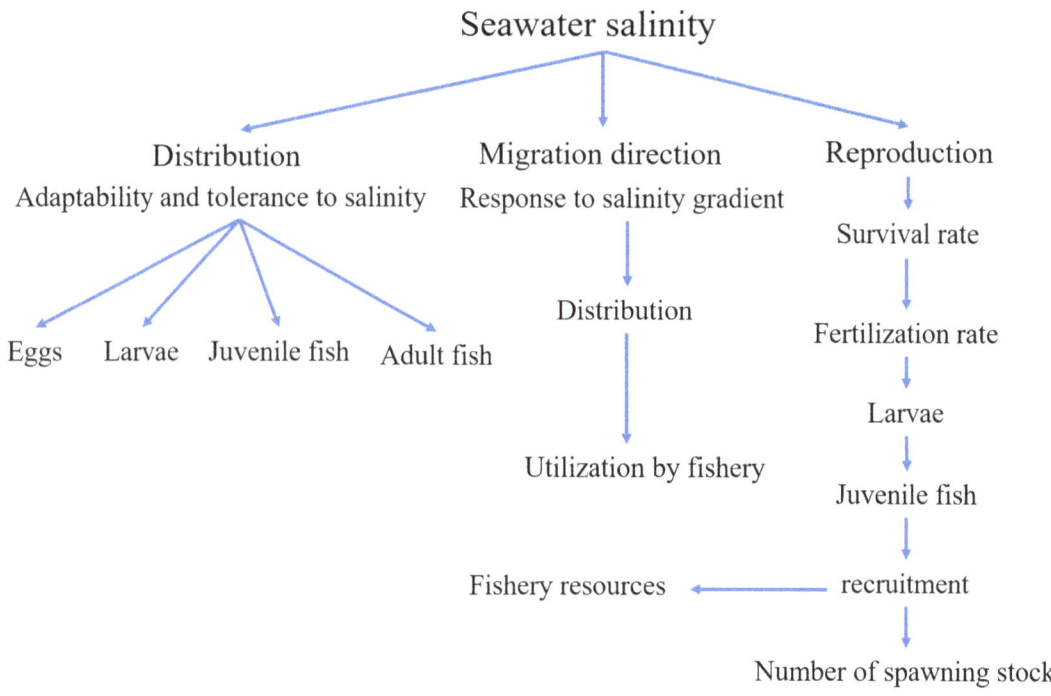

Fig. 3.12 Relationship of salinity with reproduction and migration of marine economic species (Chen 2014)

a result, *P. crocea* shoals that usually came migrated to this ground to reproduce every year moved offshore, and fishers could hardly locate dense fish shoals, resulting in a rare production reduction. Another example is the Haijiao fishing ground adjacent to the Yangtze River estuary. Because the ocean currents offshore, cold-water fronts of the Yellow Sea, and coastal waters are intertwined in this area, the area has a salinity of 32–34‰ and often attracts dense shoals of pelagic fish and demersal fish, making it a good fishing ground. Changes in seawater salinity are not isolated physical phenomena; instead, salinity changes with the movement of water systems and ocean currents. As a result, the relationship between salinity and fish is indirectly affected by ocean currents and other factors. For example, a higher catch is obtained in the tuna fishing ground located near the Hawaiian Islands in years when it is affected by the California Current, with a salinity of 34.7‰, and a lower catch is obtained in years when the fishing ground is invaded by high-salinity seawater from the Western Pacific, with a salinity higher than 35.0‰.

In general, eggs and juvenile fish can tolerate relatively large salinity changes. For example, herring can complete the entire process of fertilization, egg development, and hatching in waters with a salinity range of 5.9–52.2‰. Juvenile herring can live in seawater with a salinity of 2.5–52.5‰ for more than 68 h. Normally, during the spawning season in the nearshore areas of China, most marine economic fish species tend to spawn near the coast or in estuaries with relatively low salinity. For example, *L. polyactis* of the northern stock shoal and spawn near the Yellow River estuary in Laizhou Bay every spring. The salinity of this spawning ground is 26.4–29.6‰, and very few parent fish shoals appear in the adjacent sea areas, where the salinity exceeds this range. After hatching, *L. polyactis* larvae are densely packed in coastal waters with a salinity of 25.3–28.1‰. This fully indicates that *L. polyactis* tend to shoal in low-salinity water during the spawning and larval stages.

Salinity plays a certain guiding role in forecasting fishing conditions and reconnoitering fish shoals, and therefore, it is a piece of necessary information needed to be obtained for forecasting. However, water salinity assessments are relatively complex, and accurate data are difficult to obtain in time during fishing operations. Water color has a certain interconnection with salinity and can be used to indicate salinity in some cases, making it a customary tool used by fishers to search for fish shoals and fishing grounds. In fact, in sea areas with sharp horizontal salinity gradients, i.e., areas where different water systems meet and the difference in salinity is remarkable, the water color appears discontinuous, which can be used as an indicator of salinity.

3.7 Relationship Between Fish Movement and Dissolved Oxygen

3.7.1 Dissolved Oxygen in Seawater

The dissolved oxygen content in seawater ranges from 0 to 8.5 mg/L and is critical for the survival of fish. Like other animals, fish absorb dissolved oxygen from water (usually through gills), entering the body through blood to maintain metabolism. The oxygen content in air is approximately 200 mL/L, and gas solubility in water is related to temperature and salinity. The three main sources of oxygen in the ocean are (1) dissolved oxygen from air (through waves, convection, etc.), (2) river water supply, and (3) photosynthesis of phytoplankton.

The oxygen content in surface seawater is high and often saturated, corresponding to atmospheric pressure and water temperature because surface seawater makes direct contact with air and phytoplankton on the surface undergo vigorous photosynthesis. In sea areas where phytoplankton flourish, the dissolved oxygen in water may be temporarily supersaturated and can reach a saturation of 100–140%.

The dissolved oxygen content in water below the photic layer decreases gradually due to the lack of oxygen supplementation from photosynthesis. At depths of 400–800 m, due to the change in the density gradient and the influence of thermoclines, particulate organic matter (mainly zooplankton feces, etc.) deposited from the upper layer concentrates, leading to strong decomposition activity by bacteria. In addition, respiration by fish consumes a large amount of oxygen, and the oxygen-rich water at the bottom cannot reach and replenish this layer. Therefore, the distribution of the minimum oxygen layer becomes vertical, presenting a decline in oxygen content from the normal level of 5–6 to 2–3 mg/L.

In the water layer deeper than 1000 m, the oxygen content does not continue to decrease with the increase in depth and instead starts to increase after reaching its lowest level. There are two reasons for this increase. First, low-temperature oxygen-rich water masses descending from the ocean surface in polar regions generate undercurrents in the lower ocean layer. Second, there is less biomass in the deep ocean layer, and thus, less oxygen is consumed due to respiration and decomposition.

Certainly, in addition to the abovementioned general distribution pattern, the vertical distribution curve for oxygen shows different characteristics in different oceans. For example, the oxygen content remains in the normal range at the depth of 2000–3000 m in the Atlantic Ocean due to the injection of a large amount of new oxygen-rich cold water from the poles, but it is relatively low on the northern boundary of the Pacific Ocean and the Indian Ocean due to poor deep-water circulation caused by land barriers and the long-term oxidation of organic matter. Therefore, the oxygen content in deep water depends on whether the water is "fresh cold water" from high latitudes or "still water." The vertical distribution curve for oxygen is helpful for identifying the water systems of different oceans.

3.7.2 Harm of Anoxic Water to Fish and Its Effect on Fish Movement

Because the oxygen content in seawater is saturated, marine fish generally do not lack oxygen when living in oceans, and the oxygen supply is sufficient even for organisms in the deep sea. Therefore, oxygen is not a decisive factor in the distribution and movement of most marine organisms. However, seawater anoxia might occur under some special circumstances, e.g., in inner bays that do not connect to offshore areas, when there is no convection between the upper and lower water layers, the surface water is heated and there is no wind in the summer or the inflow of freshwater causes strong stratification of seawater. The presence of hydrogen sulfide when near-bottom water is anoxic kills all organisms, and the rise of the anoxic layer will affect fish movement.

In anoxic sea areas, the development of fish eggs is suppressed. Fish will move away from certain sea areas that lack oxygen. In summer in Mobile Bay in the United States, for example, a large number of demersal fish, shrimp, crabs, and other estuarine organisms rush to the coast, leading to an excellent fishing season along the northeast coast due to anoxic water in the trench along the eastern coast moving toward the shore at the night with high tide and strong easterly winds.

Upwelling that causes anoxic deep water to rise also has a great impact on fish. There are two types of anoxic deep water, one in the tropical zone and the other caused by massive oxygen consumption of organic sediment on the seafloor. Both tropical and subtropical zones have a remarkable anoxic layer, which is more prominent on the east side of the ocean than the west side. Under normal conditions, the anoxic layer is generally distributed at the depth of 100–150 mm with a clear top boundary. When anoxic deep water upwells along the continental shelf in tropical zones, near-demersal fish are forced to swim to shallow-water areas or near the surface.

3.7.3 Relationship Between Dissolved Oxygen and Tuna Distribution

Dissolved oxygen in water is an indispensable environmental factor for aquatic organisms; in particular, it is a pivotal environmental factor for tuna, which have strong swimming abilities. Very

few studies have been conducted on this subject due to tremendous difficulties in measuring the oxygen content. In particular, essentially no studies have been conducted to investigate the oxygen content in the water layer where potential fishing targets appear and the tolerable oxygen level of fish.

Based on theoretical studies, to maintain their ability to swim at high speeds, bigeye tuna, yellowfin tuna, and longfin tuna require an essential oxygen level of 0.5, 1.5, and 1.7 mL/L, respectively, when their body length is 50 cm and 0.7, 2.3, and 1.4 mL/L, respectively, when their body length is 75 cm. Hypoxia tolerance tests with tuna have showed that the limiting oxygen concentration for asphyxia in tuna is 1 mL/L. Studies on the catch rate for longline fishing and oxygen content distribution have shown that the oxygen tolerance of bigeye tuna is more than 1 mL/L and that almost no fish are caught in sea areas with an oxygen content below 1.0 mL/L. In summary, the lowest oxygen level that tuna can tolerate varies based on species and body length.

Research has been conducted to explore the effects of marine environmental factors, including water depth (100–250 m), suitable temperature (10–15 °C), and oxygen tolerance (1 mL/L), on the distribution of bigeye tuna caught by longline fishing in the Pacific Ocean. The results showed that the highest catch amount was achieved in sea areas where the hook depth was consistent with the depth of the water layer with a suitable temperature and dissolved oxygen content more than 1 mL/L. Research has revealed that water temperature and oxygen content are marine environmental factors that affect the distribution of bigeye tuna. Bigeye tuna are distributed in the water layer with a suitable temperature of 10–15 °C. However, even when the temperature is suitable to live, the water layer will not contain bigeye tuna if the dissolved oxygen content is less than 1 mL/L. Therefore, for a high catch rate for bigeye tuna, longline hooks should be set in the water layer with a suitable temperature of 10–15 °C and a dissolved oxygen content greater than 1 mL/L.

3.8 Relationship Between Fish Movement and Other Factors

In addition to the constantly changing environmental factors mentioned above, some marine geographical environmental factors that have relatively small fluctuations, such as water depth, seafloor topography, and substrate, as well as biological factors also affect fishing grounds. Although these factors do not exert significant impacts on fish behavior, understanding the relationship between them can help minimize the search range for fishing grounds, thus playing a certain role in reconnoitering fish shoals and identifying central fishing grounds.

3.8.1 Relationship Between Fish Movement and Water Depth and Seafloor Topography

Water depth has a close relationship with seafloor topography; therefore, the water depth distribution can be used to investigate the general terrain of the sea floor, which is difficult to directly observe. Seawater depth directly affects various hydrological elements in sea areas, in particular spatiotemporal changes in temperature, salinity, water color, transparency, water system distribution, flow direction, velocity, etc. Consequently, it indirectly affects the distribution of organisms and the shoaling of fish. Sea areas with different water depths have their own characteristic hydrological distribution and changes; the deeper the water is, the more drastic the changes are.

The distribution of fish varies with seafloor topography. Steep slopes with large inclinations are not suitable for long-term fish habitation, but relatively flat basin areas and valleys on the seafloor are good places for fish to gather. For example, the deep central area of the Yellow Sea serves as the overwintering grounds for many economic fish species and fishing grounds for cold-water fish species. On a sea floor that is uneven with occasional undulations, most fish gather in deeper depressions. Therefore, fish shoals often gather densely in small deep ditches or low-lying pits

and valleys and sparsely on convex hills or scarps. However, seafloor uplifts in the latter terrains (i.e., convex hills or scarps) cause the upwelling of deep seawater that drives epipelagic fish to gather in surface water, which is a common experience in purse seine fisheries.

The requirement for water depth varies among different life stages of marine fish species based on their physiological and survival needs. The main economic fish species in China are mostly distributed within the offshore continental shelf, and their spawning grounds are mostly in sea areas with a depth between 30 m and 80 m. For example, in the central area of the Yellow Sea, northwest and southwest of Jeju Island and east of Zhoushan are the overwintering grounds for most fish that migrate in the Bohai Sea, Yellow Sea, and East China Sea. *L. polyactis* and largehead hairtail generally are not distributed beyond the 100 m isobath and mostly concentrate in areas with depths of 40–80 m, except during spawning season, when they gather in sea areas shallower than 30 m. Areas with a water depths between 40 and 80 m are also favored by many other demersal fish shoals, making these areas good fishing grounds for bottom trawling. The following are descriptions of the relationships of fish distribution with water depth and seafloor topography:

1. Fish of the same species are distributed at different depths during different life stages and different seasons. For example, *P. crocea* in the coastal area of Zhejiang inhabits water at depths of 5–20 m during the spawning period, 20–40 m (rarely exceeding 50 m) during the foraging period, and 40–80 m in winter.

2. There is a certain relationship between fish distribution and seafloor substrates. The adaptation and selection of fish to the properties and color of seafloor substrates vary with species. Most fish do not make contact with the sea floor often or even during their entire life; their distribution seems to have little or no relation to seafloor substrates. However, some marine fish species are often close to or inhabit the sea floor; some species do not make contact with the sea floor, but their distribution

has a certain relationship with seafloor substrates during certain periods. Therefore, the role of seafloor substrates cannot be ignored in the behavioral study of fish. The following are types of adaptations of marine fish to seafloor substrates:

(a) Buried or lurking on the seafloor. Most of these fish are flat and slow. To avoid predators or hunt for food, they often inhabit the sea floor and become demersal fish. Species such as lefteye flounder, righteye flounder, and sole belong to this type of fish, and they adapt to or prefer relatively finer silty and sandy argillaceous or argillo-arenaceous substrates composed of a mixture of sand and mud.

(b) Lurking on the seafloor at certain times to forage. Many fish species fit this classification. They are referred to as lower-layer or near-demersal fish when they are close to or make contact with the seafloor. These fish can ascend and descend freely and move quickly. The bait they chase and eat is mostly demersal organisms. There are various types of seafloor substrates they prefer because their preference toward substrate properties is closely related to the distribution of demersal organisms.

(c) Migrate to areas with certain seafloor substrates to breed. Fish must migrate to places suitable for spawning and hatching during the breeding period. Once reaching gonadal maturity, fish that lay sinkable or sticky eggs or bury their eggs must migrate to a place suitable for egg attachment or burying for reproduction. For example, Pacific herring, which lay sinkable sticky eggs, often choose nearshore areas with rocky reefs and seagrass as spawning grounds. Squid lay eggs on the sea floor under rocky reefs, and their eggs stick to algae or other objects.

3. Relationship between seafloor topography and fishing grounds. The formation of fishing grounds is related to special seafloor topography. Based on related research, there is a strong relationship between the catch amount of bigeye tuna and the isobath in August in the Agulhas shoal of South Africa. Good fishing grounds may occur in sea areas near sandbars, shoals, and steep slopes of the continental shelf.

The main reason underlying the formation of the above good fishing grounds is that areas with upwelling and downwelling currents formed due to seawater disturbance and resultant eddy turbulence under the influence of seafloor topography. As a result, bait organisms reproduce and gather in these areas, attracting large fish and forming a good fishing ground.

3.8.2 Relationship Between Fish Movement and Bait Organisms

The relationship between fish and biological environmental factors mainly refers to the relationship of fish with animals and plants living in the same water bodies. In the ocean, the biological environmental factors related to fish mainly include marine organisms that are direct or indirect biological bases of fish forage and marine organisms that are predators. The following provides descriptions of such organisms.

Bait Organisms

There are many types of bait organisms for fish in the ocean, including plankton, demersal organisms, and swimming animals.

Plankton

Due to their small size, large number, and wide distribution, plankton play an important role in the circle of aquatic life and serve as a bait base for fish. Generally, fish and shrimp all eat plankton. Some fish, such as mackerel, the family Carangidae (*Decapterus maruadsi* and *Trachurus japonicus*), herring, anchovy, the genus *Coilia*, and *L. polyactis*, consume zooplankton as their main food. Some fish, such as sardine, lizard fish, and *Liza haematocheila*, consume phytoplankton as their main food. Some fish, such as *Penaeus orientalis* and *Etrumeus teres*, eat both zooplankton and phytoplankton. Most fish in the larval and juvenile stages feed on plankton and then, when adults, feed on large animals; examples of such fish are *P. crocea*, largehead hairtail, cods, basses, Spanish mackerels, sharks, and rays. Therefore, changes in the distribution and quantity of plankton can directly or indirectly affect the behavior of fish species, and the impact is particularly prominent during the foraging period of fish.

Based on the results of previous investigations, zooplankton that fish like to eat, such as *Euphausia pacifica*, *Calanus pacificus*, *Labidocera euchaeta*, opossum shrimp, *Acetes*, arrow worms, and *Pseudeuphausia latifrons*, are widely distributed in the inshore waters of China. In particular, a number of areas with dense phytoplankton populations are located in and near sea areas where warm offshore currents, coastal water systems, and cold-water masses of the Yellow Sea mix. The distribution of these phytoplankton-concentrated areas is closely related to the emergence of good central fishing grounds, e.g., the Dasha, Yangtze River estuary, Zhoushan, Yushan, Wentai, Shidao, Yanwei, and Haiyangdao fishing grounds, for several species,

Because fish behavior is closely related to plankton, the change in plankton quantity can predict changes in the catch. Data from relevant investigations indicate that the biomass of zooplankton in the northern South China Sea is also closely related to the catch of *Decapterus maruadsi*. The spring fishing season of the Pearl River estuary fishing ground occurs from November to March of the next year. The total biomass of zooplankton in the area also increases from November and is maintained at a high level until March of the next year before it begins to decrease beginning in April. The corresponding relationship between the high-biomass period and the fishing season of *Decapterus maruadsi* in the Pearl River estuarial area indicates that the emergence of the fishing season is closely related to the abundance of bait.

Demersal Organisms

Demersal organisms include those living on the sea floor for a lifetime or during a certain stage of life and those living mostly near the sea floor but capable of moving short distances. Demersal or near-demersal fish, such as yellow porgies, *Paerargyrops edita*, golden thread (*Nemipterus virgatus*), and cod, often prey on other demersal organisms. During the foraging period, the distribution of these fish species is often closely related to demersal flora. Therefore, demersal organisms that are closely related to the target fish species can be used as an indicator for reconnoitering fish shoals.

Different demersal or near-demersal fish species prefer different types of demersal organisms as food. For example, demersal fish species, including cod, snapper, flounder, and rays, all feed on demersal invertebrates such as Lamellibranchia, crustaceans, annelids, and echinoderms. However, cod favor *Crangon affinis* and hermit crabs, while *Cleisthenes herzensteini* (*C. herzensteini*) prefer to eat *Ophiura sarsii* (*O. sarsii*) and *Crangon affinis*. Most near-demersal fish have a relatively complicated diet because they often swim to the pelagic layer and demersal organisms can only be part of their forage. For example, researchers found that demersal shrimp and crabs only account for only approximately 1/4 and cephalopods and fine-legged insects accounted for even a smaller proportion of the stomach contents of largehead hairtail. Yellow porgies mainly feed on demersal organisms as well as some plankton and swimming animals, and their main bait includes Mysidacea, brittle stars, Macrura, Amphipoda, Brachyura, and fish species. *Paerargyrops edita* mainly feed on demersal organisms as well as demersal plankton, and its main bait includes brittle stars, Macrura, Polychaetes, Amphipoda, etc.

Field investigations have determined that the high-density distribution areas of demersal organisms, especially in winter, are roughly consistent with dense areas of economic fish species in the Yellow Sea and East Sea of China. For example, the slope on the sea floor in areas between the 50 m and 80 m isobaths in the southern part of the Yellow Sea and the northern part of the East Sea of China is densely distributed with demersal organisms and a good fishing ground where *L. polyactis*, white croaker, largehead hairtail, *P. crocea*, flounder, gurnard species, and other fish species gather and shoal for overwintering. Most obviously, the areas with the highest benthic biomass are the areas where the *C. herzensteini* is concentrated, while the areas with the lowest benthic biomass are less abundant.

The use of the distribution of certain demersal organisms as indicators is effective for exploring fishing grounds. For example, the distribution of brittle stars in echinoderms is related to the dynamics of *C. herzensteini* and *L. polyactis* shoals. Based on the experience gained in practice, sea areas where *O. sarsii* is dominant are excellent fishing grounds for *C. herzensteini*, and most of sea areas where Gorgonocephalidae grow thickly are the places where *L. polyactis* gather in shoals.

Swimming Animals

Many economic fish species are carnivorous fish that mainly feed on swimming animals. Generally, economic fish species feed on tiny and less active plankton during the larval stage, begin to eat larger plankton when they mature, and then feed on larger swimming animals or demersal organisms and even the larvae of various animals (including a certain proportion of fish larvae) as their body gradually grows to full size. Although a few fish species feed on plankton all their lives, the adult fish of most species also feed on plankton, demersal organisms, and swimming animals or fish that are smaller than them. That is, fish also include fish on their bait list. Small and weak fish species that large fish species prey on are called forage fish. Many carnivorous species of economic fish inhabit the coastal areas of China, and the adult fish often swallow or prey on forage fish; accordingly, fish shoals in a certain life stage migrate with forage fish. For example, largehead hairtail, *P. crocea*, Spanish mackerel, bass, and cod all consume fish that are smaller than them as their main food, and some even eat their congeners or their own larvae. Understanding the foraging habits of fish, in combination with knowledge regarding the distribution and dynamics of forage fish in areas at the same time, helps

in the identification of central fishing grounds. For example, largehead hairtail in the coastal area of Zhejiang consume anchovy, *Channa asiatica*, *Collichthys lucidus*, *Harpadon nehereus*, *Setipinna tenuifilis*, *Harengula zunasi*, *L. polyactis* juveniles, etc. as their main food; therefore, before the largehead hairtail fishing season in the Shengsi fishing ground, fishers often use the distribution of the abovementioned forage fish as an indicator to reconnoiter the fishing grounds. Japanese Spanish mackerel that enter the Bohai Sea feed intensively and often shoal to hunt small fish such as anchovy for their main food immediately after spawning. Therefore, the distribution and activity pattern of small fish such as anchovy are important reference indicators for finding the central fishing ground of Japanese Spanish mackerel.

Predators

Predatory Fish and Animals

Carnivorous fish species often prey on fish shoals and are enemies of the fish species they hunt. In addition, some other predatory marine animals also feed on fish. For example, largehead hairtail, *P. crocea*, *L. polyactis*, cod, mackerel, *Trachurus japonicus*, and Japanese Spanish mackerel all prey on anchovy; cod and largehead hairtail prey on herring; *Seriola quinqueradiata* and *Katsuwonus pelamis* prey on sardines; and sharks, stingrays, rays, dolphins, cetaceans, and seabirds also often prey on fish. Through investigating the predator/bait relationship and the pattern of predator/bait activities, the behavioral dynamics of fish shoals can be better understood. Some anchovies live in deep water (20 m below the water surface) during the day, rise to the surface or swim to shallow water at night, and leave the surface or the coast to return to the deeper water layer around midnight, presenting simultaneous horizontal and vertical movements. The reason for their regular movement is mainly that as a prey target of some predatory fish or animals, they must move as a defense. During daytime, when they might be easily attacked by predators on the water surface, they hide in the deeper water layer and disperse into small shoals to avoid threats from enemies. However, anchovies are epipelagic fish and cannot stay in deep water for a long time. When there is less threat from predators in the dark, anchovies move to the upper layer or suitable nearshore areas to forage. Therefore, based on this pattern, the status of predatory animals can be used as an indicator to explore the dynamics of anchovy shoals. Notably, the presence of predatory fish searching for food does not indicate the existence of massive fish shoals; therefore, only the predatory fish attacking their prey can be used as an indicator. Generally, predatory fish will not arrive immediately after fish shoals arrive in an area; once predatory fish are detected, it can be concluded that a fish shoal has arrived.

When fishing for epipelagic fish, the movements of seabirds flying in groups serve as an effective indicator of fish shoals because seabirds that find fish shoals often scream and swoop in groups over the shoal. The dynamics of seabirds can be used to determine the presence of fish shoals from a distance. A larger number of seabirds that chase a fish shoal indicate a larger fish shoal in the water. A flock of seabirds flying at high altitude indicates a fish shoal that is in deep water and has not yet ascended to the surface; low flying seabirds indicate that a fish shoal is swimming in shallow water; and constant swooping indicates that a fish shoal is near the surface of the water. In addition, the phenomena of seabirds staying on the surface of the water and constantly looking at the water or sometimes changing locations or flying over the water in unison indicate that a fish shoal has already gathered in the deep-water layer. The phenomenon of seabirds flying in the same direction indicates a fish shoal swimming in front of the flock of birds. The movement of a flock indicates that a fish shoal is moving; in this case, the direction of flight of the flock indicates the direction of fish movement, and a higher flying speed of the birds indicates that the fish shoal is swimming fast.

Red Tide Organisms

Red tides refer to the phenomenon of the explosive reproduction of certain plankton causing an abnormal water color change due to the change in marine environmental conditions. The seawater color during a red tide is not necessarily red; instead, the color depends on the phytoplankton species that form the red tide. At present, there are more than 260 species that can cause red tides in China's coastal waters, of which 78 are known to be toxic. Since the twentieth century, especially since the 1960s, due to the increasingly severe pollution of coastal waters, red tides have occurred successively in the coastal waters of many countries in Asia, America, and Europe, and the frequency has increased year by year. The cause of red tides has not been fully identified. Physiochemical analyses of environmental changes preliminarily revealed that the occurrence of red tides is related to many factors, including climate, sea temperature, salinity, nutrients, and environmental pollution.

A severe red tide can result in massive deaths of fish, shrimp, and shellfish in an area, causing tremendous fishery loss. For example, in May 1958, a large-scale red tide in the coastal waters of Zhejiang greatly reduced the production of *P. crocea*; the red tide that occurred along the coast of the Bohai Sea from May 5, 1952, resulted in the death of many fish and shrimp and significantly reduced the catch amount by the local fishermen, largely reducing the production of the fishery; and a large area of "stinky water" appeared on the east side of Haijiao and Langgang in Zhejiang from later September to early November in 1972 and caused a poor fishing season for epipelagic fish.

References

Chen XJ (2014) Fisheries resources and fisheries oceanography. Ocean Press. (In Chinese)
Chen XJ (2016) Theory and method of fisheries forecasting. Ocean Press. (In Chinese)
Chen XJ (2021) Fisheries oceanography. Science Press. (In Chinese)
Shelford VE (1911) Physiological animal geography. J Morphol 22:551–618
Sun LY (1992) Principles of animal ecology. Beijing Normal University. (In Chinese)
Zumholz K (2005) The influence of environmental factors on the micro-chemical composition of cephalopod statoliths. University of Kiel, Kiel, Germany

Basic Theories of Formation of Fishing Ground

4

Xinjun Chen

Abstract

Forecasting fishing ground is one of the important studying contents of fisheries forecasting. In the vast oceans, there are abundant fish and other marine living resources, but there are not all kinds of dense fish which can be caught in the ocean. Some sea areas have the value of exploitation and utilization, while others do not. Therefore, mastering the formation mechanism of fishing ground and its environmental characteristics will provide a theoretical basis for accurate prediction of fishing ground. Of course, in the ocean, the distribution of fishing grounds is not static but has the basic characteristics of dynamic change, that is, fishing grounds will change in response to factors such as climate change, changes in environmental conditions, or excessive fishing intensity; as a result, the original fishing ground will disappear or change. Also some new fishing ground will be developed because of the discovery of new fishing target, the improvement of fishing technology, and the discovery of the value of fishing ground. To this end, this chapter introduces the basic concept of fishing ground, types of fishing ground, and the basic conditions of forming fishing ground. The basic principles of the formation of five excellent fishing grounds, such as the current boundary fishing ground, the upwelling fishing ground, the Eddy fishing ground, the continental shelf fishing ground, and the reef bank fishing ground, as well as their distribution in the global sea area, are described in detail. These fishing grounds are located are rich in nutrient salts, have high primary productivity, and are abundant in bait organisms. Most of them are good places for fish and other marine animals to breed, grow, and live; therefore they provide most of the world's ocean catch.

Keywords

Fishing ground · Basic principle of forming fishing ground · Five excellent fishing grounds

4.1 Basic Concepts and Types of Fishing Grounds

4.1.1 The Concept of Fishing Grounds and Their Characteristics

Fishing Grounds

Extremely abundant fish and other marine biological resources are held in store in the vast seas and oceans, but dense schools of fish available for fishing are not distributed everywhere in the seas and oceans. Because the fish and other marine animals in the seas and oceans are not evenly distributed in various waters, different

X. Chen (✉)
College of Marine Sciences, Shanghai Ocean University, Lingang Newcity, Shanghai, China
e-mail: xjchen@shou.edu.cn

states of distribution are presented due to biological characteristics or to the influence or actions of external environmental factors. Therefore, some sea areas have a high density of fish, while others have a low density, and some sea areas have development and utilization value, while some sea areas do not have development value. For this reason, what we usually call a marine fishing ground generally refers to a place with a certain area that has development and utilization value where commercial marine fish or other commercial marine animals are more concentrated and operations can be carried out by utilizing fishing tools.

Basic Characteristics of Fishing Grounds

Fishing grounds are not immutable but have the basic characteristics of dynamic change. That is, fishing grounds will change with environmental conditions, the restriction of some factors, and factors such as excessive fishing intensity, causing changes, such as disappearance or transition, to occur to the original fishing grounds. For example, the Newfoundland fishing ground is among one of the most famous in the world. Located in the sea area along the shore of the Newfoundland peninsula in Canada (Newfoundland's Grand Banks), this fishing ground is formed by the confluence of the Labrador cold current with the warm waters from the Gulf of Mexico. Historically, the fisheries' resources at this fishing ground are very rich, and the output, especially of Atlantic cod (*Gadus morhua*), is very high, even providing fish for Europe. For several hundred years, the Atlantic cod had flourished without decline. After World War II, fishery technology developed rapidly, with large-scale mechanized trawlers becoming very pervasive and large-scale trawl nets that could catch 200 tons of cod in an hour. By the end of the 1960s, the Canadian federal government found that its fisheries' resources had been reduced by 60% compared to the golden age, with the near absence of Atlantic cod. In 1992, the amount of Atlantic cod resources had decreased to 2% of that 20 years prior; therefore, the federal government declared a fishing ban, that is, the operation of trawlers in

the Newfoundland fishing ground was permanently banned. In 2003, 11 years after the fishing ban, an investigation revealed that there was almost no recovery of Atlantic cod resources in the sea areas of Newfoundland. As of now, Atlantic cod is only allowed to be caught from May to December each year in this fishing ground.

In addition, some new fishing grounds have developed due to the discovery of new fishing objects, enhancements in fishing capacity, an increase in the utilization value of fishing objects and other factors. In fact, people's understanding of fishing grounds has been continuously enhanced through the development of fishery production processes. In ancient times, our ancestors only engaged in simple fishery activities in intertidal zones and shoals or in the vicinity of islands; through production practices, they gradually discovered, understood, and mastered the density of various schools of fish and aquatic animals and the seasonal change patterns, subsequently generating the concept of fishing grounds. Furthermore, due to advancements in science and technology and the continuous development of fishery production, fishing production tools improved and were enhanced, and the ability to fish offshore and in distant waters increased; therefore, fishing grounds developed not only in intertidal zones but also in shallow seas, open seas, deep seas, and oceans, with the continuous development of new fishing grounds. All of these fishing grounds benefited from enhancements in fishing capacity.

Several Basic Issues Studied in the Science of Fishing Ground

In addition to being affected by physiological features and ecological habits, the clustering, distribution, and migration of fish in the seas and oceans are also closely related to external environmental factors. Therefore, in research on the formation principles of fishing grounds, it is necessary to study the physiological features and ecological habits of the relevant commercial fish and commercial marine animals and their interrelationship with the surrounding environmental factors to find the patterns and general principles of the formation of fishing grounds. Therefore, one must master and understand the following basic issues when studying fisheries oceanography (Chen 2016):

1. The physiological features and ecological habits of commercial fish and commercial marine animals: the physiological features mainly include growth, reproduction, feeding, and population.
2. The fishing ground environment (including biotic and abiotic environment) and its changes: biotic conditions refer to the relationship between food organisms and symbiotic organisms as well as other various biological species, whereas abiotic conditions refer to ocean currents, water systems, water temperature, salinity, water depth, substrates, landforms, meteorology, and so on.
3. The relationship between the environmental factors, and their changes, at fishing grounds and the action of fish, and mastery of the main environmental indicators that affect the distribution, migration, and clustering of the fish.
4. Fishing conditions and their patterns of change, which compose the basic principles and main indicators of fisheries' forecasting and their patterns of change: researchers must investigate and ascertain the basic patterns of change in fishing conditions by studying the action status of fishery biological resources (clustering, distribution, migratory movement, and so on) and the interrelationship between it and the surrounding environment and utilize models to predict central fishing grounds, fishing seasons, possible fishing yield, etc.

Basic Conditions for the Formation of Fishing Grounds

Although fish and other commercial marine animals can be seen everywhere in the seas and oceans, such as in estuaries, bays, shallow seas, oceans, etc., fishing grounds do not occur everywhere, and fishing seasons do not occur at the same times during the year. Fishing grounds are often confined to a certain water layer in a certain sea area or even confined to a certain time period. Such limitations mainly depend on the density of the schools of fish and the amount of time those schools spend in that location, in addition to changes in the biological characteristics and ecological habits of the fish (commercial marine animals) and the environmental conditions; therefore, the following basic conditions are necessary for the formation of fishing grounds:

1. There must be large numbers of schools of fish that migrate to, pass, or cluster in a habitat. The main fishing objects of marine fishery production are dense populations of fish or commercial animals that carry out activities such as migration, reproduction, feeding, or overwintering, especially reproducing populations, which have a high density and are stable. Most schools of fish cluster by body length or age, which is particularly apparent for fish species such as salmon and trout. Therefore, when carrying out fishing operations, if the fishing objects have not reached the required fishing specifications (such as young or sexually immature juvenile fish), then the loss will inevitably outweigh the gain, seriously affecting the amount of resources in the coming years and even leading to a decline in fisheries' resources.
2. The environmental conditions must be suitable for the clustering and habitation of fish. If there are external environmental conditions (including biotic and abiotic conditions) for fish and other commercial animals to migrate, reproduce, feed, and overwinter during a certain period in a certain sea area, then they can cluster or inhabit together, thereby creating conditions for the formation of fishing grounds. External environmental conditions mainly include biotic conditions and abiotic conditions. Biotic conditions refer to the relationship between food organisms and symbiotic organisms as well as other various biological species, whereas abiotic conditions refer to ocean currents, water systems, water temperature, salinity, water depth, substrates, landforms, meteorology, etc. The relationship between fish and each of its abiotic and biotic environmental factors does not exist in isolation but is part of a unified and indivisible interactive system. Therefore, when

conducting fishing ground studies and analyses, a comprehensive and systemic viewpoint should be utilized to investigate the issue, and the main marine environmental factors for the formation of fishing grounds must be noted, in addition to identifying the close relationship with other environmental factors.

3. There must be suitable fishing gear and fishing methods. Even though there may be large numbers of schools of fish and other commercial aquatic animals and the marine environmental conditions are suitable for fish to remain in a location, fishing ground development is also dependent on appropriate fishing gear and fishing methods, through which fishery productivity be achieved to the utmost extent. The selection and application of appropriate fishing gear and fishing methods are indispensable key issues.

In short, among the three aforementioned basic conditions, large numbers of schools of fish are the prerequisite. Next, there must be suitable environmental conditions; otherwise, it is impossible for schools of fish to migrate to, pass, or stay in a habitat. Therefore, when selecting or determining fishing grounds, the aforementioned two conditions should be organically combined in accordance with the basic principle of unifying fish distribution with the marine environment. Finally, the ability to conduct fishing operations with the use of appropriate fishing tools and obtain a certain output are secondary conditions for establishing fishing grounds. As long as schools of fish exist and there are suitable marine environmental conditions, with advancements in science and technology, appropriate fishing gear and fishing methods can be developed and utilized responsibly.

4.1.2 Types of Fishing Grounds

Because the formation of fishing grounds is the result of the unity between the marine environment and the biological characteristics of fish and because fish and other fisheries resources are extremely abundant, with great variety, people divide fishing grounds in accordance with actual production and fisheries' management needs. The types of fishing ground divisions are multifarious. In general, multifarious divisions are established in accordance with differences in the distance between the fishing ground and the fishery base, the water depth of the fishing ground, the geographical location, environmental factors, the habitat distribution of fish in different stages of life, the mode of operation, and the fishing objects. Usually, fishing grounds are divided in the following ways:

1. Based on the distance from the fishery base and the water depth at the fishing ground, fishing grounds can be classified as follows:
   (a) Coastal fishing ground: a fishing ground that is generally distributed close to the coast, with a water depth shallower than 30 m.
   (b) Inshore fishing ground: a fishing ground that is generally distributed not far from shore, with a water depth between 30 and 100 m.
   (c) Offshore fishing ground: a fishing ground that is generally distributed farther from shore, with a water depth between 100 and 200 m.
   (d) Deep sea fishing ground: a fishing ground that is distributed in waters with a water depth deeper than 200 m.
   (e) Distant water fishing ground: a fishing ground that is distributed more than 200 nautical miles away from the home country; these grounds can be classified as either an oceanic fishing ground, i.e., a fishing ground distributed in the high seas, or a transoceanic fishing ground, i.e., a fishing ground within the 200-nautical-mile special economic zone of another country.

2. Based on differences in geographical locations, fishing grounds can be classified as follows:
   (a) Fishing ground in a bay: a fishing ground that is distributed in a bay near land

(b) Fishing ground in an estuary: a fishing ground that is distributed in the vicinity of an estuary

(c) Continental fishing ground: a fishing ground that is distributed within the scope of the continental shelf

(d) Fishing ground in a reef bank: a fishing ground that is distributed in the vicinity of a reef bank

(e) Polar fishing ground: a fishing ground that is distributed within the sea area of the two polar circles

(f) Fishing grounds according to specific geographical names: for example, the Yanwei fishing ground, i.e., the fishing ground that is distributed in the inshore area in the vicinity of Yantai and Weihai, the Zhoushan fishing ground, i.e., the fishing ground that is distributed in the inshore area in the vicinity of Zhoushan, the Beibu Gulf fishing ground, i.e., the fishing ground that is distributed in the sea area of the Beibu Gulf, etc.

3. Based on differences in oceanographic conditions, fishing grounds can be classified as follows:

(a) Fishing ground along a current boundary: a fishing ground that is distributed in the vicinity of the confluence of two different types of water systems

(b) Upwelling fishing ground: a fishing ground that is distributed in the upwelling waters

(c) Fishing ground in the eddy area: a fishing ground that is distributed in waters in the vicinity of an eddy

4. Based on differences in the stage of life of the fish, fishing grounds can be classified as follows:

(a) Spawning fishing ground: a fishing ground that is distributed in the sea area of fish spawning grounds

(b) Feeding fishing ground: a fishing ground that is distributed in the sea area of fish feeding grounds

(c) Overwintering fishing ground: a fishing ground that is distributed in the sea area of fish overwintering grounds

5. Based on differences in the mode of operation, fishing grounds can be classified as follows:

(a) Trawling fishing ground: a fishing ground where trawling is used

(b) Purse seine fishing ground: a fishing ground where purse seine gear is used

(c) Gillnet fishing ground: a fishing ground where gillnets are used

(d) Jigging fishing ground: a fishing ground where jigging is used

(e) Stationary gear fishing ground: a fishing ground where stationary gear is used

In seas and oceans, the waters with sufficient nutrient salts, high primary production, and abundant food organisms are mostly good places for fish and other marine animals to reproduce and inhabit, and they often form excellent fishing grounds. Among the various aforementioned major types of fishing grounds, upwelling fishing grounds, fishing grounds along current boundaries, fishing grounds in eddy areas, continental fishing grounds, and fishing grounds in reef banks are considered excellent fishing grounds. Notably, a certain sea area may be classified as a continental fishing ground as well as a fishing ground along a current boundary, eddy area, or reef bank. For example, the Peru fishing ground is both an upwelling fishing ground and a continental fishing ground.

4.1.3   Fishing Period (Fishing Season)

The fishing period refers to the time period when fish and other commercial aquatic animals are highly concentrated in a certain sea area and have a certain fishing scale and production value. Due to the habits of fish and other commercial aquatic animals and the effects of their water environment, regular cluster migrations, such as spawning, feeding, and overwintering, form, generating a fishing season. The formation of a fishing season is based on population density and time: "initial fishing season," i.e., fewer population numbers; "peak fishing season," i.e., the population numbers are the densest, and the output is the highest; and "end fishing season," i.e., the population numbers decrease progressively and the output declines gradually. Fishing seasons can be divided based on season: spring fishing season, summer fishing season, autumn fishing season, and winter fishing season. Fishing seasons can be based on the fishing object, e.g., the large yellow croaker fishing season, the largehead hairtail fishing season, the prawn fishing season, etc., and on the fishing object and the integration of various factors, e.g., the Zhoushan largehead hairtail winter fishing season, etc. The length of a fishing season is related to fish biology, the location of the fishing grounds, environmental changes, etc., and these factors can cause the duration of a fishing season to be long or short or for the peak period to shift to an earlier time or a later time.

The increase or decrease in yield for a fishing season is mainly related to the amount of fish resources and other commercial aquatic animals and to changes in the water environment. The disappearance of the fishing season for some traditional commercial species (such as the large yellow croaker fishing season) in the inshore sea areas of China is also due to changes in the water environment. The earliness or lateness of a fishing season is also closely related to water temperature. Research has shown that the amount of catch during the *Coilia nasus* fishing season at the Yangtze River estuary is closely related to water temperature and tide; when the water temperature increases to approximately 12 °C, the peak fishing season begins (the water temperature corresponding to the maximum daily catch each year is between 13 °C and 14.5 °C, with an average of 13.7 °C). The most obvious feature of a fishing season is its seasonal alternation (change); for example, the start of the average fishing season for *Scomber australasicus* in the central and southern parts of the East China Sea is on July 12, that is, in mid-July. The duration of the fishing season is nearly 2 months, generally ending in early or mid-September, and can be delayed to late September in individual years. The peak period for catch yield is mainly in late July and in August. The fishing season output and regional scope for different years may also change greatly, which is related to changes in

the amount of resources and the amount of fishing effort. For example, from 2003 to 2008, the scope of the fishing grounds for *Scomber australasicus* in the central and southern parts of the East China Sea was greater in 2005, 2006, and 2008; the degree of concentration in the central fishing grounds was higher in 2003, 2004, and 2008, and the annual output had greater volatility. Therefore, accurate mastery of fishing seasons is an important tool for attaining high yield in fishery production and is also an important condition for enhancing fishery production efficiency.

4.1.4 Evaluation of the Value of Fishing Grounds

Whether a certain fishing ground has development and utilization value is usually scientifically evaluated by utilizing the following standards: (1) the standing stock of fisheries resources, which can be reflected in fishing yield; (2) the density of the schools of fish in the fishing ground, that is, is it a sea area where one can carry out operations and obtain a certain output; (3) the duration of the fishing period, that is, the time in which a certain output can be obtained; (4) the degree of suitability for fishing, which is mainly manifested in the fishing gear and fishing methods; and (5) the distance from the base, which involves the maritime operating cost of the fishing vessels and other costs such as replenishment by transport vessels and hauling the catch to shore, which mainly consider economic efficiency. Among them, the amount of resources is the most important.

4.2 General Principles for the Formation of Excellent Fishing Grounds

Based on the formation conditions for and habitat environment of fishing grounds, it is generally believed that fishing grounds along current boundaries, upwelling fishing grounds, fishing grounds in eddy areas, continental fishing grounds, and fishing grounds in reef banks are the five major excellent fishing grounds. The basic principles of the formation of each fishing ground are described as follows.

4.2.1 Fishing Grounds Along Current Boundaries

Concept of a Current Boundary

The surface of discontinuity at the intersection of two water masses, water systems, or ocean currents with significantly different properties is called a current boundary or ocean front. Drastic changes occur on both sides of a current boundary, including changes in water temperature, salinity, dissolved oxygen, nutrient salts, and other quantities of oceanographic elements as well as the quality and quantity of biological phases; in particular, at the confluence of a cold and a warm current, changes in various oceanographic elements are more significant. Along its line of discontinuity, local eddy current, divergence, and convergence phenomena are more obviously generated. These hydrological conditions along current boundaries are conducive to the reproduction, growth, and aggregation of fish, and these environmental conditions for food organisms and fish and other populations are often favorable at confluence areas; therefore, good fishing grounds are often formed.

Reasons for the Formation of Fishing Grounds Along Current Boundaries

The reasons for the biological aggregation of fish along current boundaries mainly involve biology, hydrology, etc.

1. At the confluence of two ocean currents with different properties, due to divergence and counterclockwise eddy currents, the nutrient salts and organic nitrate detritus deposited in the deep layer that have not been fully utilized are brought to the upper layer, allowing phytoplankton to rapidly reproduce under photosynthesis and thus providing enriched nutrients for fish food organisms and forming a high-production sea area, where fish are able to aggregate. For example, in the divergence zone between the North Equatorial Current and the Equatorial Countercurrent, the ascendence of lower-layer seawater presents a dome shape or mountain chain form and is a good fishing ground for tuna.

2. At the interface of confluence areas, significant changes occur to the water temperature and salinity of the two different types of water systems (masses), and a greater gradient emerges, which can be regarded as a barrier or boundary for the distribution of organisms from different biospheres. The plankton and fish within different water systems that have followed the currents encounter the "barrier" and cannot go beyond, thus clustering in the vicinity of the current boundary and thereby forming a good fishing ground.

   For example, the Pacific saury (*Cololabis saira*) in the northeastern sea area of Japan migrates southward with the Oyashio Current in November to the Tokiwa area to spawn every year. In some years, the Tokiwa inshore warm water mass controls the formation of a warm water barrier, and the cold current cannot extend to the south; due to this water barrier, schools of Pacific saury stop moving southward and congregate in dense groups. As another example, every year, schools of skipjack tuna in the inshore waters of Japan migrate northward toward the Tokiwa area in May to feed. The surface water temperature in the vicinity of the Kuroshio Current front changes acutely in early June, with the north side below 17 °C, forming a cold-water barrier that prevents the skipjack tuna (*Katsuwonus pelamis*) from moving northward and causes them to cluster in the waters of the Kuroshio Current on the south side of this Kuroshio Current front. If a warm water area (or cold-water area), i.e., a narrow Oyashio Current (or Kuroshio Current), integrates at the barrier and a barrier channel is formed, then Pacific saury (or skipjack tuna) moves rapidly southward (or northward) along the channel.

3. In the mixing zone of two different types of water systems, the food organisms include

biological populations from locations with two different water system properties. Such zones have species from both a high-temperature, high-salinity water system and a low-temperature, low-salinity water system, thereby forming an abundant comprehensive food organism population generated from two types of water systems, providing fish and shrimp with food conditions that one type of water system cannot produce alone. Convergence and clockwise eddy currents make the surface seawater converge and sink. As a result, various types of organisms in the vicinity of the current boundary, from plankton and small fish to large fish, converge here, that is, they congregate at the center of the convergence zone, forming a good fishing ground.

Kitahara's Law of Fishing Conditions

In seas and oceans, current boundary areas at the confluence of two types of water masses or ocean currents with different properties provide important conditions for the formation of a good fishing ground, information that has been known by coastal fishermen for generations. However, the Japanese scholar Tasaku Kitahara was the first to theoretically summarize their patterns and propose insights regarding laws of nature. He carried out an analysis in accordance with the production report of a whaling ship from 1910 to March 1912 combined with data from many years of survey and research; in 1913, he concluded that "the place where tuna, Pacific saury, sardine (*Sardina pilchardus*), and whales aggregate in large groups is in the vicinity of the conflict line (confluence) between two ocean currents" (Fig. 4.1). After this, he carried out further surveys and research and proposed the three "Kitahara's Laws of Fishing Conditions" in 1918: (1) fish all aggregate in the vicinity of the confluence between two ocean currents; (2) because offshore ocean currents approach along the coast, the currents can drive away concentrated schools of fish; and (3) schools of fish will aggregate in the channel area connecting two ocean currents because of the approach of ocean currents flowing from both sides.

In accordance with the "First Law of Kitahara's Fishing Conditions," Michitaka Uda conducted further marine surveys of fishing grounds along current boundaries and published a paper in 1936, "Relationship Between the Center of a Fishing Ground and the Current Boundary in the Northeastern Sea Area," which discussed the distribution of the schools of fish in fishing grounds along current boundaries: "schools of fish generally cluster in the vicinity of a current boundary, and schools of fish are more concentrated in places where the current boundary has large convex and concave twists." From this, Michitaka Uda developed and enriched the "First Law of Kitahara's Fishing Conditions" (Michitaka 1963).

Method for Judging Current Boundaries

Current boundaries have generally undergone processes of occurrence to development to attenuation. During the development stage, they often form better fishing grounds. Rip currents or fronts are generally used as current boundaries, referring to the convergence phenomenon of surface flow in local waters. In the sea areas along rip currents, there is generally substantial aggregation of floating substances or foam, sea fog, and irregular waves. However, rip currents can also form in water masses with similar properties and are often confused with current boundaries by observation; therefore, current boundaries should be additionally identified by surface temperature, salinity, water color, and other features on both sides of the boundary.

Ocean fronts can be determined by calculations, visual observation, or instrumental observation. For small-scale fisheries, they are often identified by the line of aggregation of floating objects on the sea or the different water color and transparency boundary lines in the seawater. As a result of the development of marine telemetry, ocean satellite images and telemetry data are also used to observe ocean fronts. The calculation method is based on the maximum horizontal gradient of temperature and salinity. The general basis for judging the position of ocean fronts by water temperature and salinity is as follows:

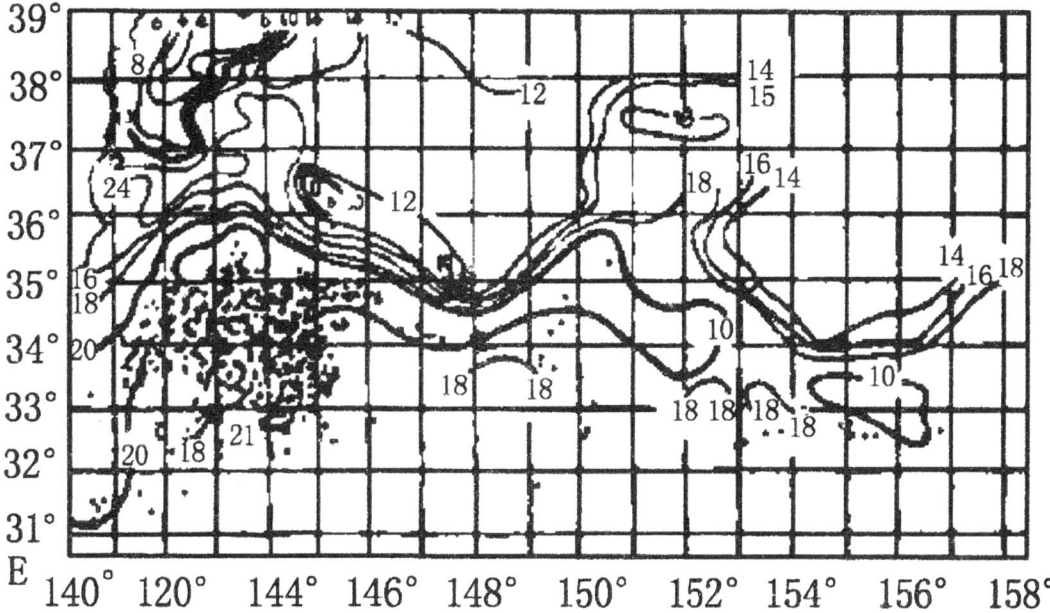

Fig. 4.1 Fishing grounds for albacore tuna (*Thunnus alalunga*) in confluence areas (Chen 2014, 2021)

$$\Delta WT/\Delta X \geq 1°C/20 \text{ nautical miles or } \Delta WS/\Delta X$$
$$\geq 0.4‰/20 \text{ nautical miles}$$

where ΔT is the amount of change in temperature, ΔS is the amount of change in salinity, and ΔX is the horizontal distance (Chen 2014, 2021).

Marine current boundaries can form in any waters in oceans and along coasts. In oceans, the representative current boundaries include the Kuroshio Current front, the Oyashiro front, the subtropical convergence line, and the Antarctic convergence line, among others. In coastal sea areas, coastal fronts are formed by the confluence of coastal waters and offshore waters near the continental margins, and current boundaries also form between estuarine continental runoff and coastal waters, i.e., estuarine fronts; in addition, current boundaries form in the vicinity of islands, reefs, capes, etc., as a result of the topography.

Amount of Aggregation of Floating Substances in Fishing Grounds Along Current Boundaries

Only the aggregation of substances such as plankton at the convergence of ocean currents is explained here. The role of clockwise eddy currents (northern hemisphere) in the aggregation of substances such as plankton in fishing grounds within eddy areas is set forth. In current boundary areas, a convergence phenomenon often exists, and the amount of aggregated plankton and other floating substances on the surface of the sea can be expressed by using the following formula:

$$A = -\int \delta\left(\frac{\partial u}{\partial x} + \frac{\partial v}{\partial y}\right) dt = \bar{\delta}KT$$

where μ and ν are the components of the horizontal current in the directions of the x and y; K is the degree of convergence and $K = -\left(\frac{\partial u}{\partial x} + \frac{\partial v}{\partial y}\right) > 0$; and $\bar{\delta}$ is the average density of the floating substances in time T (Chen 2014, 2021).

The above formula shows that in time T, the amount of aggregation A of floating substances is directly proportional to the degree of convergence K and, therefore, an important factor for ascertaining the conditions of a fishing ground. However, the relationship between the convergence of ocean currents and the aggregation of fish is not a simple statistical relationship; in fact, the following three relationships are observed:

(1) with a weaker degree of convergence, the flow rate is lower than the critical velocity for the positioning of fish; at this time, zooplankton and fish gradually accumulate in convergence zones; (2) with a moderate degree of convergence, there is less accumulation of zooplankton, and the fish can maintain positioning in the current and generally aggregate upstream of the convergence zone; and (3) with a strong degree of convergence, fish swim in the top current and drift with the current, generally aggregating downstream of the convergence zone; due to the high flow rate, zooplankton cannot possibly aggregate, and in this case, fishing grounds are not formed.

Distribution of Fishing Grounds Along Current Boundaries

The world's fishing grounds along current boundaries are mainly distributed in the following sea areas: (1) at the confluence of the Gulf Stream with the Labrador cold current offshore of New-foundland in the northwestern part of the Atlantic Ocean, an area that has high cod production; (2) at the confluence of the North Atlantic warm current with the Arctic cold current in the Northeast Atlantic Ocean from Iceland to the Spitzbergen archipelago, Bear Island, and inshore of Norway, an area with high cod and herring production; (3) at the confluence of the Kuroshio warm current with the Oyashio cold current in the north-western part of the Pacific Ocean from the inshore of eastern Japan and the Kuril Islands to Kamchatka Peninsula and the Aleutian Islands, an area that is teeming with Pacific saury, whales, skipjack tuna, tuna, herring (*Clupea pallasi*), etc.; (4) at the confluence of the East Australian Current with the West Wind Drift at the east coast of Australia and along the coast and offshore of New Zealand, an area with high tuna production; (5) at the confluence of the Brazil warm current with the Falkland cold current in southeast South America, an area with high cod, tuna, sardine, and squid production; (6) at the confluence of the Agulhas Current with the West Wind Drift in South Africa; and (7) the Antarctic whale fishing ground in the Antarctic ice belt-Antarctic Convergence Zone.

4.2.2 Fishing Grounds in Eddy Areas

Eddy currents are generated in the waters along current boundaries (water systems of different temperatures and salinities) or in places with irregular topography, such as islands, reefs, and so on. Eddy currents of various scales cause the mixing of water from the upper and lower layers, promoting the mass reproduction of food organisms, thereby forming good feeding locations for fish and shrimp. For example, the topographical eddy current in the northeastern inshore of the Tsushima Islands in Japan is a good environment for the formation of a fishing ground for Pacific mackerel (*Scomber japonicus*). In reef banks in shallow waters, sunlight penetrates to the seabed, promoting the mass reproduction of algae and providing fish with a good habitat. Eddy currents can be classified based on the mechanism of formation: a mechanically generated eddy, a topographically generated eddy, and a compound eddy.

Mechanical Eddy

Due to the unstable fluctuations caused by the instability of shearing, the difference in the relative flow rates on both sides of a current boundary generates an eddy current, and such eddy currents are called mechanical eddies. According to the circulation theory of Bjerknes, vorticity can be expressed using the following formula:

$$\xi = \frac{u_1 - u_2}{L} = \frac{\int \frac{gdp}{\rho_2} - \int \frac{gdp}{\rho_1}}{2\omega L^2 \sin \phi}$$

$$= \frac{\rho_1 - \rho_2}{\rho_1 \rho_2} \times \frac{1}{2\omega L^2 \sin \phi} \int gdp$$

where ρ_1 and ρ_2 and u_1 and u_2 are the density and velocity of the two water systems, respectively, and L is the distance between the two water systems (Chen 2014, 2021).

Because density is inversely proportional to water temperature, vorticity is directly proportional to the density gradient or water temperature gradient of the two water systems. In other words, the biomass clustered in an eddy current area has a directly proportional relationship with the water temperature gradient, and the greater the temperature difference, the greater is the possibility of forming a good fishing ground.

If a difference occurs in the relative velocity in the movement of the two water masses based on a difference in location (different latitudes), then changes will also occur to the corresponding angle, thus creating conditions for the formation of many eddy currents between the two water masses, and therefore, eddy currents often exist in the waters along current boundaries. In the Northern Hemisphere, there are two types of eddy currents: clockwise eddy currents, in which the central surface seawater converges and sinks, and counterclockwise eddy currents, in which the lower layer of the seawater in the central portion ascends and spreads. Furthermore, divergence and convergence will also be generated in the waters along current boundaries. In divergence zones, the surface seawater spreads, and the seawater from the lower layer ascends. In convergence zones, the surface seawater converges and sinks (Chen 2014, 2021).

Topographically Generated Eddy

Eddy currents formed due to topographical factors, such as islands, peninsulas, and reefs, are called topographical eddy current systems. For islands that exist within a current or peninsulas that protrude from a current, back eddy currents are formed on the back side of islands or peninsulas below the current, and local convergence occurs. Eddy currents are also generated in reefs and piles that extend up from the bottom of the sea or ocean. Topographically generated eddies of various scales cause the mixing of water from the upper and lower layers, become good feeding locations for fish by promoting the production of bait.

For example, the fishing ground for whales in the vicinity of South Georgia Island in the Antarctic Sea was formed due to the eddy current on the east side of this island when the Bellingshausen seawater system and the Weddell Drift passed through South Georgia Island, thereby promoting the mass reproduction of Antarctic krill (*Euphausia superba*) in this sea area, becoming an important fishing ground for Antarctic krill. Additionally, it is also a good fishing ground for the blue whale and the fin whale.

Compound Eddy

Eddy currents generated due to the combined action of both mechanical and topographical actions are called a compound eddy. For example, the sea area in the vicinity of the Tsushima Strait is a typical compound eddy system. During spring and summer in the Sea of Japan, the middle water layer and below are entrenched by a cold-water mass, with the Tsushima warm current above it. The topography affects the route of travel of the Tsushima warm current and makes it become serpentine, the cold-water mass trapped in this sea area experiences large protrusions of large isobath curves. An ascending cold-water eddy appears, and cold-water eddy current groups are correspondingly generated in the middle of cold-water eddy current groups, resulting in a good fishing ground for sardine, Pacific mackerel, etc. The changes in and movements of these eddy current groups are related to the distribution of the schools of fish. When the Tsushima Current is strong, the counterclockwise eddy current of the water layer is significant; additionally, a current boundary also develops, and fishery development thrives (Fig. 4.2).

4.2.3 Upwelling Fishing Grounds

General Principles for the Formation of Upwelling Fishing Grounds

Upwelling sea areas are some of the most fertile sea areas among seas and oceans of the world. Although their area accounts for only 1/1000 of the total marine area, the fishing yield accounts for about half of the total marine fishing yield of the world. In 1906, after A. Nathansohn studied a large number of fishery production materials and practices, he put forward the assertion that "upwelling waters are generally high in productivity and thus form excellent fishing grounds"; we refer to this as Nathansohn's Law (Chen 2014, 2021).

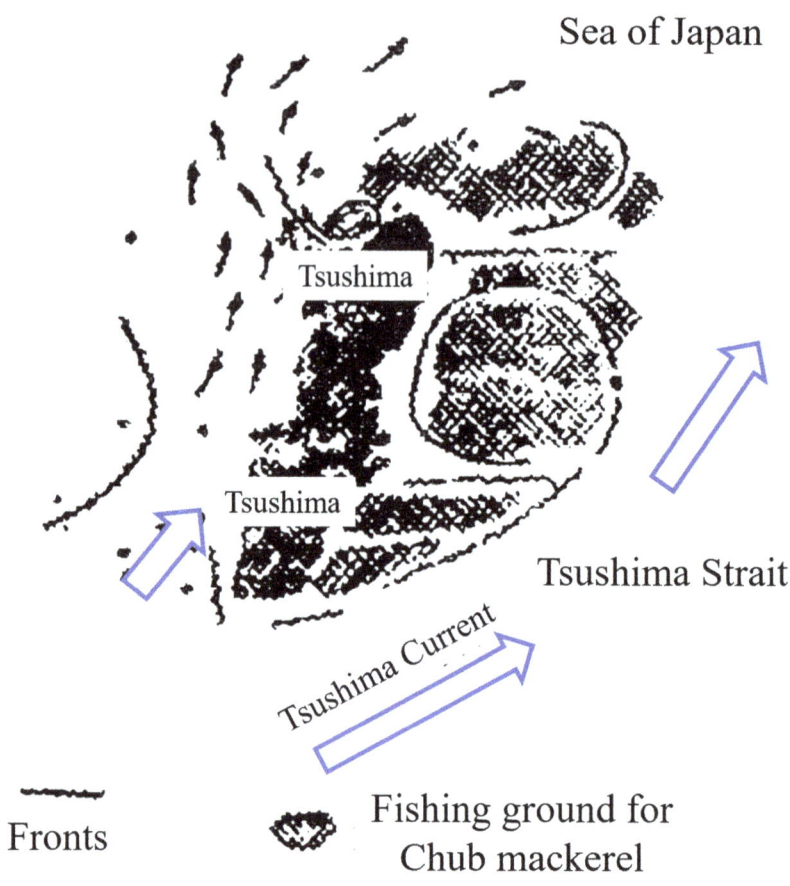

The principles of their formation are as follows. In the upper layer of seas and oceans, the photosynthesis by phytoplankton is comparatively strong, the nutrient salts (phosphates, nitrates, etc.) contained in the seawater are consumed, and the stock on hand gradually decreases. In contrast, in the sediments in the deep layers and seabed of seas and oceans, the remains of organic substances are decomposed and reduced by bacteria, continuously generating nutrients. These nutrients must be drawn to the surface through the ascending movement of seawater and generate organic substances through photosynthesis. The process that causes seawater to ascend is upwelling.

In upwelling areas, cold water in the lower layer ascends, the water temperature decreases, the salinity increases, nutrient salts are continuously replenished and enriched, promoting the mass reproduction of phytoplankton, and the transparency of the seawater decreases. Water in the lower layer contains less oxygen, and when it ascends to the surface, due to the low temperature of surface water, the oxygen in the atmosphere dissolves in large amounts for replenishment. Therefore, locations with large amounts of water ascending from lower layers that contain abundant nutrient salts are places with high productivity; thus, food organisms are abundant, thereby forming a good fishing ground. In upwelling areas, the productivity of marine organisms is generally higher. From the Somali sea area to the Gulf of Oman in the India Ocean, primary production can reach 5 g of carbon per square meter per day, and the productivity in the Peru sea area is also very high. Compared with a water mass of the same depth, upwelling water masses have a low temperature, high salinity, and low oxygen content, are rich in nutrient salts, and contain abundant plankton.

Types of Upwelling

Upwelling is generally caused by the continuous blowing of relatively stable winds along the coast and the divergence of the winds in the equatorial zone in the regression zone and subtropical zone. In principle, upwelling is caused by the divergence of marine surface water flow, and such divergence is also formed by the existence of a certain specific wind field or coastline or other special conditions. Therefore, the term upwelling, in a broad sense, includes other marine processes such as divergence and vertical circulation. The vertical convection process is limited to being generated by the cooling and sinking of surface water in middle- and high-latitude waters during winter.

The types of upwelling are generally divided into (1) wind-induced upwelling caused by prevailing winds along the coast of a continent; (2) general upwelling caused by the divergence in the confluence area of two currents and the open ocean sea area; and (3) upwelling generated by the induction of counterclockwise circulation (northern hemisphere; clockwise in the southern hemisphere). In addition, there are also local upwellings formed by special tectonic landforms such as islands, headlands (capes) that protrude into the sea, reefs, and seamounts. Among the different types, wind-induced upwelling has the greatest potential force.

Wind-Induced Upwelling

Upwelling is generated along coasts due to the effects of wind and ocean currents. For a coast in the northern hemisphere, with the wind direction parallel to or forming an angle with the coastline (Fig. 4.3), the transport direction of surface seawater is to the right of the wind direction, thus generating offshore currents. Because of fluid continuity conditions, the seawater at lower layers will replenish and ascend to the sea surface; this process is called upwelling. The depth involved in such upwelling is generally comparatively shallow, approximately 200 meters. The vertical flow rate for upwellings is generally very slow, approximately 0.1–3.0 m/day, and plays an important role in the production of

marine organisms. If the rate of ascendence is slower, then primary production is greater. The spread of ascending water masses is of important significance in terms of the formation of fishing grounds. The strength of an upwelling is closely related to wind speed, the angle between the wind direction and coastline, and topography. Based on research results, the coastal upwelling areas are good locations for sardine, anchovy (Engraulidae), Pacific mackerel, and other pelagic fishes, and fishing grounds for tuna and other large pelagic fishes can form at the outer margins of the upwelling waters.

There are four famous upwellings along continental coasts of the world: (1) the upwelling formed due to the California Current inshore along the west coast of the North American continent; (2) the upwelling formed due to the Peru Current inshore along the west coast of South America; (3) the upwelling formed due to the Canary Current inshore along the northwest coast of Africa; and (4) the upwelling formed due to the Benguela Current inshore along southwest Africa. The aforementioned four upwellings occur along the west coast of the continents, that is, in the eastern part of the ocean.

Divergent Upwelling

Upwelling caused by divergence in the confluence area of two ocean currents has already been discussed in the section on fishing grounds along current boundaries. The upwelling generated by the equatorial current system is discussed here. In the equatorial current system, the equatorial current flows from east to west. The North Equatorial Current is situated at 8°–18°N, and the South Equatorial Current can flow along the equator and extend to 5°N. An Equatorial Countercurrent exists between the two equatorial currents, flowing from west to east and situated between 3° and 10°N. There are several divergent systems in this current system, and these divergent systems are situated along the edge of the North and South Equatorial Currents. Figure 4.4 is a map of the vertical distribution of temperature, salinity, flow rate, and dissolved oxygen between 10°S and 20°N in the Pacific Ocean. As seen in

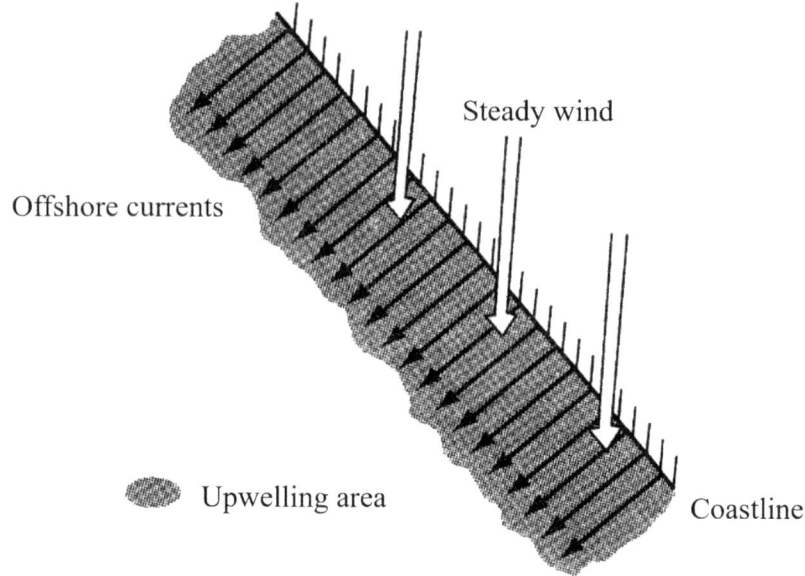

Fig. 4.3 Schematic diagram of upwelling generated by wind (Chen 2014, 2021)

the Fig. 4.4, there is a descending current at the southern boundary of the Equatorial Countercurrent and an upwelling at the northern boundary; correspondingly, between the equator and the countercurrent, a descending current is generated near the boundary of the countercurrent, and an upwelling is generated near the boundary of the equator, thereby forming two vertical circulation systems, that is, divergence is generated on the northern boundary of the countercurrent and along the equator, and convergence is generated on the southern boundary of the countercurrent. The upwelling and descending current generated by divergence and convergence are superimposed within the main ocean current, presenting a spiral movement. Due to the ascent of seawater, seawater from the lower nutrient salt-rich layer is brought to the surface layer; therefore, the two divergent sea areas at the equator and the northern boundary of the countercurrent are rich in plankton and are high in productivity.

Anticline Structure of Water Temperature

The divergence of upper layer waters, which causes the ascent of cold water from the deep layer, causes the cline to bulge (Fig. 4.5); this water temperature distribution is called the anti-cline structure of water temperature. In the eastern

part of tropical oceans, especially in the eastern part of the tropical Pacific Ocean, the Equatorial Countercurrent bifurcates at the continental shelf and enters northward and southward into the North and South Equatorial Currents, respectively. At a depth of 150 m at the equator, there is an Equatorial Undercurrent that flows from west to east, above which the nutrient salts are abundant. This undercurrent generates divergence due to the seawater along the equator and the ascent of cold water; therefore, the distribution of isotherms presents mountain-like bulges, forming a ridge-like water temperature structure.

In upwelling sea areas with an anticline water temperature structure, the thermocline rises to the vicinity of the sea surface, reducing the water layers for pelagic fish habitats, and schools of fish are more densely grouped, forming good fishing grounds where productivity is generally very high. Survey results have shown that the depth distribution pattern for the thermocline in the equatorial waters of the oceans progressively increases from east to west. For example, the depth of the thermocline in the eastern part of the equatorial Pacific Ocean is very shallow, with the shallowest part being only 10 m or 15 m and, overall, generally shallower than 50 m, but the depth gradually increases from

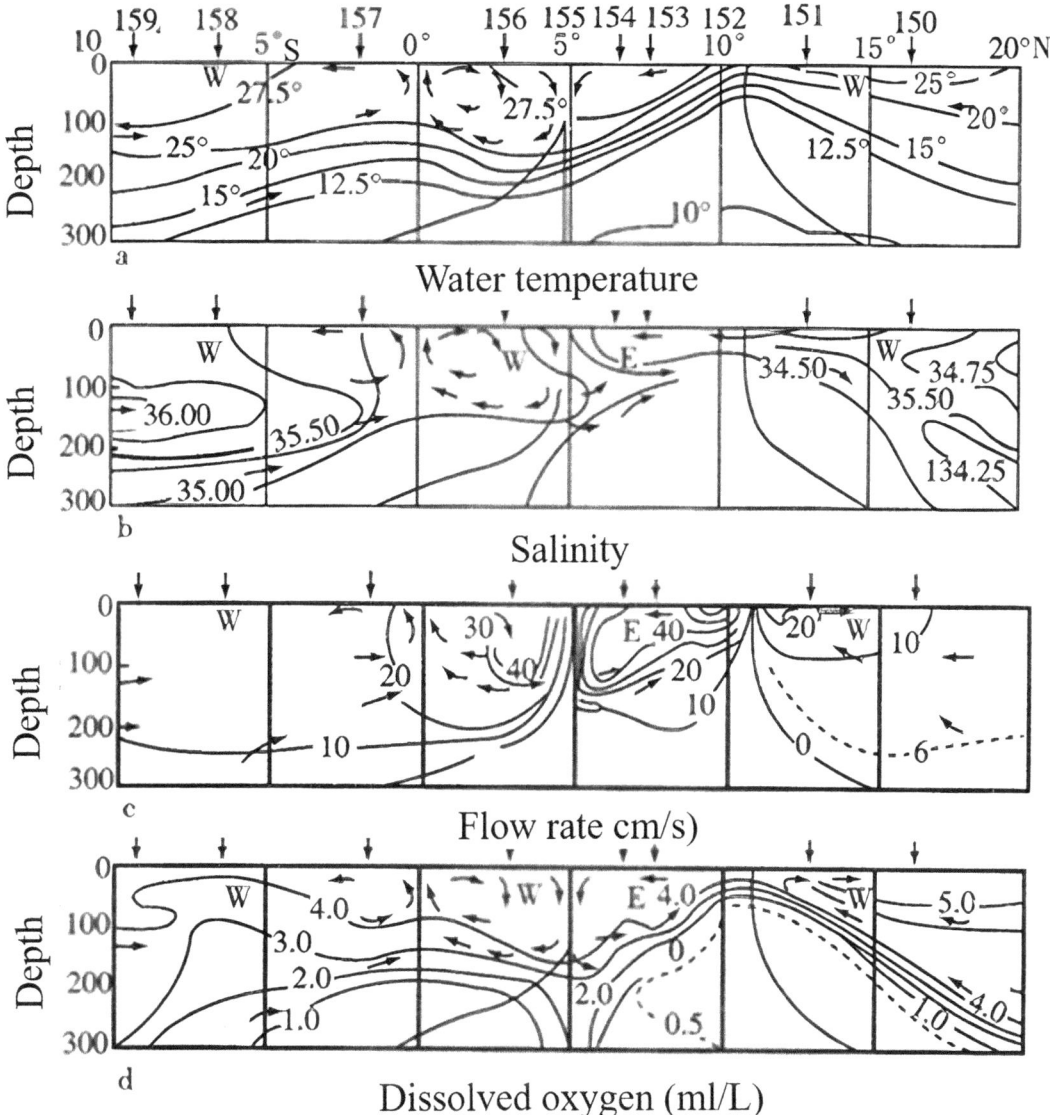

Fig. 4.4 Map of the vertical distribution of temperature, salinity, flow rate, and dissolved oxygen between 10°S and 20°N in the Pacific Ocean (Chen 2014, 2021)

east to west, ranging from 150 to 200 m in the western part of the Pacific Ocean. The gradual increase, from east to east, in the distribution of the surface water temperature of equatorial waters generates strong divergence in the eastern part of tropical and equatorial waters.

Cold-Water Dome

In the offshore waters of Costa Rica in the eastern part of the tropical Pacific Ocean, where the Equatorial Countercurrent reverses, which causes a counterclockwise circulation and induces the ascent of cold water from the lower layer, the isotherms present a rounded dome-like bulge

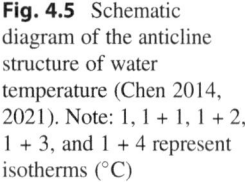 **Fig. 4.5** Schematic diagram of the anticline structure of water temperature (Chen 2014, 2021). Note: 1, 1 + 1, 1 + 2, 1 + 3, and 1 + 4 represent isotherms (°C)

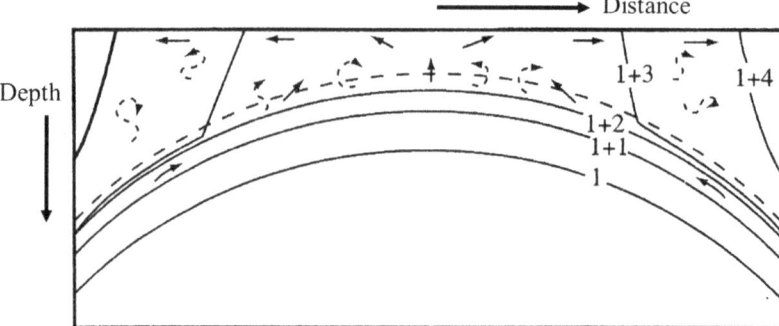

and form a dome-like water temperature structure, also called a cold-water dome, and more specifically, the famous Costa Rica Thermal Dome. The sea area surrounding this cold-water dome is often an important fishing ground for various types of fish, such as tuna and jumbo flying squid.

Upwelling waters provide prime conditions for high-yield fishery production, but there are exceptions. In actual fishery production, because of the low oxygen content of water in deep layers in upwelling areas, schools of fish can dissipate. For example, along the coast of Cochin in the Indian Ocean, the oxygen content of the upwelling waters is very low, approximately 0.25 mL/L (oxygen saturation below 5%), causing demersal fish and lobsters to dissipate and decreasing the success of trawling operations; additionally, in the Gulf of Aden, because the oxygen content of the water at the bottom layer at upwellings is below 2 mL/L, the fishing yield from trawling has significantly decreased.

Distribution of Upwelling Fishing Grounds

In upwelling areas, especially in deep sea areas far from continents, a surge of nutrient salt-rich seawater from a deep layer will result in good fishing grounds. The main upwelling fishing grounds in the world are shown in Fig. 4.6: (1) the California Current area, which produces sardine, mackerel, and albacore tuna; (2) the Peru Current area, which produces anchovy, tuna, and hake; (3) the Benguela Current area, which produces sardine; (4) the northwest African Canary Current area, which produces sardine, Pacific mackerel, cod, tuna, octopus (*Octopoda*), squid, and demersal fishes; and (5) the

northwestern current area of the Indian Ocean, which produces tuna and purpleback flying squid (*Symlectoteuthis oualaniensis*).

4.2.4 Continental Fishing Grounds

The continental shelf, especially the shallow sea near land, has comparatively sufficient sunlight penetration from the sea surface to the seabed, and various nutrients are brought from continental runoff and from the deep layer of offshore waters; it is a good location for the reproduction, feeding, and overwintering of fish, shrimp, shellfish, and other commercial animals. Various fish and other fishing objects migrate and cluster in sea areas of the continental shelf, thereby forming excellent fishing grounds.

Main Water Systems That Affect Continental Fishing Grounds

Generally, the main water systems that affect continental fishing grounds are coastal water and open ocean water (ocean water). Usually, the body of water between the coastline and the 200 meter isobath is called coastal water, and the body of water beyond the 200 meter isobath is called ocean water. Due to the differences between coastal water and ocean water in temperature, salinity, and water color transparency, a frontal surface forms in their confluence, providing very good conditions for the formation of fishing grounds. The main factor that affects coastal water is continental runoff, especially in sea areas such as estuaries and bays. Additionally, because coastal waters, tides, tidal currents, and

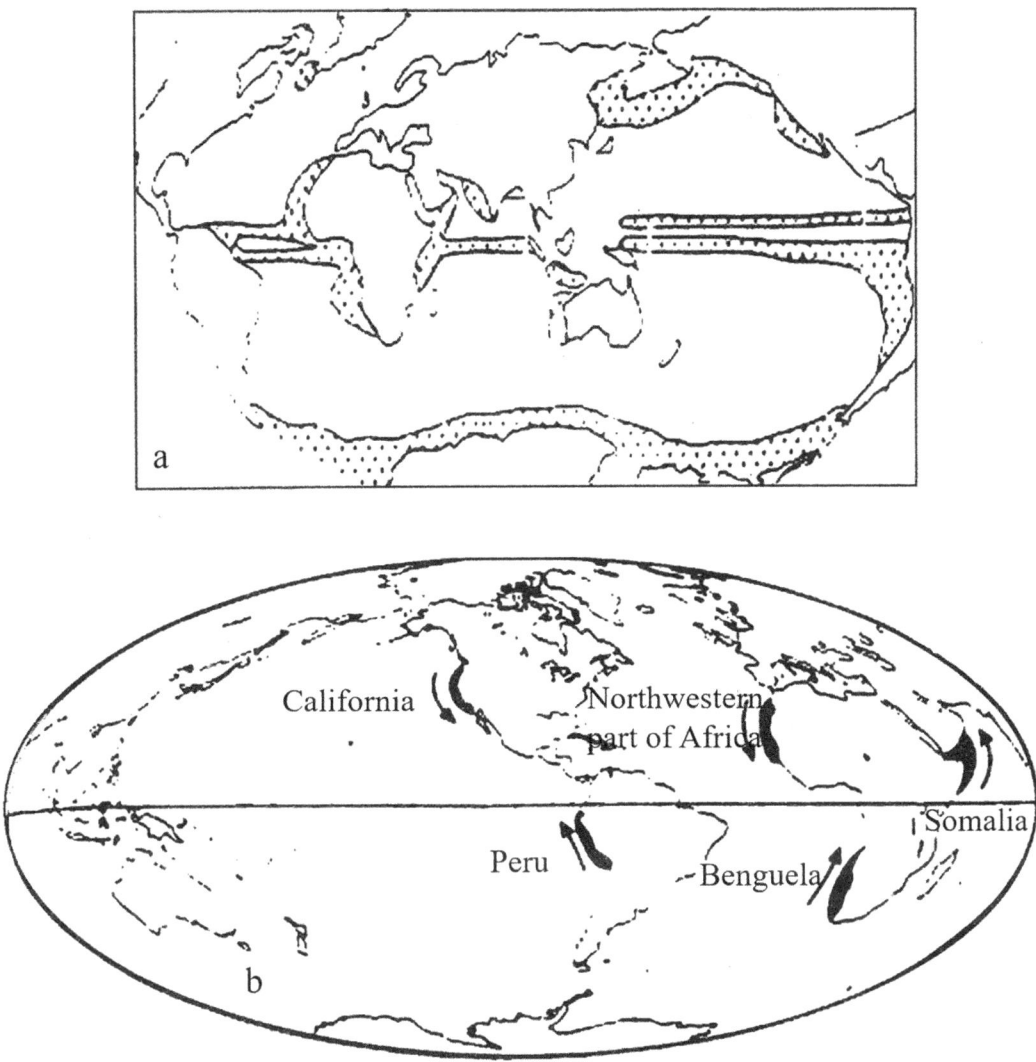

Fig. 4.6 Schematic diagram of the distribution of the main upwellings in the seas around the world (Chen 2014, 2021). (**a**) Distribution of upwellings; (**b**) distribution of main upwellings in sea areas along coasts

waves, among other factors, have comparatively large effects, these bodies of water are fully mixed.

Conditions for the Formation of Continental Fishing Grounds

The continental shelf provides good fishing grounds with the highest development rates; with only 7.6% of the area, it accounts for 90% of the world's fishing yield. Based on existing data analyses, among the worldwide seas and oceans, the fishery output is approximately 12.5 liters per square kilometer in shallow seas with a water depth within 100 m, 5.4 liters per square kilometer in sub-shallow sea areas at depths between 100 and 200 m, and only 1 liter per square kilometer in sub-deep seas with a water depth of 300 meters.

The following characteristics lead to the formation of continental fishing grounds:

(1) Rivers input a large amount of nutrients.
(2) The water is shallow, the bodies of water are fully mixed under the action of wind, tides, and convection, the bottom layer replenishes the top layer, and the nutrients in the entire body of water are plentiful.
(3) Photosynthesis is sufficient, and there is mass reproduction of phytoplankton; generally speaking, the photosynthesis layer stops at 60–150 meters.
(4) The water is shallow; therefore, the material cycle is fast, and the primary production is high.
(5) Due to abundant food organisms, the continental shelf is generally a spawning ground, and bays are locations for fish maturation.
(6) The water depth of the continental shelf is appropriate, and the seabed is comparatively flat, which is suitable for fishing.
(7) At the margins of the continental shelf, due to the offshore action of the upper layer current, lower layer inshore water is introduced, producing upwelling.
(8) In addition, a frontal surface is generated along the shore by the coastal water system and the inshore water system in the vicinity of the margins of the continental shelf.

A large number of rivers enter the sea, but the nutrients they transport and the upwelling action are negligible. The estuary waters of several famous major rivers in the world, such as the Yangtze River, the Mississippi River, and the Amazon River, are all excellent fishing grounds, but they are not world-famous major fishing grounds. In contrast, there are no large rivers in the sea areas with developed upwellings, such as the inshore waters of Peru, the inshore waters of the southwest coast of the United States, and the east coast of North America, but they are world-famous major fishing grounds, i.e., the productivity of these upwelling sea areas is comparatively higher.

Distribution of the Main Continental Fishing Grounds Around the World

At present, most of the fishing grounds that have been developed and utilized around the world are distributed on the continental shelf, among which the following are famous fishing grounds:

1. Fishing grounds for demersal fish, shrimp, crab, and pelagic fishes in sea areas such as China's seas (including the Bohai Sea, the Yellow Sea, the East China Sea, and the South China Sea), the Sea of Okhotsk, and the Bering Sea
2. Fishing grounds for flounder, cod, sardine, and Pacific mackerel in the North Sea in Europe, the inshore waters of Norway, and the Barents Sea
3. Fishing grounds for cod, tuna, sardine, squid, and crab in sea areas from Brazil on the southeast coast of South America to the inshore waters of Argentina
4. Fishing grounds for demersal fish, shrimp, crab, sardine, and other pelagic fishes along the coast of Guinea in West Africa
5. Fishing grounds for demersal fish, shrimp, sardine, and other fish in the inshore waters of India, Saudi Arabia, and Iran
6. Fishing grounds for demersal fish, tuna, and squid in the inshore waters of Australia
7. Fishing grounds for demersal fish along the coast of Alaska in the United States to Canada
8. Fishing grounds for cod, herring, flounder, salmon, and other fish from the Atlantic coast of Canada to the vicinity of Newfoundland.

4.2.5   Fishing Grounds in Reef Banks

Types of Sea Areas with Reef Banks and Their Distribution

The following types of sea areas have reef banks:

1. The continental slope, the margins of islands, and the area surrounding rocky reefs, as well as rocky reef sea areas that are not exposed out of the water surface, have reef banks. There are steep slopes within these areas, and the ocean

currents or tidal currents that flow from offshore along the steep slopes bring unutilized nutrient salts from the bottom layer of the deep sea to the middle and upper layers and even to the surface layer. Additionally, the waves that often impact the steep slopes or rocky reefs also, to a certain extent, mix the seawater from the upper and lower layers. Therefore, the seawater in the sea areas of the continental slope, the margins of islands, and around rocky reefs is more fertile, and the food organisms are more abundant, thereby creating conditions for the formation of fishing grounds, such as the Shengshan fishing ground and the Haijiao fishing ground, which are famous in China.

2. Sea areas such as estuaries, bays, straits, channels, and capes have reef banks. In these sea areas, in addition to the upwelling and the convective mixing of seawater from the upper and lower layers, back eddy currents caused by topography also occur frequently, spreading the seawater that contains abundant food organisms and thereby forming a good environment for certain fish to gather. At the mouths of large rivers, the water depth is comparatively shallow, and the water is brackish, containing both organic substances brought in by fresh water and large amounts of nutrient salts brought in with tidal currents, which are conducive to the mass reproduction of plankton. Therefore, the main spawning grounds for many fish are in the vicinity of estuaries. There are varying degrees of mixed zones outside estuaries, which are generally places for the reproduction and maturation of many types of fish; they are also excellent fishing grounds. For example, the estuary of the Yangtze River is both the spawning ground for fish such as *Coilia nasus* and the spawning ground for fish such as largehead hairtail, silver pomfret, and Chinese herring.

3. Sea areas with sea domes, ocean ridges, and submarine ridges have reef banks. The bulging of the seabed in these areas can lead to the ascent of seawater along the slope, forming an upwelling sea area and making food organisms abundant; these are prime conditions for good fishing grounds, for example, the Emperor Mountain fishing ground in the central part of the North Pacific Ocean and the Yamato Tai fishing ground in the Sea of Japan.

4. Depressions in the seabed can contain reef banks. Depressions in the seabed of the continental shelf change the flow rate of ocean currents and tidal currents that pass through. The water current near the bottom layer is calmer, and the hydrological conditions are more stable, accommodating the habitation, feeding, and movement of clustered fish. Fishers with practical experience know that these deep pools and trench bottoms in which the depth changes suddenly are often good places for dense schools of fish.

Distribution of Fishing Grounds in Reef Banks

In sea areas with submarine ridges with reef banks, excellent fishing grounds are formed due to the emergence of upwelling. The following are typical fishing grounds in reef banks around the world: (1) the Zunan–Ogasawara–Mariana Islands fishing ground and the Satsunan–Ryukyu fishing ground, which produce skipjack tuna and tuna; (2) the reef bank and submarine ridge fishing grounds in the offshore waters of the North and South Pacific Ocean, which produce tuna; and (3) the reef bank and submarine ridge fishing grounds in the offshore waters of the North and South Atlantic Ocean, which produce demersal fish (Grand Banks) and tuna.

4.2.6 Distribution of the Main Fishing Grounds Around the World

The main fishing grounds around the world provide the vast majority of the marine fishing yield around the world. These grounds are mainly distributed in sea areas along boundary currents in the central and western parts and boundary currents in the eastern part of seas and oceans. The primary excellent fishing grounds are upwelling fishing grounds, fishing grounds along current

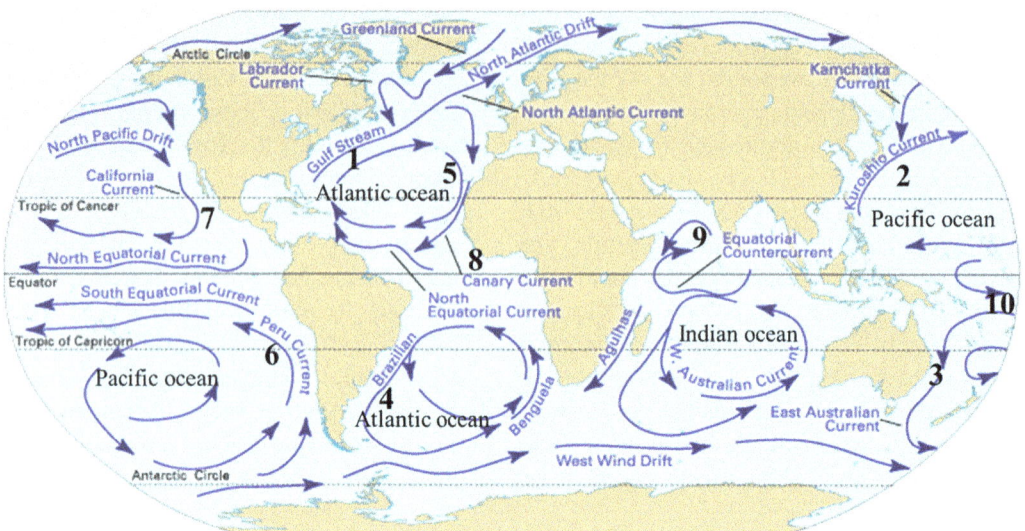

Fig. 4.7 Schematic diagram of the distribution of the world's main fishing grounds. Note: (1) Newfoundland fishing ground; (2) Hokkaido fishing ground; (3) Australia-New Zealand fishing ground; (4) Argentine fishing ground; (5) North Sea fishing ground; (6) Peru fishing ground; (7) California fishing ground; (8) Canary fishing ground; (9) Arabian Sea fishing ground; and (10) Western Central Pacific Ocean fishing ground

boundaries, fishing grounds in eddy areas, continental fishing grounds, and fishing grounds in reef banks. The sea areas where these fishing grounds are located have sufficient nutrient salts, high primary production, and abundant food organisms, and most of them are good places for the reproduction, growth, feeding, and habitation of fish and other marine animals.

Divided based on the geographical location where they are, the main fishing grounds in the world usually include the Newfoundland fishing ground, the Hokkaido fishing ground, the Argentine fishing ground, the Australia-New Zealand fishing ground, the North Sea fishing ground, the Peru fishing ground, the California fishing ground, the Canary fishing ground, the Arabian Sea fishing ground, and the Western Central Pacific Ocean fishing ground (Fig. 4.7), and these fishing grounds are mostly distributed in the sea areas of the continental shelf. (1) The Newfoundland fishing ground is formed by the confluence of the Gulf Stream with the Labrador cold current, located along the coast of Newfoundland in the northwestern part of the Atlantic Ocean. It is a current boundary fishing ground and

a continental fishing ground, and it is teeming with cold-water demersal fish, such as cod, as well as salmon, skipjack tuna, largehead hairtail, herring, plaice, mackerel (*Scomberomorus niphonius*), drums (Sciaenidae), squid, and tuna. (2) The Hokkaido fishing ground is formed by the confluence of the Kuroshio warm current with the Oyashio cold current, located in the sea area surrounding Hokkaido in Japan in the northwestern part of the Pacific Ocean. It is a current boundary fishing ground and is teeming with pelagic fishes such as pollock, Pacific saury, Japanese flying squid (*Todarodes pacificus*), squid, skipjack tuna, Oriental sardine (*Sardinops sagax*), and herring. (3) The Australia-New Zealand fishing ground is formed by the confluence of the East Australian Current with the West Wind Drift, located in the sea area surrounding Australia and New Zealand in the southwestern part of the Pacific Ocean. It is a current boundary fishing ground, an upwelling fishing ground, and a continental fishing ground, and it is teeming with species such as tuna, squid, and cod. (4) The Argentine fishing ground is formed by the confluence of the Brazil warm current with the

Falkland cold current, located in the sea area of the Patagonian continental shelf in the southwestern part of the Atlantic Ocean. It is a current boundary fishing ground and a continental fishing ground, and it is teeming with species such as cod, Argentine shortfin squid (*Illex argentinus*), drum, and Pacific mackerel. (5) The North Sea fishing ground is formed by the confluence of the North Atlantic warm current with the East Greenland cold current, located between the island of Great Britain, the Scandinavian Peninsula, the peninsula of Jutland, and the Holland-Belgium lowlands. It is a current boundary fishing ground and a continental fishing ground, and it is teeming with species such as cod, herring, and capelin (*Mallotus villosus*) (6). The Peru fishing ground is formed by the strong Peru cold current passing by the coast of Peru, blown by the prevailing westerlies and southeasterlies year round, and the occurrence of surface seawater deviating from the coast and the upwelling of cold water from the lower layer. It is a current boundary fishing ground, and it is teeming with species such as anchovy, jumbo flying squid (*Dosidicus gigas*), and Chilean jack mackerel (*Trachurus murphyi*) (7). The California fishing ground is formed by the extensive upwelling caused by the California cold current passing by the coast of California in the northeastern Pacific Ocean. It is a current boundary fishing ground, and it is teeming with species such as sardine, mackerel, and albacore tuna (8). The Canary fishing ground is formed by the extensive upwelling caused by the Canary cold current passing by the coast of northwestern Africa. It is a current boundary fishing ground, and it is teeming with species such as sardine, Pacific mackerel, tuna, octopus, squid, and demersal fishes (9). The Arabian Sea fishing ground is formed by the extensive upwelling generated by monsoons in the Arabian Sea in the northwestern part of the Indian Ocean. It is a current boundary fishing ground, and it is teeming with species such as tuna, skipjack tuna, sardine, wahoo (*Acanthocybium solandri*), Pacific mackerel, mackerel, purpleback flying squid, and cuttlefish (10). The Western Central Pacific Ocean fishing ground in the western central Pacific Ocean is formed by the extensive upwelling and current boundary generated by the North Pacific Equatorial Current, South Pacific Equatorial Current, and Equatorial Countercurrent, located in the sea area in the vicinity of South Pacific island nations. This fishing ground is teeming with species such as skipjack tuna and Pacific mackerel.

References

Chen XJ (2014) Fisheries resources and fisheries oceanography. Ocean Press. (In Chinese)

Chen XJ (2016) Theory and method of fisheries forecasting. Ocean Press. (In Chinese)

Chen XJ (2021) Fisheries oceanography. Science Press. (In Chinese)

Michitaka U (1963) Fisheries oceanography. Kouseisha Kouseikaku Co., Ltd., Tokyo. Publisher (In Japanese)

Basic Principles and Methods of Fisheries Forecasting

5

Xinjun Chen, Wei Yu, and Jintao Wang

Abstract

Fisheries forecasting is a comprehensive application of the basic principles and methods of fisheries oceanography in fishery production and is one of the main tasks of marine fishery production and service. It is necessary to master and understand the basic principles and methods of fisheries forecasting in order to carry out scientific research on fisheries forecasting. The basis of fisheries forecasting is the relationship and laws between the movement and biological status of the fish and the environmental conditions, as well as data on various fishing and oceanographic conditions, such as fish catch, resource status, marine environment, etc., obtained through various real-time pre-season surveys, and its main task is to forecast the fishing ground, fishing period, and possible catch. In this chapter, the types and research contents of fisheries forecasting are introduced, and the basic process and data of fisheries forecasting are put forward. A reasonable forecast model of fishing condition should include the basis of fishing ground, data model, and forecast model. The part of fishing ground mainly includes the law of fish shoaling and migration, the effect of environmental conditions on fish behavior, and the effect of short-term and long-term environmental events on fishery resources. The part of data model mainly includes the methods of collection, processing, and application of fishery data and environmental data and the influence of these different data on forecasting model. In the part of forecasting model, it mainly includes the theory foundation and method of establishing the forecasting model; the estimation, optimization, and verification of the model parameters; and the uncertainty analysis. Finally, the application of marine remote sensing, Geographical Information System (GIS), and habitat theory and method in fisheries forecasting is introduced in detail. Through the introduction and study of this chapter, we can understand the current situation of fisheries forecasting research in the world and provide a basis for the future research work.

Keywords

Fisheries forecasting · Marine remote sensing · GIS · Habitat suitability index

X. Chen (✉) · W. Yu · J. Wang
College of Marine Sciences, Shanghai Ocean University, Lingang Newcity, Shanghai, China
e-mail: xjchen@shou.edu.cn; wyu@shou.edu.cn; jtwang@shou.edu.cn

5.1 Overview of Fisheries Forecasting

5.1.1 The Concept of Fisheries Forecasting

Fisheries forecasting, which can also be referred to as fishing condition forecasting, is the comprehensive application of the basic principles and methods of fisheries science in fishery production and is also one of the main tasks of marine fishery production services. Fisheries forecasting refers to making forecasts on various elements of the aquatic resource status in a certain period in the future and within a certain range of waters, such as fishing period, fishing grounds, quantity and quality of the fish stock, and fish catch quota, among other elements. The basis of fisheries forecasting is the relationship and laws between the movement and biological status of the fish and the environmental conditions, as well as data on various fishing and oceanographic conditions, such as fish catch, resource status, marine environment, etc., obtained through various real-time pre-season surveys (Chen 2014, 2016, 2021).

The main task of fisheries forecasting is to predict the fishing grounds, fishing period, and possible catch, that is, to answer questions such as at what time, in what place, which fish species, how long of fishing (operation time), when to fish (beginning and end of the fishing season and the best fishing season), where to fish (location of the central fishing ground), and how much the catch (the possible catch for the entire fishing season).

In the ocean, the priority of fishing vessel is given mainly to pursuing and catching the main commercial fish species in the migration process, such as squid and tuna. Therefore, accurate fisheries forecasting can provide a scientific basis for fishery authorities and production units regarding how to carry out production plan and production management in the next fishing season; furthermore, it can also provide a basis for fishery management departments to predict the amount of resources.

5.1.2 Types and Content of Fisheries Forecasting

Divided Based on the Forecast Period

There are different methods for dividing fisheries forecasting, but it is mainly divided in accordance with the forecast period. For example, Fei and Zhang (1990) divided fisheries forecasting into prospective fisheries forecasting, long-term fisheries forecasting, mid-term fisheries forecasting or semi-long-term fisheries forecasting, and short-term fisheries forecasting. (1) Prospective fisheries forecasting refers to the prediction of the status of fishing conditions for several years or even several decades, for example, determining the scale of development and utilization of a certain type of resource. (2) Long-term fisheries forecasting refers to annual forecasting, which involves predicting the status of fishing conditions in the following year or a longer time based on the data from previous years, including the location of fishing grounds, migration routes, and so on; it is established based on the forecasting of oceanographic conditions. (3) Mid-term fisheries forecasting refers to seasonal forecasting or fishing season forecasting, which involves predicting the future status of fishing conditions during the entire fishing season and primarily focuses on the current fishing season for the location of fishing grounds, the lateness or earliness of the fishing period, clustering status, etc. (4) Short-term fisheries forecasts can be divided into the initial fishing season period, peak fishing season period, and late fishing season period, and it refers to special forecasting on the fishery development status in a certain stage of the fishing season. Fei and Zhang (1990) stated that prospective and long-term forecasts, as advanced stages of forecasting, are fundamental, strategic forecasts that mainly provide references for fishery authorities and production units when formulating development plans; in contrast, mid- and short-term forecasts, as low-level stages of forecasting, are practical, tactical forecasts that mainly provide references for production departments when arranging for production.

The Japan Fisheries Information Service Center (JAFIC) divides fisheries forecasting into two categories, that is, mid- to long-term forecasting and short-term forecasting (Chen 2016, 2021). (1) Mid- to long-term forecasting refers to the utilization of the relationship and laws between the movement and biological aspects of fish and the marine environment and in accordance with the information collected in terms of biology and oceanography, especially through surveys on the number of juvenile fish for the target fish species in the early period of the fishing season, to thereby provide forecasts regarding the amount of resources of and fish catch for the target fish species in the coming year. Such long-term forecasting is in fact more academic and provides a service to fishery management departments and research institutions. (2) Short-term forecasting, also referred to as a quick reports on fishing grounds, refers to forecasting the changes in and development trends for fishing grounds in combination with the real-time water temperature, salinity, and other environmental factors; such forecasts are extremely time dependent and directly serve fishery production.

As seen in the aforementioned analyses, the division of fisheries forecasting types is mainly based on the length of the forecast time; different types of forecasting require different basic data, forecast periods, and objects of use. In this book, fisheries forecasting is generally divided into three categories, that is, full fishing season forecasting, fishing season stage forecasting, and short-term forecasting or near real-time forecasting (Chen 2016).

1. Full fishing season forecasting – The effective time of the forecast is the entire fishing season, and the contents include the amount of resource replenishment, the start and end times of the fishing period, and possible fish catch, among other content. This type of forecast is published at an appropriate time before the fishing season and can provide a reference for fishery management departments and production units. The basic data and survey data that this forecasting requires are large-scope (scale) data on the marine environment and changes therein, surveys on the number of juvenile fish for the target fish species before the fishing season, trends in the strength of ocean currents, and so on, and the development trends for and overall overview of the annual fishing season are analyzed from a macro-perspective.

2. Fishing season stage forecasting – The entire fishing season is generally divided into three stages—the initial period, peak period, and late period—for forecasting. These forecasts supplement full fishing season forecasting and provide a scientific basis for production departments in a timely and more accurate way for scheduling production. Forecasts are published on the eve of each production stage, and the time dependence is strong. The basic data and survey data that this forecasting requires are the development and changing trends for the marine environment by stage as well as production survey data for the target fish species.

3. Short-term forecasting – This type of forecasting is also referred to as a quick report on fishing conditions; it predicts the location of central fishing grounds, the direction of movement of schools of fish, and the possibility of a bloom in the next 24 h or within the next few days. Forecasting on fishing grounds is carried out by obtaining real-time information on the marine environment. This type of forecasting has the strongest time dependence, and the data on oceanographic conditions obtained are generally published on the same day. The basic data that this forecasting requires are fishery production and survey data in the last few days, including individuals caught and size composition as well as changes in water temperature, and weather conditions (such as typhoons, low atmospheric pressure, etc.), among other data.

Divided Based on Fisheries Forecasting Content

Based on different forecasting content, fisheries forecasting can be divided into three types, that is, forecasting regarding the resource status, forecasting regarding time, and forecasting regarding space. The emphasis of each type of forecasting is different, and the corresponding forecasting principles and models also differ.

1. Forecasting regarding the resource status, that is, forecasts on the quantity and quality of the fish stock as well as the fish catch under certain fishing conditions, is mainly mid- to long term. Accurate mid- to long-term forecasting has important significance to fishery management and production; not only can fishery management departments use the forecasting results as reference information for formulating fishery policies, but fishery production enterprises can also reasonably plan their fishing effort based on these forecasts, which can be an advantage during times with intense competition for fish. At present, forecasting models regarding the fishery resource status mainly use fish population dynamics as a basis, and mathematically, statistical regression, artificial neural networks (ANNs), and time series analysis are mainly used.

2. Forecasting regarding time mainly includes forecasting the time of appearance and the duration of the fishing period. This type of forecasting not only requires the forecaster to have a very good understanding of the migration and clustering status of the target fish species but also a certain means of observation to understand, in real time, the weather, ocean currents, water temperature structure, and the food organism situation in the target area; forecasting is carried out in combination with the experience of fishers and fishery researchers. With changes in marine fishery production modes, fisheries forecasting researchers have separated from the front line of fishery production; therefore, currently, this category of forecasting mainly gives priority to the qualitative field analysis of experienced fishery producers. Its principles are very difficult to clearly and quantitatively interpret, and existing quantitative studies have generally only used simple linear regression.

3. Forecasting regarding space, that is, forecasting the location of fishing grounds or the spatial distribution status of fish resources, is commonly referred to as fishing ground forecasting. Due to the gradual scarcity of fishery resources and the continuous increase in fuel and fishing access costs, forecasts regarding the location of fishing grounds have become increasingly more important in the fishery production process, and enterprises have increasingly higher real-time and accuracy requirements for these forecasts. Therefore, research on forecasting models for the location of fishing grounds is quite active, and most fisheries forecasting models in China and abroad are forecasting models for identifying the location of fishing grounds.

5.2 Techniques and Methods for Fisheries Forecasting

5.2.1 Basic Process of Fisheries Forecasting

Specialized research institutions, research centers, or commercialized companies are generally responsible for fisheries forecasting research and the daily publishing of forecasts. These institutions possess the sources of data for two conditions—fishing conditions and oceanographic conditions—and their network information systems and sources of data are multifaceted. For example, in terms of oceanographic conditions, the main sources are ocean remote sensing, fishery survey vessels, fishery production vessels, transport vessels, and buoys, among others. In terms of fishing conditions, the main sources are fishery production vessels, fishery survey vessels, wharfs, production command departments, aquatic product markets, etc.

Based on actual survey and research results, fisheries forecasting institutions quickly process, forecast, and publish data regarding

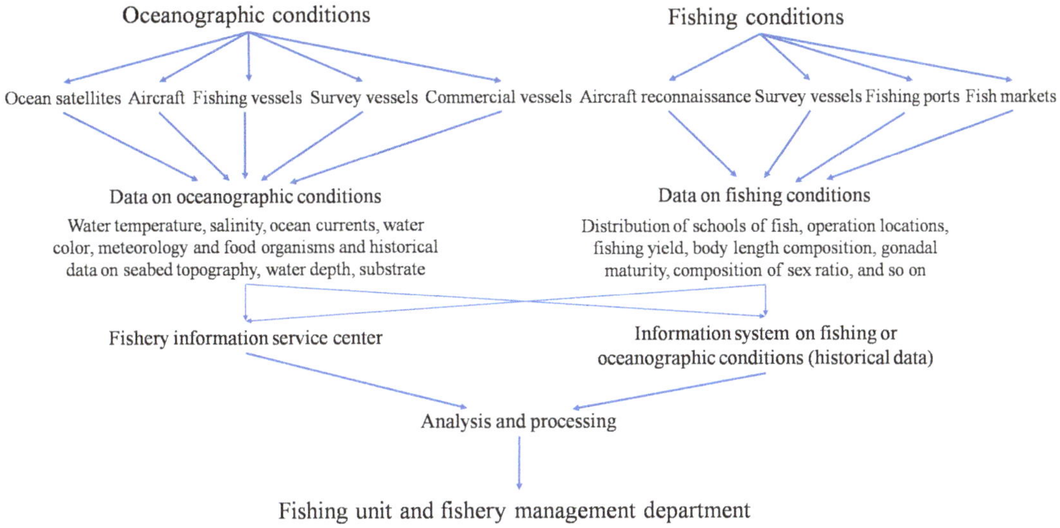

Oceanographic conditions

Fishing conditions

Ocean satellites Aircraft Fishing vessels Survey vessels Commercial vessels Aircraft reconnaissance Survey vessels Fishing ports Fish markets

Data on oceanographic conditions

Water temperature, salinity, ocean currents, water
color, meteorology and food organisms and historical
data on seabed topography, water depth, substrate

Data on fishing conditions

Distribution of schools of fish, operation locations,
fishing yield, body length composition, gonadal
maturity, composition of sex ratio, and so on

Fishery information service center

Information system on fishing or
oceanographic conditions (historical data)

Analysis and processing

Fishing unit and fishery management department

Fig. 5.1 Schematic diagram on the process of fisheries forecasting technology (Chen 2016)

oceanographic conditions, fishing conditions, and so on, serving fishery production at every opportunity. The form of publishing can be diverse, such as meteorological fax, networks, etc. While fisheries forecasting institutions publish various analyses and data on fishing conditions and oceanographic conditions, they also must hold training classes to familiarize the users with the related basic knowledge in order to facilitate full use of the various data published to effectively engage in fishery production. To analyze and forecast the fishing conditions and oceanographic conditions, a complete fishery information network is usually established to carry out work such as data collection, processing, and analysis, forecasting, and publishing. A schematic diagram of forecast processing is shown in Fig. 5.1.

5.2.2 Composition of a Fisheries Forecasting Model

A reasonable fisheries forecasting model should consider content from three respects, that is, the basis of fisheries science, data models, and forecasting models. Fisheries science mainly includes the clustering and migration laws of fish, the effects of environmental conditions on fish behavior, and the effects of short-term and long-term environmental events on fishery resources. Data models mainly include the methods for the collection, processing, and application of fishery data and environmental data as well as the effects of these methods on forecasting models. Forecasting models mainly include the theoretical basis and methods for establishing a fisheries forecasting model and the estimation, optimization, and verification of parameters for the corresponding model, as well as an uncertainty analysis (Chen et al. 2013).

Basis of Fisheries Science

The distribution of fish in the seas and oceans is jointly determined by their own biological characteristics and external environmental conditions. First, marine fish generally have clustering and migration habits, and their clustering and migration laws determine the general distribution of fishery resources in time and space. Second, the behavior of fish is closely related to the external environment in which they live. The external environment in which fish live includes two categories—biotic factors and abiotic factors. Biotic factors include harmful organisms, food organisms, and population relationships. Abiotic factors include water temperature, ocean currents, salinity, light, dissolved oxygen, meteorological conditions, seabed topography, water quality factors, and so on. Finally, various types of sudden, staged, or even long-term and slow marine environmental events, such as red tides, the El Nin o phenomenon, and global warming, also generate short-term and long-term effects on fishery resources, which then cause oscillations in fishery resources in terms of time, space, quantity, and quality. Only by comprehensive consideration of the effects of factors in these three respects can a reasonable fisheries forecasting model be established.

Data Models

Fisheries forecasting requires fishery data and marine environmental data, and the collection, processing, and application of these data have important effects on fisheries forecasting models. When constructing a fisheries forecasting model, to unify the temporal and spatial resolution of fishery data and environmental data, it is generally necessary to resample the data. Because the place of operation for commercial fishing is not random, combined processing in terms of space and time will cause the model to generate different deviations; marine phenomena such as eddy currents and frontal surfaces that are closely related to the formation of fishing grounds have stronger variability, and taking the average of marine environmental data, in terms of spatial and temporal scales, will weaken or even cover up these phenomena. Therefore, when constructing a fisheries forecasting model, the appropriate spatiotemporal resolution should be selected to reduce model deviation and enhance prediction accuracy. In addition, the construction of a fisheries forecasting model should also fully take into account the particularities of the fishery data itself. For example, fishery data are a type of "presence-only" data; that is, importance is attached to recording the places where fish are caught, but no importance is attached to recording the places where there is no fish catch. Finally, appropriate strategies should also be selected for the application of low-resolution historical data, spatial location information, and other data.

Forecasting Models

Fisheries forecasting models can mainly be divided into three types (Chen et al. 2013), that is, empirical/phenomenon models, mechanism/process models, and theoretical models. Generally, existing fisheries forecasting models still give priority to empirical/phenomenon models. There are two common development bases for these types of models. One uses the ecological niche or the resource selection function (RSF) as the theoretical basis, and the ecological niche of the target fish species or the resource response function for key environmental factors is analyzed mainly through frequency analysis, regression, and other statistical methods to thereby establish a fisheries forecasting model. The other is based on the idea of knowledge discovery; that is, fisheries forecasting models are established by discovering the laws for the formation of fishing grounds in the data through various machine learning and artificial intelligence methods with the use of fishery data and marine environmental data as the basis.

Generally, statistics-based fisheries forecasting models are regression-centered, their model structure is preset, and the model coefficients are estimated mainly from existing data; then, these models are used to predict fishing grounds. These are often referred to as "model-driven" models. However, prediction models based on machine learning and artificial intelligence methods are centered on model learning; they are "data-driven" models that extract the rules for the formation of fishing grounds from data through various data mining methods and then use these rules to carry out fishing ground forecasting. In the last few decades, great changes have occurred to traditional statistics and calculation methods, and the distinction between statistical methods and machine learning methods has already become blurred.

Drawing lessons from research on biological distribution prediction models, the process of establishing a fisheries forecasting model can be divided into four steps: (1) research on the mechanism for the formation of fishing grounds; (2) establishment of a fisheries forecasting model; (3) model calibration; and (4) model evaluation and improvement.

The construction of a fisheries forecasting model should be based on the biology of the target fish species and fisheries science research, striving to fit the model with the realities of fisheries science. If there is clarity on the clustering and migration characteristics of the target fish species as well as a mechanism for the formation of fishing grounds, a mechanism/process model or a theoretical model can be selected for use to quantitatively describe these characteristics and mechanisms. In contrast, if the understanding of these characteristics and mechanisms is incomplete, then an empirical/phenomenon model can be selected to carry out an averaged description of the process for the formation of fishing grounds in accordance with basic ecological principles. In addition, regardless of which type of prediction model is constructed, the characteristics of the data that are used in the model should be fully considered, a factor that is especially important for statistics-based models.

Model calibration refers to the estimation of model parameters and the adjustment of a model after a forecasting model equation is established. The different forecasting models also have different methods for estimating the model parameters. For example, for various statistical models, the parameters are estimated mainly by using methods such as minimum variance or maximum likelihood estimation; however, for ANN models, the weight coefficients are obtained by iterative calculation of the model until convergence. For fisheries forecasting models, in addition to estimating and adjusting the model parameters and constants, model calibration also includes the selection of independent variables. When utilizing marine environmental elements to carry out fisheries forecasting, determining which environmental factors to select is a comparatively important and very difficult task. When Zhou (1987) was utilizing a regression model to carry out a study on forecasting the fishing period of

Japanese Spanish mackerel (*Scomberomorus niphonius*), it was thought that forecasting using a combination of multiple factors was more accurate than forecasting using a single factor. Research has shown that to increase the accuracy of prediction models, it is best to not have too many individual independent variables. In addition, for certain models, model calibration also includes the transformation of independent variables and the selection of a smoothing function, among other tasks.

Model evaluation mainly involves assessing the performance and actual effect of a prediction model. There are two primary methods for model evaluation. One uses the same data for model evaluation and model calibration, and the variation coefficient method or the self-help method is used to evaluate the model; the other method is the use of completely new data for model evaluation, and the evaluation criterion is generally the degree of fit of the model or a certain distance parameter. Because the main objective of fisheries forecasting models is to forecast, the latter method is generally used for model evaluations; that is, the degree of conformity between the predicted fishing situation and the actual fishing situation is tested.

5.2.3 Introduction of the Main Fisheries Forecasting Models

Statistical Models

Linear Regression Models

Early or traditional fisheries forecasting mainly uses methods such as regression analysis, correlation analysis, discriminant analysis, and cluster analysis, which give priority to classical statistics. The most representative among them is general linear regression models. By analyzing the relationship between sea surface temperature (SST), chlorophyll-a (Chl-a) concentration, and other marine environmental data and historical fish catch, catch per unit effort (CPUE), or fishing period, a regression equation is established:

$$\text{Catch (or CPUE)} = \beta_0 + \beta_1 \cdot \text{SST} + \beta_2 \cdot \text{Chl} - a \\ + \cdots + \varepsilon$$

where β_0, β_1, β_2, and so on are regression coefficients and ε is the error term. In general linear regression models, the least-square method is used to estimate the coefficients, and then these equations are utilized to forecast the fishing period, fish catch, or CPUE.

General linear regression models have a stable structure and simple operation method, and they have achieved certain effects in early practical applications. However, general linear models also have limitations. There is fuzziness and randomness in the relationship between the formation of fishing grounds and marine environmental elements, and it is generally very difficult to establish a regression equation with a very high correlation coefficient. Additionally, actual fishery production and marine environmental data generally do not satisfy the assumptions of general linear models for data, then leading to comparatively poor prediction results of the regression equation. Currently, the application of general linear regression models in fisheries forecasting is comparatively rare, and they have been gradually replaced by more complex models such as segmented linear regression, polynomial regression, exponential (logarithmic) regression, and quantile regression.

Generalized Regression Model

Generalized linear models (GLMs) carry out certain transformations of response variables through a contiguous function and integrate regressions based on the exponential family of distributions with a general linear regression; the regression equation is as follows:

$$g(E(Y)) = \beta_0 + \sum_{i=1}^{p} \beta_i \cdot X_i + \varepsilon$$

GLMs can transform the independent variables themselves and can also add a function term that reflects the interrelationship between the independent variables, thereby realizing

nonlinear regression in a linear form. The transformation of independent variables includes many forms; for example, the GLM equation in polynomial form is as follows:

$$g(E(Y)) = \text{LP} = \beta_0 + \sum_{i=1}^{p} \beta_i \cdot (X_i)^p + \varepsilon$$

Generalized additive models (GAMs) are a nonparametric extension of GLMs. The equation is as follows:

$$g(E(Y)) = \text{LP} = \beta_0 + \sum_{i=1}^{p} f_i \cdot X_i + \varepsilon$$

The regression coefficients β_i in GLMs are replaced by a local scatter point smoothing function f_i. Compared with GLMs, GAMs are more suitable for processing nonlinear problems.

Starting in the 1980s, GLMs and GAMs have been successively applied to the prediction of the spatial distribution of fishery resources, such as the American lobster in the Gulf of Maine. GLMs and GAMs can process nonlinear problems to a certain extent; therefore, they have better prediction accuracy. However, their application is comparatively complicated and requires researchers to have a deeper understanding of the error distribution and transformation of predictive variables in fishery production data; otherwise, effects on the prediction results can be easily generated.

Bayesian Method

The Bayesian statistical theory is based on Bayes' Theorem; that is, posterior probability is calculated through the prior probability and the corresponding conditional probability. Among them, prior probability refers to the total probability of the formation of a fishing ground, conditional probability refers to the probability that an environmental element satisfies a certain condition when a fishing ground is "true," and posterior probability is the probability of the formation of a fishing ground under the conditions of the current environmental elements. In the Bayesian method, the prior probability and conditional probability are obtained through the frequency statistics of the historical data; after the posterior probability is calculated, the forecast is completed by way of a similarity checklist. Existing research has shown that the Bayesian method has a better forecast accuracy rate; when a fisheries forecasting model for Western Pacific tuna was established using the Bayesian statistical method, the comprehensive forecast accuracy rate was 77.3%.

A significant advantage of the Bayesian method is that it is easy to integrate; it can be integrated together for application with almost any existing model, and a commonly used method is to calculate and modify the prior probability with different models. At present, Bayesian models applied in fisheries forecasting use the simple Bayesian classifier; this method assumes that the effects of environmental conditions on the formation of fishing grounds are independent of each other. This assumption obviously does not conform to the realities of fishing grounds. It is believed that the Bayesian belief network model, which considers the joint probability of each predictive variable, should also have a larger application space in terms of fisheries forecasting.

Time Series Analysis

Time series refers to a set of numerical series with a time sequence. The processing and analysis of time series have the unparalleled advantages of static statistical processing methods. With the development of computers and numerical calculation methods, a complete set of analysis and prediction methods has been generated. Time series analysis is mainly applied in terms of fish catch prediction in fisheries forecasting. For example, time series analysis models are used to predict the annual yield of commercial fishing for brown shrimp in the northwestern part of the Gulf of Mexico (Grant et al. 1988). The time series analysis, an ANN, and a Bayesian dynamic model have been used to predict the yield of *Loliginidae* and the yield of *Ommastrephidae* in the sea area of Greece, and the results showed that the time series analysis method had very high accuracy (Georgakarakos et al. 2006).

Spatial Analysis and Interpolation

The basis of spatial analysis is the spatial autocorrelation of the geographic entity; that is, the closer the geographic entities are, the higher the degree of similarity, and the farther the distance away the geographic entities are, the greater the difference. Spatial autocorrelation is referred to as the "first law of geography," and ecological phenomena also satisfy this law. Spatial analysis is mainly used to analyze the correlation and heterogeneity of fishery resources in a spatiotemporal distribution, such as changes in the center of gravity of fishing grounds, the spatiotemporal distribution pattern of fishery resources, etc. However, there are also some scholars who use geostatistics-based interpolation methods (such as kriging interpolation) to interpolate fish catch data and estimate the total amount and spatial distribution of fishery resources. For example, geostatistical methods have been used to predict the spatial distribution of the fin whale in the northwestern part of the Mediterranean Sea. It is necessary to explain that fisheries have very strong dynamic change features, and geostatistical methods are essentially static methods; therefore, there are strict requirements on these methods regarding fishery data collection.

Machine Learning and Artificial Intelligence Methods

The prediction of fishing grounds in terms of space can also be viewed as a type of "classification," that is, the process of dividing each grid in space into "fishing ground" and "non-fishing ground." Such a classification process is generally a type of supervised classification; that is, the rules for the formation of fishing grounds are extracted from sample data through different methods, and then these rules are used to classify actual data and divide each grid point in the sea area into two types, i.e., "fishing ground" and "non-fishing ground." There are many methods for extracting classification rules, primarily being machine learning methods. Machine learning involves computers simulating or implementing human learning behavior to acquire new knowledge. There are similarities among machine learning, artificial intelligence, and data mining, but each has its respective focus, which are not elaborated upon here. There are numerous machine learning and artificial intelligence methods; at present, ANNs, rule-based expert systems, and case-based reasoning (CBR) methods are most applied for fisheries forecasting. In addition, decision trees, genetic algorithms, maximum entropy methods, cellular automata, support vector machines, classifier aggregation, relevance analysis and cluster analysis, fuzzy reasoning, and other methods have all started to be applied to the analysis and forecasting of fishing situations.

ANN Models

ANN models are generated from simulating the biological nervous system. They consist of a set of interconnected nodes and directional chains. The main parameter of ANNs is the weighted value of connecting each node, which is generally obtained by iterative calculation of the sample data until convergence; the principle of convergence is to minimize the error sum of squares. The process of determining the weighted values in a neural network is referred to as the neural network learning process. The learning of a neural network with a complex structure is very time-consuming, but the speed for predictions is very fast. ANN models can simulate very complex nonlinear processes and have already been widely applied in marine and aquatic product disciplines. In fisheries forecasting applications, ANN models have been successfully applied for spatial distribution predictions and yield predictions.

ANNs do not require the fishery data to satisfy any assumptions, nor does the interrelationship between the response function of fish to environmental conditions and various environmental conditions need to be analyzed; therefore, ANNs are convenient to apply, and there are no significant differences compared with other models in terms of application effects. However, there are many types of ANNs and much structural variability; relative to other models, the application of an ANN is more difficult and requires the modeler to have abundant experience. In addition,

ANN models are implicit in the expression of knowledge and, thus, equivalent to a black box model; this aspect makes ANN models perform adequately in high-dimensional situations but unable to clearly explain prediction principles. Of course, there are also methods to test the degree of contribution of a single input variable to the model output in the ANN model.

Rule-Based Expert System

An expert system is a type of intelligent computer program system; it contains the knowledge and experience of human experts in a specific field and can utilize the problem-solving methods of human experts to process complex problems in that field. In fisheries forecasting applications, this expert knowledge and experience generally manifest as the rules for the formation of fishing grounds. At present, the most common expert system methods in fisheries forecasting are the environmental envelope method and the habitat suitability index (HSI) model.

The environmental envelope method is the earliest and one of the most widely applied fisheries spatial forecasting models. Fish have a suitable range for environmental elements, and it is assumed in the environmental envelope method that schools of fish will appear in suitable environmental conditions but will not appear when environmental conditions are unsuitable. When this type of model is implemented, the grids that satisfy a single environmental condition are usually calculated first, and then, spatial superposition analysis is carried out to calculate the results of different environmental conditions to obtain a final prediction result; therefore, this method is also often referred to as the spatial superposition method. The spatial superposition method can fully utilize expert knowledge in the fishery field, and the model is simple in structure and easy to implement, which is especially suitable for use with environmental grid data obtained by ocean remote sensing retrieval; therefore, it has attained considerably wide application in the field of fisheries forecasting.

The HSI model is a framework model proposed by the US Geological Survey's Wetland and Aquatic Research Center for use in describing the quality of a habitat for fish and wild animals. Its basic concept and implementation method are similar to those of the environmental envelope method, but there are also some differences. First, the prediction result of the HSI model is a habitat suitability index similar to "fishing ground probability," rather than a binary result of "fishing ground" and "non-fishing ground" provided by the environmental envelope method. Second, in the HSI model, the adaptability of fish to a single environmental element is not described with the use of an absolute numerical range but is expressed by an RSF. Last, when describing the comprehensive action of multiple environmental factors, the HSI model can use numerous expressions, such as continued product, geometric mean, arithmetic mean, hybrid algorithm, etc. The HSI model is widely applied in fish habitat analysis and fisheries forecasting. However, as an averaged index, the habitat suitability index does not have a strict correlation with real-time fishing grounds; therefore it is necessary to be very cautious when utilizing the HSI model to predict fishing grounds.

CBR

CBR simulates problem-solving; that is, when a new problem is encountered, the problem is analyzed first, a similar case to the problem is found in stored memory, and then the information and knowledge related to the case are slightly modified to solve the new problem. In the CBR process, the new problem faced is the target case, and the case in stored memory is referred to as the source case. CBR is a strategy in which the source case is obtained by prompting from the target case, and the source case guides the solution for the target case. This method simplifies knowledge acquisition and enhances the efficiency of problem-solving by way of the direct reuse of knowledge, and the quality of the solution is comparatively high; it is suitable for use in noncomputational derivation and is widely applied in fishing ground forecasting. The principle of CBR is simple, and the resulting models manifest as fishing ground rules; therefore, it can be very easily applied to an expert system.

However, the CBR method requires sufficiently numerous sample data to establish the case database, and the extracted cases are still mainly a summary of historical data, making it difficult to predict new fishing grounds.

Mechanism/Process Models and Theoretical Models

The two types of models mentioned previously are both empirical/phenomenon models. Empirical/phenomenon models are static, averaged models, and they assume that there is a certain equilibrium between fish behavior and the external environment. Different from empirical/phenomenon models, mechanism/process models and theoretical models focus on considering the dynamics and randomness in the actual process of the formation of fishing grounds. In this process, the behavior of fish is always affected by various instantaneous and random elements, and the assumptions may not necessarily be able to reach equilibrium with the external environment. The formation of a fishing ground is a complex process, and the model used is different with different understandings of this process. Some of the models draw support from numerical calculation methods to reproduce dynamic processes, such as fish migration, clustering, and population changes. Some common models are the biomass equilibrium model, the advection-diffusion interaction model, and the physical-biological coupled model based on the three-dimensional hydrodynamic numerical model. Li et al. (2017) utilized the Finite Volume Coastal Ocean Model (FVCOM) to establish the physical-biological coupled model of the early life history process of mackerel in the East China Sea. Some other models look at the behavior of individual fish and study group behavior and changes through the choices of individuals. For example, the artificial life model, based on the genetic algorithm and a neural network, has been utilized to study the movement process of tuna, and the individual-based ecological model (individual-based model, IBM) has also been widely applied to study the transport process of fish eggs and larval fish.

5.3   Application of Ocean Remote Sensing and Geographic Information Systems (GIS) in Fisheries Forecasting

5.3.1   Application of Remote Sensing in Fisheries Forecasting

The marine environment is a necessary condition for the survival and activities of marine fish, and each change in environmental parameters generates important effects on the migration, distribution, movement, clustering, and quantitative changes in fish. Fishing ground analysis and forecasting require a certain timeliness. Remote sensing provides a tool for collecting marine ecosystem and environmental data over a large area quickly and dynamically that is able to acquire extensive, synchronous, real-time, and effective high-precision environmental information on fishing grounds, and it can greatly enrich the means of studying and analyzing fishing grounds Therefore, the response relationship between such spatiotemporal distribution and behavior and changes in the environment can be sought by utilizing remote sensing data, and corresponding models can be established to thereby make scientific forecasts of the fishing situation.

The development in the use of ocean satellite remote sensing through satellite observation of the marine environment can be roughly divided into three stages. The first stage is exploratory experiments (1970–1978), primarily manned spacecraft tests and the utilization of meteorological satellites (Television Infrared Observation Satellite (TIROS)-N, Defense Meteorological Satellite Program (DMSP) series satellites, and Geostationary Operational Environmental Satellite (GOES) series satellites, among others), and land satellites (e.g., the Landsat series) are used to obtain oceanographic information. During this stage, scholars in ocean remote sensing started to use the data acquired by meteorological satellites and land satellites to analyze marine environmental information and to use them in research on marine fishing ground analysis and

forecasting. However, meteorological satellites and land resource satellites have their own characteristics and cannot completely replace ocean satellites. The second stage comprised experimental research (1978–1985). During this stage, the United States launched an ocean satellite (SeaSat-A) and a Nimbus satellite (NIMBUS-7) loaded with the Coastal Zone Color Scanner (CZCS), which enriched marine environmental information, and the interest of scholars in oceanography circles regarding the utilization of ocean satellite remote sensing to study oceanography and marine biological resources was further enhanced. In 1983, the US Sea Grant Marine Advisory Service and the Graduate School of Oceanography at the University of Rhode Island (URI) utilized SST data retrieved by the Advanced Very High Resolution Radiometer (AVHRR) to carry out research and analysis of the temperature of the entire sea area, the temperature of sea areas of interest, and the horizontal temperature gradient of the entire sea area and made product images to distribute to fishers, thus reducing the time required for fishing vessels to find fish. The third stage (1985 to the present) comprises research and application, during which many ocean satellites have been launched, including ocean topography satellites (e.g., Geodetic Satellite (GEOSAT), Geosynchronous (GEO)-1, and Ocean Topography Experiment (TOPEX)/Poseidon), ocean dynamic environment satellites (e.g., European Remote-Sensing Satellite (ERS)-1, ERS-2, and Radar Satellite (Radarsat)), and ocean water color satellites (e.g., SeaSat, Republic of China Satellite (ROCSAT), and Korean Multi-Purpose Satellite (KOMPSAT)).

In recent years, with the continuous development of remote sensing technology in the direction of hyperspectral remote sensing and high spatial resolution remote sensing, the precision of ocean remote sensing data retrieval has greatly increased, providing more abundant marine environmental information, such as SST, ocean water color (e.g., Chl-a concentration), sea surface salinity (SSS), and the dynamic topography of the sea surface (e.g., sea surface height (SSH)), and a broad application space for research on marine fishing grounds and the analysis of fishing situations.

SST

Water temperature is one of the most important environmental factors affecting fish activities, and the distribution, migratory movement, and clustering of fish are directly or indirectly limited by the environmental temperature. Marine fish have a certain suitable temperature interval and a most suitable temperature; therefore, water temperature is the most important and commonly used environmental element in analyzing the marine environment and the living habits and resource abundance of fish. SST has a greater effect on the fishing ground distribution of pelagic fish inhabiting the mixed layer of seas and oceans. At present, the technology for utilizing ocean satellite remote sensing to retrieve the SST is comparatively mature, and its accuracy is 0.5–0.8 °C. Using SST data, abundant physical oceanographic information can be obtained, such as the spatial distribution of surface temperatures, temperature fronts, temperature anomalies, surface water masses, and the El Nino phenomenon, and these water temperature indicators can characterize the distribution of fishing grounds from different perspectives. For example, in a study of the central fishing ground of the silver pomfret (*Pampus argenteus*) in the Gulf of Mexico, SST remote sensing images by the National Oceanic and Atmospheric Administration (NOAA)/ AVHRR sensor every April and May from 1985 to 1987 indicated that there was a certain spatial relationship between the central fishing ground and a front formed by the warm offshore water with a low chlorophyll concentration and cold water with a high chlorophyll concentration in the shelf break area; additionally, the silver pomfret catch was higher in the frontal area than in the areas far from the shelf break where the frontal surface had gradually weakened or dissipated.

In the SST field, the narrow zone containing the maximum value within the horizontal temperature gradient is usually the transition area where cold and warm water masses converge, thereby forming a temperature frontal surface. From the temperature contour map generated by the SST data, flow gaps can be intuitively identified, with narrow and long zones with denser contours representing the temperature frontal surface. Upwelling will usually form in the vicinity of temperature frontal surfaces, and the abundant nutrient salts carried in these waters provide conditions for the reproduction and growth of plankton, thereby forming a high-productivity area. For example, in September 2001 and from April to May 2005, by comparing the SST temperature field forecasted in the eastern sea area of Japan and the distribution of *Skipjack tuna* fishing grounds recorded in fleets' fishing logs, it was determined that the horizontal temperature gradient of the operating area was 0.1 °C and that the preferred temperature interval for *Skipjack tuna* differed with the seasons and as sea conditions changed. Large-scale variation anomalies in the spatial field of marine water temperature can often indicate important marine events, such as the El Nino-Southern Oscillation (ENSO) and La Nina, among other phenomena. When the ENSO phenomenon occurs, the weakening of the southeaster trade winds causes a large amount of warm water from the sea area of the equatorial Pacific Ocean to flow toward the equatorial East Pacific Ocean, thereby causing a decrease in water temperature in the western part of the Pacific Ocean and an increase in water temperature in the eastern part of the Pacific Ocean. The large-scale changes in the warm water layer and the changes in climatic conditions accompanying the ENSO phenomenon generate an important effect on the amount of resources in fishing grounds and the distribution of fishing grounds. Guo and Chen (2005) utilized sea surface temperature anomaly (SSTA) in the Nino 3.4 area as an ENSO indicator and carried out a cross-correlation analysis on the longitudinal center of gravity for annual average yield between 1990 and 2001 and the changes in ENSO index years, the results of which showed that the largest negative correlation between a high-yield longitudinal center of gravity and average longitude presented 1 year after the ENSO index.

Ocean Water Color

The utilization of remote sensing to acquire information on ocean water color aims to detect spectral radiation information on biotic and abiotic parameters (such as Chl-a concentration, suspended substances, soluble organic matter, and pollutants) related to ocean water color through sensors onboard aircraft or onboard satellites. After undergoing atmospheric correction, biological and optical characteristics are used to retrieve the sea water chlorophyll concentration, soluble organic matter, and other marine environmental information. At present, inshore water color information can be obtained from the CZCS sensor; the Sea-viewing Wide Field-of-view Sensor (SeaWIFS) and the Moderate Resolution Imaging Spectroradiometer (MODIS) can provide water color information for all water areas in the world and are the two most widely used sensors in ocean water color remote sensing at present. The utilization of remote sensing to retrieve ocean water color concentration, especially Chl-a concentration, can reflect the distribution conditions of zooplankton and phytoplankton in the seas and oceans. Research results have indicated that sea areas with a Chl-a mass concentration of 0.2 mg/m^3 or higher have an abundant stock of plankton and that fishing grounds for fishing operations can form in these areas. Information on marine dynamic environmental features, such as flow field and flow pattern, can be extracted with the use of remote sensing images of Chl-a concentration through artificial visual interpretation, which can also indicate the distribution of marine fishing grounds.

For example, researchers used remote sensing technology to analyze the habitat features of *Skipjack tuna* in the Northwest Pacific Ocean and, through a GAM, evaluated various environmental factors in the habitat, such as the SST, sea surface chlorophyll (SSC), sea surface height anomaly (SSHA), and eddy kinetic energy (EKE) as well as the interaction effect between various factors. They concluded that SST, followed by the SSC, is the most important indicator affecting the migration of *Skipjack tuna* and pointed out that the oligotrophic side of the Kuroshio frontal surface and the Kuroshio Extension are important features of the *Skipjack tuna* habitat in the Northwest Pacific Ocean and that the mesoscale eddy current is also an important factor in the formation of the *Skipjack tuna* habitat.

Not only can the ocean Chl-a concentration indicate the stock of plankton and dynamic marine environmental features, it can also be combined with lighting conditions to estimate the primary production of seas and oceans through related remote sensing retrieval algorithms. The amount of primary production of seas and oceans can reflect the photosynthesis rate of marine phytoplankton; therefore, the amount of primary production of seas and oceans is a fundamental factor that determines the stock of, distribution of, and changes in marine organisms. Assessment of primary production is conducive to understanding marine organisms, especially the role played by marine fish in the dynamic system of marine ecology. For example, Lehodey et al. (1998) combined net primary production and data such as ocean currents and used a coupled dynamical biogeochemical model to predict the potential distribution of feed in the *Skipjack tuna* fishing grounds in the Western Central Atlantic Ocean; the simulation results coincided more with the plankton distribution actually observed and changes in its spatiotemporal sequence, and it was noted that combining temperature, dissolved oxygen, and other environmental elements to simulate the potential tuna fishing ground environment will be closer to the real fishing ground habitat environment, thus greatly benefiting the establishment of a large-scale dynamic model for the tuna population.

5.3.2   Application of GIS in Fisheries Forecasting

GIS is an emerging marginal science that integrates disciplines such as computer science, space science, information science, surveying, mapping, remote sensing, environmental science, and management science into one field. GIS started in the 1960s, and it has become a basic platform for multidisciplinary integration and has been applied to various fields; as a basic means and tool for geospatial information analysis, it has also been widely applied in the study of fisheries forecasting.

Development History of Fisheries GIS

GIS is a computer system used to input, store, query, analyze, and display geographic reference data. The basic operations of GIS are summed as spatial data input, attribute data management, data display, data analysis, and GIS modeling. In the 1980s, GIS started to be applied to the management of fisheries in inland water areas and to the selection of farms. In the late 1980s, GIS was gradually applied to marine fisheries. However, compared with land, the application of GIS to marine fisheries is still greatly limited.

There are three primary reasons that impede the rapid development of fisheries GIS. First, in terms of funding support, the collection of data regarding the biology, physical chemistry, and topography of aquatic organisms requires substantial funds, especially for resource and environmental surveys that require a long time. The second reason is the complexity and dynamics of the aquatic system; the aquatic system is more complex and dynamically variable than the terrestrial system and requires different types of information. The aquatic environment is usually unstable and usually requires the use of three or even four dimensions (three dimensions + time) to express. The third reason is that because many commercial software developers usually use terrestrial information as a basis, these software programs cannot directly and effectively process fisheries and marine environment data.

Although the development of marine fisheries GIS technology has faced many challenges, due to the rapid development in computer technology and in the means of acquiring marine data and to meet the needs of the development of marine fisheries science itself, after 2000, marine fisheries GIS technology developed considerably and has been increasingly more valued by scientific researchers and international organizations.

In 1999, the First International Symposium on GIS in Fisheries Science was held in Seattle, United States, and it has been held once every 3 years subsequently. The content of the symposiums has included the application of GIS technology in remote sensing and acoustic surveys, habitat and environment, marine resource analysis and management, mariculture, geostatistics and models, artificial fishing reefs and marine conservation areas, and other fields in marine fisheries as well as the development of GIS systems, such as Marine Explorer, researched and developed by the Environmental Simulation Laboratory in Saitama, Japan, the Arc Marine or ArcGIS Marine Data Model, researched and developed by research institutions such as Ohio State University, Duke University, and NOAA in the United States and others in Denmark, and fishery Analyst for ArcGIS 9.1, researched and developed by the company Mappamondo GIS.

Utilization of GIS to Study the Relationship Between Fishery Resources and the Marine Environment

Marine fishery resources are closely related to the marine environment; this relationship is the most basic issue in marine fisheries GIS research and usually involves content such as GIS mapping and modeling. As a type of spatial analysis tool, GIS can be used to explain the differences between different regions. GIS modeling involves the application of GIS in the process of establishing a model with spatial data, and GIS can integrate different data sources to establish various models, such as binary models, index models, regression models, and process models. Index models and regression models are commonly used for fisheries, and they are commonly used in habitat suitability analysis and vulnerability analysis.

In addition, it is very important to determine the key habitats of fish for fishery resource management. The primary characteristics of a habitat include the presence of a set of biotic and abiotic parameters that adapt to support and maintain all life history stages of fish populations. Due to the significant spatiotemporal changes in the key

Table 5.1 Application of GIS in research on the relationship between fishery resources and the marine environment

Research objective	Research cases and their content	Reference
Relationship between resource distribution and the environment	Relationship between cephalopod resources and the environment Spatial distribution of fattening grounds for *Solea solea* Relationship between the spatial distribution of fattening grounds for juvenile plaice (*Pleuronectes platessa*) and environmental variables	Pierce et al. (1998) Eastwood et al. (2003) Stoner et al. (2007)
Habitat determination and mapping	Mapping marine benthic habitats by GIS image processing technology A new mapping method for marine benthic habitats that utilizes physical environmental data Utilizing GIS environmental modeling methods to design important fish habitats Identification of important habitats for small pelagic fish species in the Mediterranean water areas of Spain	Sotheran et al. (1997) Huang et al. (2011) Valavanis et al. (2004) Bellido et al. (2008)
Dynamic monitoring of resources	Spatial dynamic changes in the amount of replenishment for southern blue fin tuna (*Thunnus maccoyii*) The activity range and habitat use of *Paralabrax nebulifer* in the marine conservation areas of Southern California	Nishida et al. (1999) Mason and Lowe (2010)
Habitat assessment and management	Utilization of GIS and GAMs to establish a habitat model of Antarctic lantern fish (*Electrona antarctica*) Application of GIS in habitat assessment and marine resource management	Loots et al. (2007) Stanbury and Starr (2000)

habitats of fish, GIS, as an efficient spatiotemporal analysis tool, has attracted increasingly more attention and high regard from managers, and research in this respect is also increasing day by day.

Based on the current status of research in China and abroad, GIS has been widely applied in terms of the relationship between fishery resources and the marine environment. The objectives are to understand the relationship between fishery resource distribution and the marine environment and to study and determine the scope of distribution of fish habitats to thereby further master the dynamic distribution of fishery resources and to ultimately assess and manage the distribution of fish habitats or fishing grounds (Table 5.1).

Utilization of GIS to Study Fisheries Forecasting

After 2000, with continuous improvements in the means of acquiring satellite remote sensing information and the continuous deepening of the application of GIS technology in marine fisheries, the application of GIS in fisheries forecasting has also developed, enriching and developing the means

and methods of fisheries forecasting (Table 5.2). Usually, the methods for combining GIS technology to carry out fisheries forecasting involve statistical analysis and forecasting (such as linear regression analysis, correlation analysis, discriminant analysis, and cluster analysis), spatial statistical analysis and spatial modeling (such as spatial association analysis and spatial information analysis models), artificial intelligence (such as expert systems and ANNs), fuzziness and uncertainty analysis (such as the Bayesian statistical theory), and numerical calculation and simulation (such as the Monte Carlo simulation method), among other methods (Table 5.2), and very good prediction results have been obtained, as seen in the cases in Table 5.2. GIS relies on established autonomous databases and can realize functions such as the integrated management of spatiotemporal data, spatial superposition and buffer analysis, contour analysis, the exploration and analysis of spatial data, the visual display of model analysis results, and the vectorized output of maps, combining various statistical methods and fishing and oceanographic condition data to realize intelligent fisheries forecasting.

Table 5.2 Examples of the application of GIS in marine fisheries forecasting

Fisheries forecasting method	GIS application examples	Reference
Statistical analysis and forecasting	Identification of the optimal habitat and suitable fishing ground for squid (*Ommastrephes bartramii*) in the Northwest Pacific Ocean	Chen et al. (2010)
Spatial analysis and modeling	Software design and implementation of a marine fishery electronic map system	Shao et al. (2001)
Artificial intelligence	Fishery ground forecasting for the coastal water areas in the southern and central parts of Sulawesi, Indonesia	Sadly et al. (2009)
Uncertainty analysis	Fisheries forecasting for pelagic fish in the northern sea areas of Iceland based on remote sensing and GIS	Sanchez (2003)
Numerical calculation and simulation	Prediction of the distribution of food organisms for skipjack tuna in the equatorial Pacific Ocean	Lehodey et al. (1998)

Based on GIS and ocean remote sensing as well as technology such as expert systems, the Key Laboratory of Sustainable Exploitation of Oceanic Fisheries Resources of Ministry of Education at Shanghai Ocean University; the Environmental Simulation Laboratory in Saitama, Japan; Collecte Localisation Satellites (CLS) France; and other research institutions have successively developed marine fisheries forecasting service systems that can realize the real-time forecasting of central fishing grounds and predict the amount of resources for the main fishing species, thus providing technical means for the sustainable development and efficient utilization of marine fishery resources.

5.4   Application of Habitat Theories and Methods in Fisheries Forecasting

5.4.1   Basic Concept

Generally, habitat refers to the total sum of the spatial range and environmental conditions in which organisms appear in the environment, including abiotic environments and other organisms needed for the survival of individual or group of organisms. The National Research Council (NRC 1982) states that habitat refers to the environment where animals or plants usually live, grow, or reproduce. The US Fish and Wildlife Service defines habitat as a place that provides direct support to a specific species, population, or community, including all environmental characteristics such as air quality, water quality, vegetation and soil features, and water supply in that place. In research on fishery resources, a very important issue is fish habitats, and the main content of research is the effects of changes in the habitat environment of living things on biological activities. In recent years, increasingly more studies have found that by only considering the effects of abiotic factors, some phenomena that cannot be explained by the distribution of biological habitats will appear; therefore, some biotic factors have also been gradually considered.

The HSI is an index that evaluates the degree of suitability of wildlife habitats. The HSI model was first proposed in the early 1980s by the Fish and Wildlife Service of the National Wetlands Research Center of the US Geological Survey, and it was used to describe the habitat quality of wild animals. The Fish and Wildlife Service has also established HSI models for 157 species of wild birds and fish. Currently, the HSI model is widely used in species management, environmental impact evaluation, abundance distribution, and ecological restoration research. The HSI model has been widely used in fishery resource assessments, protection, and management since the early 1980s and has become one of the important tools for identifying fishing grounds and estimating the abundance of fish. The HSI model has also been utilized and applied to study the effects of the marine environment on the distribution of fish and for the forecasting of central fishing grounds, among other applications, and has also become a research hotspot in fisheries science.

5.4.2 Brief Description of Research Methods

The value range of the HSI is generally 0.0–1.0; it is a quantitative index, with 0.0 indicating an unsuitable habitat and 1.0 indicating the most suitable habitat. Habitat evaluation procedures (HEP) of the HSI have been widely used in the evaluation of habitat quality of wildlife. Usually, the development process of the HSI model includes (1) acquiring habitat data; (2) constructing a single-factor suitability function; (3) assigning weight to the habitat factor; (4) combining multiple suitability indexes to calculate the overall HSI value; and (5) drawing an HSI distribution map.

Essentially, the construction of the HSI is based on the following concept: first, simulate a suitability index (SI) for an organism to all environmental elements and then associate the various SIs together through certain mathematical methods to obtain a comprehensive HSI.

Analyzing the effect of a single factor on biological distribution is the most basic method used in HSI research. However, a habitat is a very complex ecosystem, and comprehensive consideration of the effects of multiple factors can better explain and predict the distribution of organisms. However, data collection requires substantial human power, material resources, and time, and it is impossible to take all factors into consideration. Generally, the selection of appropriate input factors for the HSI model should be based on the following criteria: (1) morphological and biochemical factors must be significantly related to the carrying capacity of the habitat or the survival or growth rate of commercially developed species; (2) there is sufficient understanding of the relationship between the factors and the habitat; and (3) these factors can be obtained or measured or sampled by practical and cost-effective methods. Therefore, the selection of

environmental factors is crucial because the factors can indicate the habitats of the species.

In fisheries science, there are many factors that affect the HSI, including abiotic and biotic factors and human influence. The selection of environmental factors is different for different research areas and research objects and their life history stages. Generally, for marine organisms, the main impact factors considered are temperature, including the surface temperature, the horizontal gradient of the surface temperature, the temperature at different water depths, etc., as well as SSS, SSHA, SSH, and Chl-a, among other factors; in an estuary, the main impact factors considered are temperature, salinity, depth, dissolved oxygen, etc.; and in rivers, the main impact factors considered are temperature, depth, water flow rate, etc. For benthic organisms, sediment types and substrates would also be considered. The sources of data for these environmental factors generally include (1) remote sensing environmental data; (2) field measurement data; (3) experimental data; and (4) indirect acquisition (obtained by calculation through mathematical models or methods on the basis of the first three data sources).

Those who develop HSI models usually assume that (1) species or populations actively choose suitable habitats for their survival and (2) there is a linear relationship between species and environmental variables, and such a linear relationship mainly comes from empirical data, expert judgment, or a combination of the two. Usually, the constructed linear functions are segmented; to simplify the SI model, many scholars also directly assign values in accordance with historical data or expert knowledge (Table 5.3). In the natural environment, such a hypothetical linear relationship almost does not exist; therefore, increasingly more researchers have started to simulate the relationship between biological distribution and environmental variables in accordance with knowledge through mathematical statistics to thereby calculate and obtain an SI curve for impact factors.

Generally, the weight of each factor is obtained based on expert knowledge. However,

Table 5.3 Construction and application of an SI

Basis	SI function construction method	Application target	Reference
Historical data and expert knowledge or a combination of the two	Empirical value assignment	*Scomber japonicus* in the East Yellow Sea *Ommastrephes bartramii* in the Northwest Pacific Ocean Greater redhorse in the Sandusky River Invertebrate *Plecoptera*, Sole (*Solea solea* and *Solea senegalensis*) in the estuary of the Tagus River in Portugal *Sander vitreus* in the Sandusky River	Chen et al. (2009a, 2010); Tomsic et al. (2007); Vinagre et al. (2006); Gillenwater et al. (2006)
	Linear function	Bigeye tuna (*Thunnus obesus*) in the Indian Ocean *Acipenser sinensis* in the Yangtze River *Odontesthes bonariensis* in Buenos Aires *Potamogeton pectinatus* *Alosa sapidissima* *Salmo trutta* *Oncorhynchus tshawytscha* *Morone saxatilis* (Walbaum)	Chen et al. (2008); Yi et al. (2007, 2008); Ban et al. (2009); Gómez et al. (2007); Lee et al. (2006); US Fish and Wildlife Service (1984, 1985, 1986)
Knowledge in mathematical statistics	Linear regression	Bigeye tuna (*Thunnus obesus*) in the Indian Ocean	Wang (2006)
	Nonlinear regression	Skipjack tuna (*Katsuwonus pelamis*) in the Western Central Pacific Ocean Manila clam (*Tapes philippinarum*) in the Mediterranean Sea	Guo and Chen (2009); Vincenzi et al. (2006)
	Normal distribution model	Squid (*Ommastrephes bartramii*) in the Northwest Pacific Ocean	Chen et al. (2009b);
	Quantile regression	Argentine shortfin squid (*Illex argentinus*) in the Southwest Atlantic Ocean Bigeye tuna (*Thunnus obesus*) in the Indian Ocean	Feng et al. (2007, 2009, 2010)
	Exponential-polynomial model	Crayfish (*Orconectes neglectus*) in the streams of the Ozark Mountains	Gore and Bryant (1990)

in many studies, there may be a lack of sufficient information to assign different weights to different environmental variables. At present, in fisheries science, most HSI applications treat the weight of each factor equally. However, some researchers believe that the weight values obtained from expert judgments are stronger than the weak links of the current algorithms because these weight values represent the common knowledge of those engaged in fishery scientific research and management; therefore, they are widely accepted.

5.4.3 Commonly Used HSI Calculation Formulas and Their Applications

In fisheries science, the following are the commonly used HSI comprehensive algorithms:

1. Continued product model (CPM)

$$HSI = \prod_{i=1}^{n} SI_i$$

2. Minimum model (MINM)

$$HSI = Min(SI_1, SI_2, \cdots SI_n)$$

3. Maximum model (MAXM)

$$HSI = Max(SI_1, SI_2, \cdots SI_n)$$

4. Geometric mean model (GMM)

$$HSI = \sqrt[n]{\prod_{i=1}^{n} SI_i}$$

5. Arithmetic mean model (AMM)

$$HSI = \frac{1}{4} \sum_{i=1}^{n} SI_i$$

6. Hybrid algorithm

In accordance with the principle of differentiating the large from the small, the minimum value for the SI for each factor is taken at different times, and the maximum value for each time period is taken at the same place, that is:

$$HSI = Max\{Min(SI_1, SI_2, \cdots SI_n)_1, \cdots$$
$$\times Min(SI_1, SI_2 \cdots SI_n)_j\}$$

7. Weighted geometric mean (WGM)

$$HSI = \left(\prod_{i=1}^{n} SI_i^{w_i}\right) 1/\sum_{i=1,\cdots,n} w_i$$

8. Weighted mean model (WMM)

$$HSI = \frac{1}{\sum_{i=1,\ldots,n} w_i} \sum_{i=1}^{n} w_i SI_i$$

In the formula, i is the i-th impact factor, n is the total number of impact factors, SI_i is the SI value of the i-th impact factor, w_i is the weight or weight number of the i-th factor, and j is the j-th life history stage or the j-th time period.

The above models are all constructed in connection with a single-factor SI, which ignores the effect of the interaction between factors on biological distribution. Wang (2006) and Feng et al. (2009) utilized the quantile regression (QR) method to study bigeye tuna, calculated the relationship between all factors and their interaction factors and the distribution of bigeye tuna, and further calculated the HSI; the output results could better predict the biological distribution. In addition, in the terrestrial ecosystem, some other models and algorithms for constructing the HSI have also emerged, such as fuzzy HSI models, fuzzy neural network (FNN) models, and binomial logistic models.

To make the output results more intuitive, researchers generally use mapping software (such as ArcGIS and Marine Explorer) or programming software (such as Matlab and R language) to enable the visualization of results, divide the HSI into different levels from 0 to 1, and give biological habitats different degrees of fitness, such as unsuitable, generally suitable, moderately suitable, more suitable, and most suitable.

The steps of model testing generally include model calibration, verification, and validation. Although the US Fish and Wildlife Service has established many HSI models for wildlife and fish, it has rarely tested its models, which has thus affected the development and application of these models to a certain extent.

5.4.4 Application of the HSI in Marine Fisheries

The use of the HSI in marine fisheries started comparatively late. There have been more studies in recent years, mainly focused on central fishing ground analysis, resource amount estimations, ecological management, etc. For example, Vincenzi et al. (2006) utilized factors such as sediments, dissolved oxygen, temperature, depth, and Chl-a to study the habitat of Manila clam (*Tapes philippinarum*) in the Mediterranean Sea. They used historical data to construct an SI equation for each factor, set the weight of each factor in accordance with expert knowledge, used the weighted geometric mean model (WGMM) to establish an HSI model, and finally utilized the piecewise function generated from the experimentally measured values to convert the HSI values into an annual potential yield estimate. This model provides managers with the potential economic yield of Manila clams in different places, and this reasonable, fast, and cost-effective method provides a basis for the fair distribution of catch among competitors and significantly improves the transparency of the decision-making process.

Based on the application of the HSI in marine fisheries, most researchers utilized production statistics to construct SI functions, utilized five basic methods of the HSI for calculation, and used the Akaike information criterion (AIC) and other standards to screen for the optimal model. Due to the presence of certain difficulties in acquiring data on the abundance of biological resources, especially in commercial fisheries, independent research is rarely conducted on the different life stages of research objects. With the development of remote sensing (RS) and GIS, the utilization of remote sensing environmental data has provided support for studying the habitat of marine organisms on a large scale.

In short, HSI theories and methods have been very well applied in fisheries science, especially with the development of GIS technology and its application in the field of fisheries. The HSI will become an important tool and means for fishery resource prediction, central fishing ground analysis, etc. However, there are still uncertainties in using the HSI model for fisheries forecasting research, as follows (Chen 2016).

1. Comprehensiveness and objectivity of the acquisition of habitat data – The choice of the number and form of environmental variables in the HSI model is key to identifying the most suitable habitat or not. Factors that do not significantly affect the spatial distribution of organisms or too many factors in the HSI model may negatively affect the establishment of an HSI model and affect the collection of data for different factors, a massive project in and of itself. Although some scholars have proposed appropriate criteria for the selection of factors, it is still possible to miss factors that very significantly affect biological distribution; therefore, it is very difficult to explain the appearance of some biological phenomenon. Generally, the required input factors can

accurately reflect the spatiotemporal distribution of organisms, include all factors that are significantly related to it as much as possible, and, at the same time, discard the irrelevant or less relevant factors.

2. Reliability of the SI curve – The acquisition of an SI curve relies on historical data, field experience, and expert judgment.

3. Representativeness of the input data – The sample must reflect the distribution characteristics of the overall data, and model verification is needed to reduce the uncertainty of the input data.

4. Structure of the model – For the same data, the results obtained with different SI and HSI models may be significantly different.

Although the HSI model has certain problems and limitations, its superiority is incomparable to other habitat models. The HSI can indicate the most suitable environmental conditions for organisms, predict the distribution of central fishing grounds, estimate the amount of resources, and so on, thereby contributing to the sustainable development and utilization of fishery resources as well as to scientific management. In summing up the current status of research and its existing problems in China and abroad, the following issues have to be considered in the process of HSI research and application (Chen 2016):

1. Fully understand the life history process and biological characteristics of the research object as well as the marine environment in which it lives.

2. In connection with different growth stages and the external environments, make full use of the development of global positioning system (GPS), GIS, and RS (3S) technology to acquire data on appropriate environmental factors and select appropriate environmental factors as well as indicators that characterize resource abundance or habitats. For example, fishing effort and CPUE are both usually used to establish fish habitat models, but studies have shown that habitat models established based on fishing effort are better than those based on CPUE (Tian et al. 2009).

3. Carry out research on the spatiotemporal criteria for suitable factors and establish norms and standards.
4. Assign weight to each factor in accordance with historical data as much as possible, and set the optimal weight value through an appropriate optimization algorithm.
5. Choose the appropriate models for different targets (central fishing ground, biomass estimate, etc.).
6. Choose the appropriate HSI model by comparing and analyzing various models.
7. Utilize measurement data and the latest data to continuously improve and revise chosen models in order to improve their accuracy.

References

Ban X, Li DM, Li D (2009) Application of habitat suitability criteria on spawn-sites of Chinese sturgeon in downstream of Gezhouba Dam. Eng J Wuhan Univ 42(2):172–177. (in Chinese)

Bellido JM, Brown AM, Valavanis VD et al (2008) Identifying essential fish habitat for small pelagic species in Spanish Mediterranean water. Dev Hydrobiol 203:171–184

Chen XJ (2014) Fisheries resources and fisheries oceanography. Ocean Press. (in Chinese)

Chen XJ (2016) Theory of and methods for fisheries forecasting. Ocean Press. (in Chinese)

Chen XJ (2021) Fisheries oceanography. Science Press. (in Chinese)

Chen XJ, Feng B, Xu LX (2008) A comparative study on habitat suitability index of bigeye tuna, *Thunnus obesus* in the Indian Ocean. J Fish Sci China 15(2): 269–278. (in Chinese)

Chen XJ, Gao F, Guan WJ et al (2013) Review of fisheries forecasting technology and its models. J Fish China 08:1270–1280. (in Chinese)

Chen XJ, Li G, Feng B et al (2009a) Habitat suitability index of chub mackerel (*Scomber japonicus*) in the East China Sea. J Oceanogr 65(1):93–102

Chen XJ, Liu BL, Tian SQ et al (2009b) Forecasting the fishing ground of *Ommastrephes bartramii* with SST-based habitat suitability modelling in Northwestern Pacific. Oceanologia Et Limnologia Sinica 40 (6):707–713. (in Chinese)

Chen XJ, Tian SQ, Chen Y et al (2010) A modeling approach to identify optimal habitat and suitable fishing grounds for neon flying squid (*Ommastrephes bartramii*) in the Northwest Pacific Ocean. Fish Bull 108:1–14

Eastwood PD, Meadena GJ, Carpentier A et al (2003) Estimating limits to the spatial extent and suitability of sole (*Solea solea*) nursery grounds in the Dover Strait. J Sea Res 50:151–165

Fei HN, Zhang SQ (1990) Fisheries resources science. China Science and Technology Press. (in Chinese)

Feng B, Chen XJ, Xu LX (2007) Study on distribution of *Thunnus obesus* in the Indian Ocean based on habitat suitability index. J Fish China 31(6):805–812. (in Chinese)

Feng B, Chen XJ, Xu LX (2009) Multivariate quantile regression on habitat suitability index of *Thunnus obesus* in the Indian Ocean. J Guangdong Ocean Univ 29(3):48–52. (in Chinese)

Feng B, Tian SQ, Chen XJ (2010) The habitat suitability index of *Illex argentinus* by using quantile regression method in the Southwest Atlantic. Trans Oceanol Limnol 124(1):15–22. (in Chinese)

Georgakarakos S, Koutsoubas D, Valavanis V (2006) Time series analysis and forecasting techniques applied on loliginid and ommastrephid landings in Greek waters. Fish Res 78(1):55–71

Gillenwater D, Granata T, Zika U (2006) GIS-based modeling of spawning habitat suitability for walleye in the Sandusky River, Ohio, and implications for dam removal and river restoration. Ecol Eng 28(3):311–323

Gómez S, Menni R, Naya J et al (2007) The physical–chemical habitat of the Buenos Aires pejerrey, *Odontesthes bonariensis* (Teleostei, Atherinopsidae), with a proposal of a water quality index. Environ Biol Fish 78(2):161–171

Gore JA, Bryant RM (1990) Temporal shifts in physical habitat of the crayfish, *Orconectes neglectus* (Faxon). Hydrobiologia 199(2):131–142

Grant WE, Matis JH, Miller W (1988) Forecasting commercial harvest of marine shrimp using a Markov chain model. Ecol Model 43(3-4):183–193

Guo A, Chen XJ (2005) The relationship between ENSO and tuna purse-seine resource abundance and fishing grounds distribution in the Western and Central Pacific Ocean. Mar Fish 04:338–342. (in Chinese)

Guo A, Chen XJ (2009) Studies on the habitat suitability index based on the vertical structure of water temperature for skipjack *Katsywonus pelamis* purse-seine fishery in the Western-Central Pacific Ocean. Mar Fish 31(1):1–9. (in Chinese)

Huang Z, Brooke BP, Harris PT (2011) A new approach to mapping marine benthic habitats using physical environmental data. Cont Shelf Res 31(2):4–16

Lee GEMV, Molen DTV, Boogaard HFPV et al (2006) Uncertainty analysis of a spatial habitat suitability model and implications for ecological management of water bodies. Landsc Ecol 21:1019–1032

Lehodey P, Andre JM, Bertignac M et al (1998) Predicting skipjack tuna forage distributions in the equatorial Pacific using a coupled dynamical bio-geochemical model. Fish Oceanogr 7:317–325

Li YS, Chen XJ, Chen CS (2017) Ecological dynamics simulation of the early life cycle of Chub Mackerel (*Scomber japonicus*) in East China Sea. Ocean Press. (in Chinese)

Loots C, Koubbi P, Duhamel G (2007) Habitat modelling of *Electrona antarctica* (Myctophidae, Pisces) in Kerguelen by generalized additive models and geographic information systems. Polar Biol 30:951–959

Mason TJ, Lowe CG (2010) Home range, habitat use, and site fidelity of barred sand bass within a southern California marine protected area. Fish Res 2010(106): 93–101

National Research Council (1982) Imparts of emerging agricultural trends on fish and wildlife habitats. National Academy, Washington, D. C, pp 1–333

Nishida T, Lyne V, Miyashita K, et al (1999) Spatial dynamics of southern bluefin tuna recruitment [C]// The first international symposium on GIS in fishery science, pp 89–106

Pierce GJ, Wang J, Bellido JM et al (1998) Relationships between cephalopod abundance and environmental conditions in the Northeast Atlantic and Mediterranean as revealed by GIS. ICES J Mar Sci 55:14–33

Sadly M, Hendiarti N, Sachoemar SI et al (2009) Fishing ground prediction using a knowledge-based expert system geographical information system model in the south and Central Sulawesi coastal waters of Indonesia. Int J Remote Sens 30:6429–6440

Sanchez EEL (2003) Remote sensing and geographic information system for pelagic fishing ground forecasting in North Iceland waters. UNU–Fisheries Training Programme, 1–56

Shao QQ, Zhou CH, Zhang MJ et al (2001) Software design and development for marine fisheries electronic mapping system. J Fish China 25(4):367–372. (in Chinese)

Sotheran IS, Foster-Smith RL, Davies J (1997) Mapping of marine benthic habitats using image processing techniques within a raster-based geographic information system. Estuar Coast Shelf Sci 44:25–31

Stanbury KB, Starr RM (2000) Applications of geographic information systems (GIS) to habitat assessment and marine resource management. Oceanol Acta 22:699–703

Stoner AW, Spencer ML, Ryer CH (2007) Flatfish-habitat associations in Alaska nursery grounds: use of continuous video records for multi-scale spatial analysis. J Sea Res 57:137–150

Tian SQ, Chen XJ, Chen Y et al (2009) Evaluating habitat suitability indices derived from CPUE and fishing effort data for *Ommatrephes bratramii* in the Northwestern Pacific Ocean. Fish Res 95:181–188

Tomsic CA, Granata TC, Murphy RP et al (2007) Using a coupled eco-hydrodynamic model to predict habitat for target species following dam removal. Ecol Eng 30 (3):215–230

U.S. Fish and Wildlife Service (1984). Habitat suitability index models and instream flow suitability curves: inland stocks of Striped Bass. Biological report

U.S. Fish and Wildlife Service (1985) Habitat suitability index models and instream flow suitability curves: American Shad. Biological report

U.S. Fish and Wildlife Service (1986) Habitat suitability index models and instream flow suitability curves: Brown Trout. Biological report

Valavanis VD, Georgakarakos S, Kapantagakis A et al (2004) A GIS environmental modelling approach to essential fish habitat designation. Ecol Model 178:417–427

Vinagre C, Fonseca V, Cabral H et al (2006) Habitat suitability index models for the juvenile soles, *Solea solea* and *Solea senegalensis*, in the Tagus estuary: defining variables for species management. Fish Res 82(1–2):140–149

Vincenzi S, Caramori G, Rossi R et al (2006) A GIS-based habitat suitability model for commercial yield estimation of *Tapes philippinarum* in a Mediterranean coastal lagoon (Sacca di Goro, Italy). Ecol Model 193(1–2):90–104

Wang JQ (2006) Study on the habitat index model of bigeye tuna in the Indian Ocean. Shanghai Fisheries University, Shanghai, pp 1–40. (in Chinese)

Yi YJ, Wang ZY, Lu YJ (2007) Habitat suitability index model for Chinese sturgeon in the Yangtze River. Adv Water Sci 18(4):538–543. (in Chinese)

Yi YJ, Wang ZY, Yao SM (2008) Habitat suitability model for evaluating Chinese sturgeon spawning sites. J Tsinghua Univ Sci Technol 48(3):340–343. (in Chinese)

Zhou BB (1987) Study on prediction of fishing season by using the method of adjusting complex correlation coefficient of degree of freedom. Acta Oceanol Sin 9(6):774–779. (in Chinese)

Case Studies of Fisheries Forecasting

6

Xinjun Chen, Jintao Wang, and Wei Yu

Abstract

Scientific fisheries forecast is an important guarantee to realize sustainable development of marine fishery, improve production efficiency, and reduce fuel cost. Because of the different life history processes and resource characteristics of different marine species, the mechanisms of resource change and fisheries formation are different, and the methods of fisheries forecasting and environmental factors are different, for example, species with short life cycles such as cephalopods have a 1-year life cycle and die after spawning, so their resource abundance and annual catches are closely related to the spawning environment; in general, the optimum environmental range of their spawning grounds can be one of the important indicators for the prediction of the next year's resources and catch. In addition, skipjack tuna, which is distributed in the Central and Western Pacific Ocean, usually inhabits waters with sea surface temperature of 20–30 °C; the previous results have been shown that the distribution, migration, and schooling of skipjack are closely related to the variation of sea surface temperature and El Niño-Southern Oscillation (ENSO) in the tropical Pacific Ocean; and the catch position of skipjack moves with the 29 °C isotherm along the longitude line at the edge of the warm pool. Therefore, in this chapter, we choose largehead hairtail *Trichiurus lepturus* in the East China Sea, Japanese horse mackerel *Trachurus japonicus* and blue-spotted mackerel *Scomberomorus niphonius* in the East China Sea and Yellow Sea, skipjack *Katsuwonus pelamis* in the Western and Central Pacific Ocean, neon flying squid *Ommastrephes bartramii* in the Northwest Pacific Ocean, Antarctic krill *Euphausia superba*, and other important economic species as an example. Based on the previous research results with the above mentioned species in the world, the fishing season prediction, the central fishing ground forecasting, and the annual catch and resource abundance index forecasting are described in detail, which will provide the basis for the scientific fisheries forecasting.

Keywords

Fisheries forecasting · Fishing season forecasting · Resources forecasting · Case study

X. Chen (✉) · J. Wang · W. Yu
College of Marine Sciences, Shanghai Ocean University,
Lingang Newcity, Shanghai, China
e-mail: xjchen@shou.edu.cn; jtwang@shou.edu.cn;
wyu@shou.edu.cn

© The Author(s), under exclusive license to Springer Nature Singapore Pte Ltd. 2022
X. Chen (ed.), *Theory and Method of Fisheries Forecasting*,
https://doi.org/10.1007/978-981-19-2956-4_6

6.1   Fishing Season Analysis and Fishing Ground Forecasting

Fishing season analysis and fishing ground forecasting are important research topics of fisheries forecasting studies. The timing of fishing seasons is directly affected by marine environmental factors, especially surface temperature and ocean currents; the timing of fishing seasons also directly affects the management of marine fishery production. The emphasis of fishing ground forecasting is to forecast the location of central fishing grounds and their movement trends, and the location of fishing grounds and their movement are directly affected by a variety of factors such as sea temperature, sea surface height, chlorophyll, and ocean currents, which are being studied and analyzed in China and abroad at present through the use of a variety of methods. Various models for fishing ground forecasting with comparatively high precision have been established to provide a scientific basis for fishery production. In this chapter, important commercial species are used as the study objects, with the goal of analyzing their fishing season and predicting the location of their expected and central fishing grounds by assessing the migration distribution of the studied species and their basic biological characteristics in combination with previous research results to provide a basis for the scientific production of fisheries (Chen 2016, 2021).

6.1.1   Fishing Season Analysis: Case Studying for *O. bartramii* in the North Pacific Ocean

Migration Distribution of *O. bartramii*

As an oceanic species, *O. bartramii* is widely distributed throughout the entire sea area of the North Pacific Ocean and has high resource abundance. *O. bartramii* is a warm-water species that

migrates seasonally in the North Pacific Ocean (Ichii et al. 2009). The early larvae of *O. bartramii* born in the winter and spring in the Northwest Pacific Ocean are generally distributed in the area of the Kuroshio Counter Current and south of 35°N and west of 155°E and grow to the juvenile *O. bartramii* stage, growing and feeding northward with the Kuroshio Current starting in May. Between May and August, immature *O. bartramii* migrate northward or northeastward, entering the confluence of the warm and cold currents of the Subpolar Front at 35°–40°N. Due to abundant food organisms in this confluence area and the northward migration being hindered by the cold water of the Oyashio Current, *O. bartramii* feed and cluster, potentially forming a central feeding ground. Generally, for *O. bartramii* in the North Pacific Ocean, they migrate northward to feed from May to October and begin migrating southward after October for spawning (Fig. 6.1).

There are two major current systems—the Kuroshio Current and Oyashio Current—in the Northwest Pacific Ocean, and their confluence and role in mixing generate numerous famous fishing grounds, such as saury (*Cololabis saira*) fishing grounds and *O. bartramii* fishing grounds. The Kuroshio Current has a high temperature (15–30 °C) and high salinity (34.5–35‰) and is derived from the North Equatorial Current. The Oyashio Current has a low temperature and low salinity and originates in the Bering Sea, flowing from north to southwest along the Kuril Islands. One branch of the Kuroshio Current continues to flow northeastward in the vicinity of 35°N, reaches 40°N, and converges with the southward Oyashio Current, merging in the eastern sea area of Hokkaido and flowing eastward after converging and mixing. Their mixed water constitutes the Subpolar Front (at approximately 40°N), with a width of approximately 2–4° of latitude, which is more apparent in the sea area west of 160°E, but the front is not apparent in the sea area east of 160°E. The extension current east of 160°E is also referred to as the North Pacific Current. A

Fig. 6.1 Migration patterns of the autumn and winter-spring cohorts of neon flying squid in the North Pacific Ocean (Ichii et al. 2009)

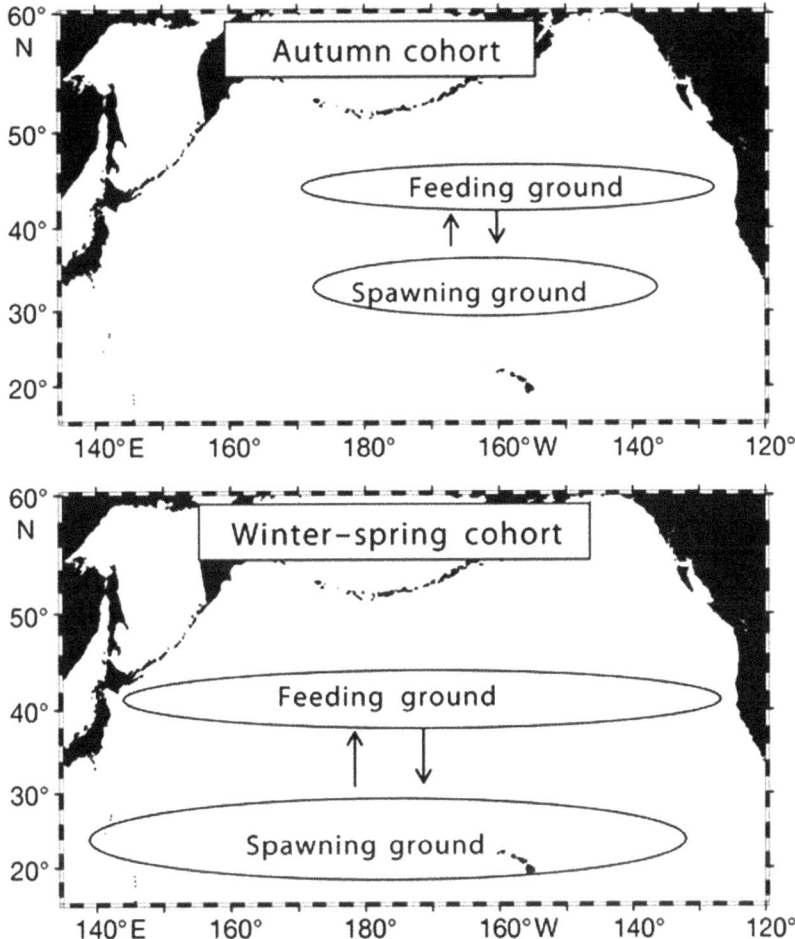

mixing area is formed in the area between the more southerly front (generally at 36°–37°N) and the more northerly front (generally at 42°–43°N) of the Subpolar Front. On the north side of the front, because the subpolar circulation is a persistent and divergent cyclonic circulation and the cold water rises, it is high in nutrient salts, and the basic phytoplankton and zooplankton production is also comparatively high. This provides the most basic guarantee for the formation of fishing grounds. In the autumn, when the North Pacific Subpolar Front is weakened and desalinated, *O. bartramii* and other marine animals cross or forage in the vicinity of the front.

Relationship Between Fishing Season Analysis for *O. bartramii* and the Marine Environment

Relationship Between the Fishing Season of *O. bartramii* and Sea Temperature

In the Northwest Pacific Ocean, sea temperature is closely related to the *O. bartramii* fishing season. Generally, if the sea temperature is high, the fishing season will occur earlier; if the sea temperature is low, the fishing season will occur later. The sea temperature distribution can be reflected in three aspects: surface sea temperature, vertical sea temperature (thermocline), and deep sea temperature (100 m or 200 m).

Studies have shown (Chen 1995, 1997; Chen et al. 2003, 2005; Wang and Chen 2005) that the surface sea temperature where *O. bartramii* is generally distributed in the sea area west of 155°E is generally at 20–23 °C and that the distribution of the 20 °C isotherm can be used as one of the indicators for the formation of an *O. bartramii* fishing season. The surface sea temperature for fishery harvesting at 155°–160°E is 17–18 °C, and the 17 °C isotherm can be used as one of the indicators for the formation of an *O. bartramii* fishing season. In the *O. bartramii* fishing ground in the sea area of 160°–175°E, the surface sea temperature is generally at 11–13 °C, and the surface sea temperature is 5–7 °C lower on average than that in the sea area west of 160°E.

In analyzing the formation of an *O. bartramii* fishing season, relying only on the surface sea temperature is not sufficient; it is also important to measure the sea temperature structure in the vertical direction in the sea area west of 160°E, and a thermocline must form within a 50-meter water layer. Studies have reported that the daily output of a single vessel is basically directly proportional to the sea temperature difference $\triangle T_1$ at 0–100 m, i.e., catch per unit effort (CPUE, daily catch, unit: kg/day) $= -1213 + 314\triangle T_1$ ($R = 0.69$), and that the relationship between the daily output and the sea temperature difference $\triangle T_2$ at 0–50 m is even closer, i.e., CPUE $= -880 + 365\triangle T_2$ ($R = 0.779$) (Chen 1995). In the sea area west of 160°E, the thermocline mainly exists within the 50-m water layer, but the thermocline in the sea area east of 160°E is not apparent or does not form. Therefore, whether a thermocline has formed is also one of the main indicators for analyzing the *O. bartramii* fishing season.

In the sea area east of 160°E, the habitat water layer for *O. bartramii* is deep, generally approximately 300–400 m in daytime, but the habitat water layer for small-sized *O. bartramii* west of 160°E is only approximately 100 m. Additionally, because the sea area east of 160°E is only the extension current after the confluence of the Kuroshio Current with the Oyashio Current, the confluence force is not strong, and the mixing of the upper and lower layers is comparatively sufficient; therefore, the deep sea temperature plays an extremely important role in the formation of the *O. bartramii* fishing season in the sea area east of 160°E. Sea areas with concentrated *O. bartramii* distribution are mainly located in the warm-water front zone of the 200-meter deep water layer, and thus, the sea temperature of the 200-m water layer can be used as an important indicator of *O. bartramii* fishing seasons, i.e., generally 10 °C in June, 8 °C in July, and 6 °C in August (Wang and Chen 2005).

Relationship Between the Formation of the *O. bartramii* Fishing Season and Ocean Currents

Generally, in the years when the Kuroshio Current is stronger and the Oyashio Current is weaker, the northward force of all branches of the northward Kuroshio Current is stronger, the surface temperature of the sea area increases rapidly after May, *O. bartramii* migrate northeastward with the current, the position of the central fishing ground shifts to the north and to the east, and the fishing season is early. In the years when the Kuroshio Current is weaker and the Oyashio Current is stronger, the surface temperature is low, the increase in temperature is slow, the central distribution area of *O. bartramii* shifts more to the south and to the west, and the fishing season is late (Wang and Chen 2005).

6.1.2   Fishing Period Prediction: Case Studying for Japanese Spanish Mackerel *S. niphonius* in the East China Sea

Overview of the Migration Distribution of *S. niphonius*

S. niphonius is a warm-temperate pelagic fish that is distributed in the waters of the Indian Ocean and the western part of the Pacific Ocean, and it is also distributed throughout inshore areas of China. *S. niphonius* is a large-sized long-distance migratory fish species, with a population of *S. niphonius* in the Yellow Sea and the Bohai

Sea and a population in the East China Sea and the South Yellow Sea (inshore area) of China.

The *S. niphonius* populations in the Yellow Sea and the Bohai Sea pass the Dasha fishing ground in the last 10 days of April and reach the eastern sea area of the Sheyang Estuary in Jiangsu (range, 33°00′ to 34°3′0′N and 122°00′ to 123°00′E) from the southeast; thereafter, the fish stock swims northwestward and enters various spawning grounds in Haizhou Bay and the southern coast of the Shandong Peninsula, with the spawning period being from May to June. Around early September, the fish stock begins to successively swim away from the Bohai Sea; in mid-September, the feeding population in the Yellow Sea is mainly concentrated in the Yantai, Weihai, Haiyang Island, and Lianqingshi fishing grounds; and in the first 20 days of October, the main population moves southeastward, passing the peripheral sea area of Haizhou Bay, converging with the stock of fish feeding in Haizhou Bay and quickly migrating southeastward in early November, passing the northwestern part of the Dasha fishing ground to return to the Shawai and Jiangwai fishing grounds to overwinter.

The *S. niphonius* population in the East China Sea and the South Yellow Sea overwinters in the offshore sea area of the East China Sea from January to March. In April, the fish stock overwintering inshore enters along the coast to spawn, and the fish stock overwintering offshore successively migrates westward or in a northwestern direction to reach the sea areas around the coastal estuaries, harbors, and islands in Zhejiang, Shanghai, and the southern part of Jiangsu to spawn; the vigorous stage of the spawning period occurs from mid-May to mid-June. After spawning, part of the spawning stock remains in the sea area in the vicinity of the spawning grounds and feeds together with the juvenile fish born that year, and the other part of the spawning stock migrates northward to feed. Important feeding grounds and good fishing seasons form in the autumn. At the end of autumn, fish stock that are feeding leave the feeding ground and migrate eastward or in a southeastern direction, returning to the overwintering ground to overwinter from December to January (Chen 2016, 2021).

Fishing Period Forecasting for *S. niphonius*

The migration route and distribution status of *S. niphonius* often vary with changes in the hydrological status of its living environment. The timing of the fishing period, the shift in the location of the fishing ground, the degree of centralization or decentralization, and the length of the residence time of the fish stock, among other parameters, are all closely related to changes in the hydrological environment. For this reason, the fishing period for *S. niphonius* can be predicted based on the relationship between changes in the hydrological environment and food organism environment and the distribution characteristics of actions by *S. niphonius*.

Sea Temperature and Fishing Period

Using the average surface sea temperature of the Yangtze Estuary in the early April over the years, the anomaly values and the fishing periods for *S. niphonius* at the Yangtze Estuary over the years can be plotted on a graph. As seen in Fig. 6.2, except for the abnormal situation in 1980, the fishing period was earlier with higher sea temperatures (early April); otherwise, it was late.

Suppose we set Y_{date} as the fishing season date (April Y_{date}), using SSTA as the sea surface temperature anomaly (SSTA); then, the following relational expression can be obtained (Wei and Zhou 1988):

$$Y_{\text{date}} = 42.6419 - 2.1187\text{SSTA}, r = -0.8196$$

For the significance test of r, using the $a = 0.05$ level, $r = 0.8196 > a_{0.05} = 0.666$, and the test is significant.

Thus, there is a close relationship between the sea surface temperature (SSTA) in early April and the timing of the fishing period; based on verification by actual forecasting work in previous years, the results of fishing period forecasting based on sea temperature as the predictor are more accurate. Therefore, the sea surface temperature (SSTA) in early April can be used as one of the main indicators for forecasting the timing of the fishing period (Wei and Zhou 1988).

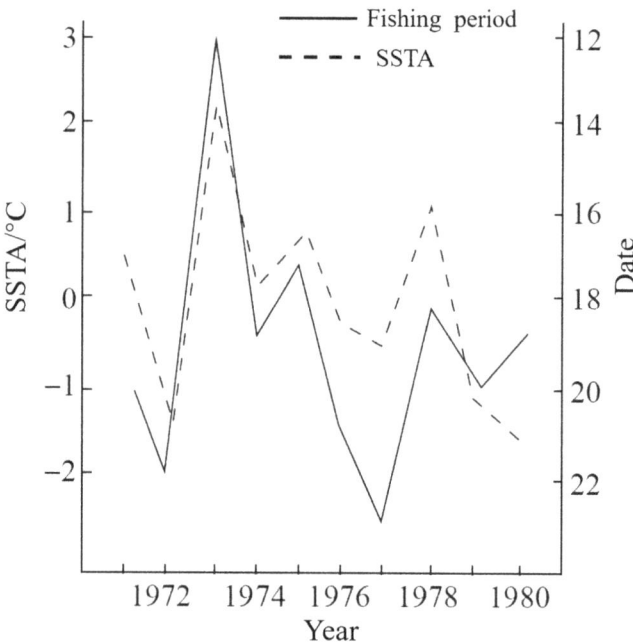

Air Temperature and Fishing Season Date

The average air temperature anomaly (ATA) values at the Yangtze Estuary in early April across the years and the fishing season date for *S. niphonius* (Y_{Date}) across the years are plotted in Fig. 6.3. Except for 1977, the fishing season date was related to high or low air temperature. The following relational expression can be obtained:

$$Y_{Date} = 34.849 - 1.4638 \text{ATA}, r = -0.6467$$

Using the $a = 0.05$ level, $r = 0.6467 > r0.05 = 0.666$; therefore, the use of air temperature in early April as a predictor has certain significance (Wei and Zhou 1988).

Because the range of variation in air temperature is larger than that in sea temperature, when the air temperature increases or decreases greatly, then the action of the fish stock is indirectly affected by sea surface temperature, but the effect is not as direct as that of sea temperature on the action of the fish stock; therefore, air temperature can be used as a reference indicator.

6.1.3   Prediction of Central Fishing Grounds

Prediction of the Central Fishing Grounds for *Katsuwonus pelamis* (Skipjack Tuna) in the Western and Central Pacific Ocean

Biological Overview of the *Katsuwonus pelamis* Fishery in the Western and Central Pacific Ocean

In the Pacific Ocean, the distribution of and central fishing grounds for *Katsuwonus pelamis* are closely related to the warm Kuroshio Current. *Katsuwonus pelamis* in the Pacific can be divided into two populations: the western population and the central population. The western population is distributed in the vicinity of the Mariana Islands and the Caroline Islands and migrates toward Japan, the Philippines, and New Guinea; the central population inhabits the vicinity of the Marshall Islands and the Tuamotu Archipelago (belonging to French Polynesia) and migrates toward the west coast of Africa and the Hawaiian

Fig. 6.3 Relationship between air temperature anomaly and fishing season date in early April across the years (Wei and Zhou 1988)

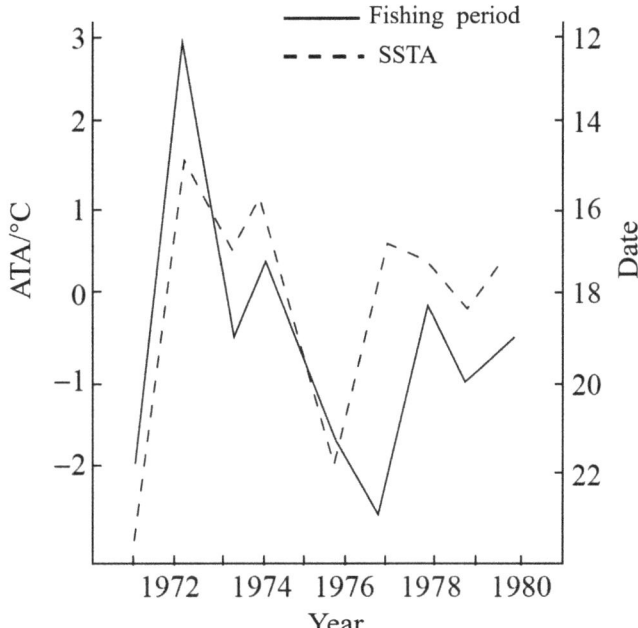

Islands. Sea areas with abundant food in the vicinity of tropical islands south of 20°N with a surface sea temperature above 20 °C are spawning grounds. In the Pacific Ocean, year-round spawning occurs in the tropical sea areas of the Marshall Islands and Central America, with the main spawning ground in the Central Pacific Ocean between 150°E and 150°W (Fig. 6.4).

Analysis of the Distribution of *Katsuwonus pelamis* Fishing Grounds in the Western and Central Pacific Ocean and Distribution Patterns

Katsuwonus pelamis is widely distributed in tropical sea areas, with the sea temperature ranging from 20 to 30 °C, and clusters in the sea areas where upwellings and cold and warm water masses converge. Additionally, these fish follow seabirds, debris floating on the water surface, sharks, whales, and dolphins and the migration of other tuna. A large number of studies have shown that the distribution, migration, and clustering of *Katsuwonus pelamis* in the Western and Central Pacific Ocean are closely related to sea temperature variations and the El Niño-Southern Oscillation (ENSO) in the tropical Pacific ocean.

Lehodey et al. (1997) carried out an analysis based on the *Katsuwonus pelamis* catch in the western equatorial Pacific Ocean by US *Katsuwonus pelamis* purse seiners from 1988 to 1995 and confirmed that the fishing ground for *Katsuwonus pelamis* moved as the 29 °C isotherm at the margins of the warm pool moved along the longitudinal lines.

Some scholars have analyzed the relationship between fishing ground movement and the environment by assessing the *Katsuwonus pelamis* catch and marine environmental factors (ocean current, sea surface temperature, Southern Oscillation Index (SOI), chlorophyll concentration, and so on), with an emphasis on the analysis of the fishing ground gravity center and the relocation of the fishing ground. The researchers have proposed the following main conclusions:

Correlation of CPUE with Surface Sea Temperature Utilizing the Kolmogorov-Smirnov (K-S) test, the results show that when the sea surface temperature (SST) is between 28 and 29 °C, it can be used as an index sea temperature for the selection of fishing grounds and the distribution of fish stocks. This temperature range is actually the

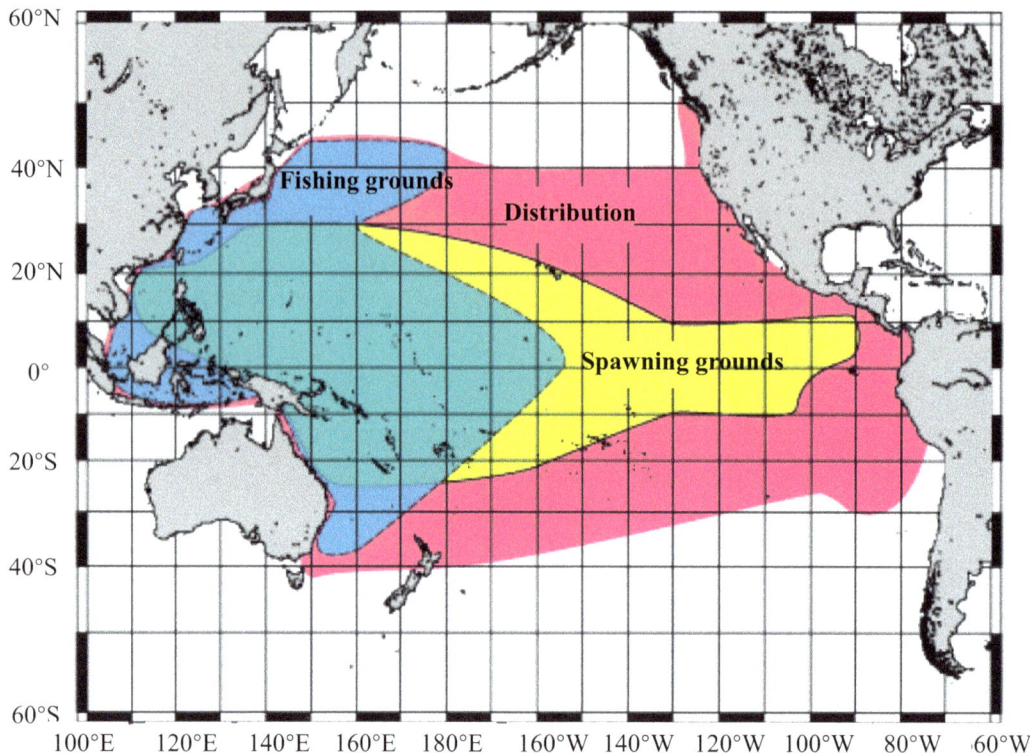

Fig. 6.4 Schematic diagram of the distribution of *Katsuwonus pelamis* fishing grounds and spawning grounds in the Western and Central Pacific Ocean (JFA 2021)

SST at the margins of the warm pool, which can be used as an indicator for the spatial distribution of stocks of *Katsuwonus pelamis*.

The linear correlation analysis of the SST and CPUE indicates a significant positive correlation in the sea area east of 160°E and a significant negative correlation in the sea area to the west. In addition, based on the correlation analysis of temporal sequences, the CPUE is positively correlated with the SST, and the CPUE is slightly delayed as the SST varies; the delay is 3 months.

CPUE Spatial-Temporal Distribution and Gravity Center Movement

The main operating fishing grounds for *Katsuwonus pelamis* purse seiners over the years have been concentrated in the sea area of the Western Central Pacific Ocean west of 180°E, with sparser activity in the sea area east of 180°E. Prior to 1996, these fish were mostly distributed in the sea area between 141° and 156°E, especially in 1994, when they were more limited between 147° and 153°E. In 1997, the fishing ground gravity center moved eastward on a large scale, and the movement extended approximately 2000 kilometers from June to August, exactly when the El Niño phenomenon occurred. At the end of 1997, the center moved westward. The El Niño phenomenon declined in 1998 and 1999, and the fishing ground gravity center returned to the sea area on the west side, with the bottom of the center at 165°E.

ENSO and the Movement of the Fishing Ground Gravity Center

Based on the monthly movement, gravity center is mainly distributed in sea area between 5°N and 5°S, and there is a greater bottom variation in longitude. Therefore, the movement of the gravity center along the longitudinal line of the fishing ground can be used to simplify the bottom displacement of *Katsuwonus pelamis* stocks. The results show that during the El Niño period (SOI has a negative value), *Katsuwonus pelamis* stocks relocate eastward on a large scale with the 29 °C isotherm, and during the La Niña period (SOI has a positive value), the stocks relocate to the West Pacific Ocean with the 29 °C isotherm. There is a delay in terms of time; the delay is approximately 3 months.

Relationship Between Fishing Ground Displacement and Sea Temperature Changes

Results indicate that the size and displacement direction of the CPUE and SST vectors are consistent and that sea temperature can be used for vector analysis to estimate the movement mechanism of *Katsuwonus pelamis* stocks.

Changes in the Gravity Center of the Central Fishing Ground for *Katsuwonus pelamis* and Establishment of a Prediction Model

Wang and Chen (2013) utilized statistical data on *Katsuwonus pelamis* production in the Western and Central Pacific Ocean from 1990 to 2010 (21 years) and sea surface temperature anomaly (SSTA) data in the Nino 3.4 region and, with a quarter as the time unit, used the clustering method of minimum spatial distance to analyze the spatial distribution patterns of *Katsuwonus pelamis* fishing grounds at a smaller time scale and establish a spatial-temporal distribution model of the central fishing ground for *Katsuwonus pelamis* to provide a basis for mid- and long-term fisheries forecasting for *Katsuwonus pelamis* in the Western and Central Pacific Ocean.

The production gravity center of each month was used to express the spatial-temporal distribution of the central fishing ground for *Katsuwonus pelamis*. Using the month as the unit, the production gravity center of each month from 1990 to 2010 was calculated, taking the 3-month mean value as the production gravity center of each

quarter. The formula for calculating the production gravity center is as follows:

$$X_{long} = \sum_{i=1}^{M}(C_i \times X_i)/\sum_{i=1}^{M}C_i$$

$$Y_{lat} = \sum_{i=1}^{M}(C_i \times Y_i)/\sum_{i=1}^{M}C_i$$

where X_{long} and Y_{lat} are the longitude and latitude of the position of the gravity center, respectively; C_i is the yield of fishing area i; X_i and Y_i are the longitudinal and latitudinal positions of the center of fishing area i, respectively; and M is the total number in the fishing area.

The quarterly ENSO index data were calculated, that is, the 3-month SSTA mean for the Nino 3.4 region is used, and linear correlation was conducted to calculate the correlation coefficient between the longitude and latitude of the production gravity center and the SSTA value for each quarter. Unary linear regression and a backpropagation (BP) neural network based on a fast algorithm were utilized to establish a prediction model for the fishing ground gravity center of *Katsuwonus pelamis* based on the quarterly mean SSTA values in the Nino 3.4 region and to compare the forecast results. The following conclusions were reached.

Correlation Between Longitude and Latitude of the Quarterly Production Gravity Centers Each Year and the SSTA for the Nino 3.4 Region

There is a significant correlation ($r = 0.35$, $p < 0.01$, $n = 84$) between the quarterly production gravity center in the longitudinal direction and the quarterly SSTA (Fig. 6.5), but there is no significant correlation ($r = 0.03$, $p < 0.01$, $n = 84$) between the quarterly production gravity center in the latitudinal direction and the quarterly SSTA (Fig. 6.6).

Relationship Between the Longitude Distribution of the Quarterly Production Gravity Centers Each Year and the SSTA for the Nino 3.4 Region

The quarterly production gravity centers each year were clustered based on the minimum Euclidean distance, and four categories were

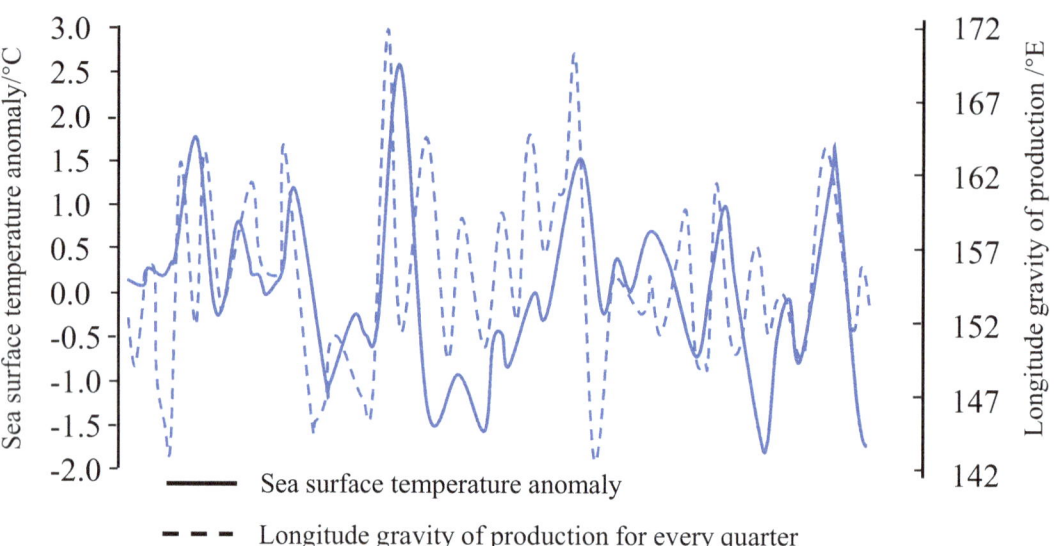

Fig. 6.5 Relationship between the quarterly yield gravity center and the quarterly SSTA in the longitudinal direction (Wang and Chen 2013)

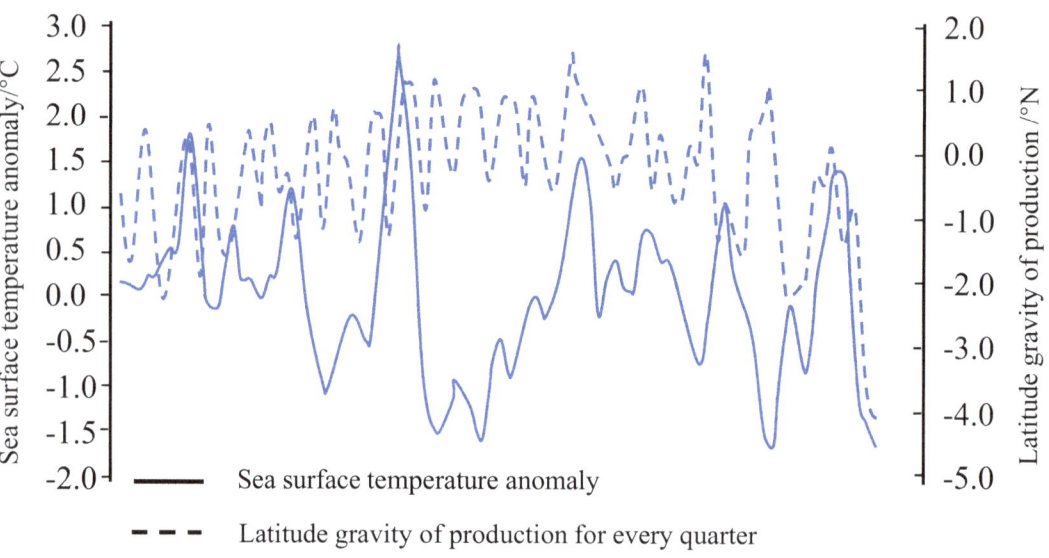

Fig. 6.6 Relationship between the quarterly production gravity center and the quarterly SSTA in the latitudinal direction (Wang and Chen 2013)

obtained (Fig. 6.13). The mean values for the four categories are as shown in Table 6.1. Based on the results in Table 6.1 and Fig. 6.13, as the SSTA increases, the longitude of the production gravity center shifts eastward, and the higher the SSTA is, the more apparent the eastward shift.

Table 6.1 SSTA for the Nino 3.4 region and the longitude of the average production gravity center (Wang and Chen 2013)

SSTA interval	Longitude of the average production gravity center
SSTA ≤ -0.5 °C	153.96
-0.5 °C $<$ SSTA<0.5 °C	153.91
0.5 °C $\leq$ SSTA<1 °C	157.12
SSTA>1 °C	160.08

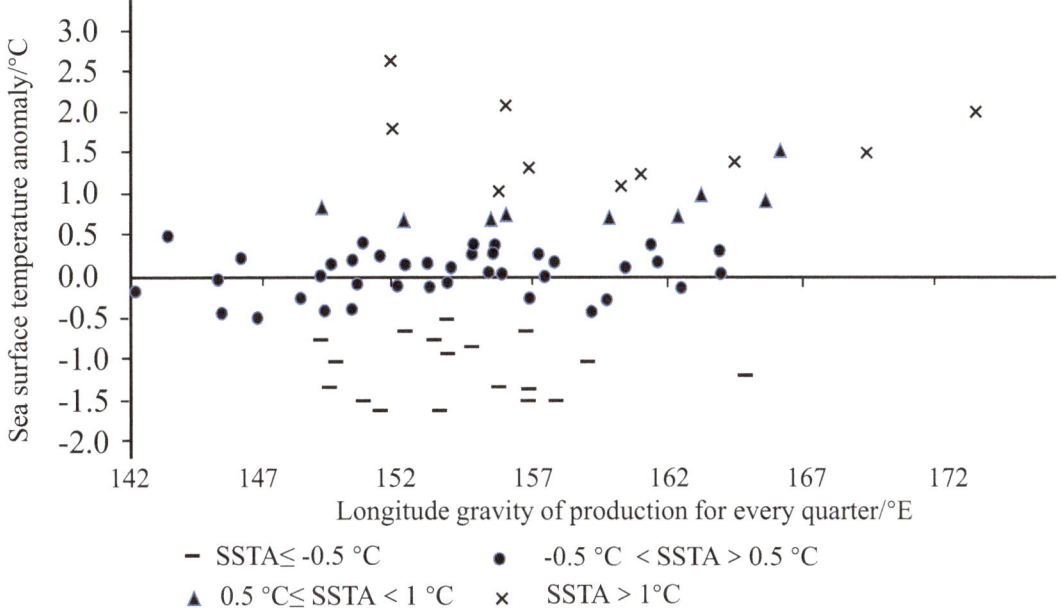

Fig. 6.7 Relationship between quarterly production gravity centers in the longitudinal direction and the SSTA (Wang and Chen 2013)

Prediction Model for the Longitude of the Production Gravity Center of *Katsuwonus pelamis* Based on the SSTA Value

When establishing the unary linear regression model, analysis of variance (ANOVA) indicated a significant difference between the SSTA value and the longitude value ($F = 8.3815$, $p = 0.0049 < 0.04$), with the following equation established (Fig. 6.7):

$$Y_E = 1.995 \times X_{SSTA} + 155.05$$

where Y_E represents the longitude of the production gravity center and X_{SSTA} represents the SSTA for the Nino 3.4 region.

When establishing the BP neural network model based on a fast algorithm, the input neuron is set to 1 (SSTA value), there is one output neuron (longitude value), there are three hidden layer neurons (their functions are sigmoid functions), and the residual error of fitting after training is $\delta = 0.0296$; the data obtained from the model are provided in Table 6.2.

There were 84 samples in total, among which 60 were used as training data to establish the model, while the remaining samples were used as test data. The accuracy rates of the prediction by the two models were compared by calculating the magnitude of the mean square error (MSE). The results were $MSE_Y = 19.25$ and $MSE_{BP} = 11.34$, indicating that the BP model was better than the unary linear model.

Table 6.2 Data obtained from the BP neural network model (Wang and Chen 2013)

Weights from the input layer to the hidden layer	Weights from the hidden layer to the output layer
0.0986	−0.2473
−1.8283	−0.705
1.2462	0.2771

Forecasting the Central Fishing Ground for *O. bartramii*

Establishing a Forecast Model for the Central Fishing Ground of *O. bartramii*

As an oceanic species, *O. bartramii* is widely distributed throughout the entire sea area of the North Pacific Ocean and has high resource abundance. It is the main fishing target for China's distant-water squid-jigging fishery. Chen et al. (2003) analyzed the central fishing ground for *O. bartramii* in the North Pacific Ocean from 150°E to 165°E from 1998 to 2000 by calculating the yield and the gravity center based on the number of voyages taken by operating vessels; in combination with the marine environmental conditions, they found the distribution pattern for the central fishing ground for *O. bartramii* year-to-year and established a forecast model, providing a basis for forecasting the central fishing ground of *O. bartramii* in the North Pacific Ocean.

The production data were obtained from data related to *O. bartramii* angling production in the Northwest Pacific Ocean in the vicinity of China from 1998 to 2000; the data included the date, longitude, latitude, yield, number of operations, and average daily yield. The temporal resolution was daily, and the spatial resolution was $0.5° \times 0.5°$.

In terms of the marine environment, the latitudes of the 20 and 15 °C isotherms distributed along the 150°E, 155°E, 160°E, and 165°E longitude lines were used as indicators of the strength of the warm water of the Kuroshio Current (20 °C) and the cold water of the Oyashio Current (15 °C), respectively.

The positions of the gravity centers for production and number of operations in the sea area from 150°E to 165°E from August to October of each year were calculated (Table 6.3).

Linear regression was utilized to establish the relational expression between the position and time (month) of the central fishing ground for *O. bartramii* and the marine environment (strength of the Kuroshio Current), resulting in the following research conclusions.

Comparison of the Distribution of the 20 and 15 °C Isotherms in Each Month

Because *O. bartramii* often cluster where cold and warm water converge, thus forming a fishing ground, analysis was carried out on the distribution status of the 20 °C (representing the Kuroshio front) and the 15 °C (representing the cold water front) isotherms from August to October each year based on the quick report chart of SST in the North Pacific, adding in understanding the relationship between operating fishing grounds and the marine environment. Regarding the distribution of the 20 °C isotherm in August, the sea area between 150° and 155°E shifted north by 1–2° of latitude compared with the sea area between 160° and 165°E. The warm water force was strongest in the sea area between 150° and 155°E in 1999, and its 20 °C isotherm shifted north by 1–1.5° of latitude compared with that in other years (Table 6.4). For the distribution of the 15 °C isotherm, the cold water in the sea area west of 160°E was relatively weak in August 1999, but the cold water in the sea area between 160°E and 165°E was stronger, shifting south by 1–3 latitudes compared with that in other years. From 1998 to 2000, the average sea temperature gradients between the 20 and 15 °C isotherms were 0.033 °C/nautical mile, 0.158 °C/nautical mile, and 0.068 °C/nautical mile, respectively, among which the horizontal temperature gradient was the largest in the sea area in the vicinity of 150°E in 1999 and 2000, at 0.500 °C/nautical mile and 0.167 °C/nautical mile, respectively.

Table 6.3 Positions of the gravity centers based on yield and number of operations between August and October from 1998 to 2000 (Chen et al. 2003)

Year	August				September				October			
	Production gravity center		Gravity center for the number of operations		Production gravity center		Gravity center for the number of operations		Production gravity center		Gravity center for the number of operations	
	Longitude E	Latitude N	Longitude E	Latitude N	Longitude E	Latitude N	Longitude E	Latitude N	Longitude E	Latitude N	Longitude E	Latitude N
1998	155°44'	42°30'	155°45'	42°30'	156°28'	42°55'	155°47'	42°43'	159°03'	43°37'	158°53'	43°34'
1999	156°52'	42°48'	156°48'	42°41'	159°08'	43°47'	158°56'	43°52'	158°06'	43°10'	158°05'	43°17'
2000	156°19'	43°37'	156°22'	43°36'	158°39'	44°28'	158°44'	44°27'	160°50'	44°09'	160°31'	44°04'
Average	156°18'	42°58'	156°18'	42°56'	158°05'	43°43'	157°49'	43°41'	159°20'	43°39'	159°10'	43°38'

Table 6.4 Distribution of the 20 and 15 °C isotherms between August and October from 1998 to 2000 (Chen et al. 2003)

Month	Isotherm	1998				1999				2000			
		150°E	155°E	160°E	165°E	150°E	155°E	160°E	165°E	150°E	155°E	160°E	165°E
August	20 °C	40°40'	41°20'	41°30'	41°00'	42°20'	43°40'	41°05'	40°50'	41°30'	42°30'	40°58'	39°40'
	15 °C	43°10'	43°00'	44°30'	45°05'	42°30'	46°00'	44°30'	41°50'	42°00'	43°55'	44°40'	43°01'
September	20 °C	41°40'	42°10'	41°00'	41°00'	43°30'	43°40'	42°20'	40°40'	42°20'	43°25'	40°50'	40°08'
	15 °C	42°05'	44°15'	46°30'	44°25'	44°25'	44°35'	45°45'	45°00'	43°35'	44°22'	45°35'	45°50'
October	20 °C	39°20'	39°35'	40°40'	40°05'	40°30'	40°25'	40°40'	40°00'	41°25'	39°52'	40°55'	39°18'
	15 °C	41°35'	43°30'	42°55'	44°00'	43°40'	43°20'	44°35'	42°10'	42°00'	43°08'	44°35'	43°52'

Table 6.5 Comparisons of the Kuroshio branch fronts and high-yield fishing areas between August and October from 1998 to 2000 (Chen et al. 2003)

Month	1998		1999		2000	
	Kuroshio branch fronts	Range of high-yield fishing area	Kuroshio branch fronts	Range of high-yield fishing area	Kuroshio branch fronts	Range of high-yield fishing area
August	First branch 150°–154°E, 42°30′N; second branch 155°–157°E, 42°N	152°–156°30′E, 42°–44°N	Second branch 155°–157°E, 44°N	155°–157°E, 43°N–44°30′N	Second branch 156°–157°30′E, 43°30′N	155°–158°E, 44°–45°30′N
September	Third branch 158°30′–160°30′E, 43°N	156°30′–160°30′E, 43°N–45°N	Third branch 159°–162°E, 42°30′N	159°–163°30′ E, 43°30′N–44°30′N	Second branch 156°–158°E, 44°30′N; third branch 161°–163°30′E, 41°45′N	156°30′–158°30′E and 161°30′E–164°E, 44°30′–45°30′N
October	Third branch 158°–160°30′E, 41–41°30′N	158°–161°30′E, within the 42°30′–45°N sea area	Third branch 158°–159°30′E, 42°N	158°–160°30′ E, 43°N–44°N	Third branch 163°–165°E, 41°30′N	162°–164°E, 44°30′–45°30′N

The distribution of the 20 and 15 °C isotherms in September from 1998 to 2000 is provided in Table 6.4. The 20 °C isotherm in the sea area between 150° and 155°E shifted north by 1–2° of latitude compared with that in the sea area between 160° and 165°E, and the 15 °C isotherm generally shifted south by 1–2° of latitude compared to that between 160° and 165°E. The average sea temperature gradient between the 20 and 15 °C isotherms in the sea area between 150°E and 155°E was the strongest, with an average between 0.04 and 0.20 °C, but it was under 0.025 °C/nautical mile in other sea areas.

The 20 °C isotherm was basically located in the sea area south of 41°N (Table 6.4). In the sea area west of 155 °E, the 20 °C isotherm varied more greatly each year, while that in the sea area east of 155°E was basically consistent. The warm water force in the sea area in the vicinity of 160°E was comparatively strong, and the warm water force in the vicinity of 165°E was relatively weak. The horizontal gradient between the 20 and 15 °C isotherms in the sea area between 150°E and 165°E was between 0.018 °C and 0.143 °C/nautical mile, and its strength was relatively lower than that in the previous period.

Warm-Water Branch Distribution of the Kuroshio Current Each Year and Its Relationship with the Gravity Center of Operating Fishing Grounds

Because *O. bartramii* is a warm-water species and its juvenile individuals feed and migrate north with the Kuroshio Current, a full understanding of the strength of the warm-water force of the Kuroshio Current and the development of its fronts has important significance for assessing the migration distribution and operating fishing grounds for *O. bartramii*. As seen in Table 6.5, the migration distribution of *O. bartramii* is extremely closely related to the marine environment. From August to October, the main operating fishing grounds for *O. bartramii* and their gravity centers are basically located in the frontal zones of the second and third warm-water branches of the Kuroshio Current. In August, the main operating fishing grounds are basically located in the frontal zone of the second branch of the Kuroshio Current, with a small portion located in the frontal zone of the first branch, whereas in September and October, they are basically located in the frontal zone of the third branch of the Kuroshio Current, with a small portion located in the frontal zone of the second branch (Table 6.5).

Forecast Model for Fishing Ground Movement

Through the aforementioned analysis, the direction of migration for *O. bartramii* from August to October is generally in a southwest-northeast direction, and their migration is affected by the strength of the force of the Kuroshio Current as well as the branches. To be able to forecast the operating fishing grounds for *O. bartramii* from August to October, we established a relational expression between the position and time (month) of the operating fishing grounds for *O. bartramii* and the various branches of the Kuroshio Current (20 °C isotherm). For cases in which the effects of marine environmental conditions were not taken into account, unary linear equations were established for the longitude and latitude of the gravity center of the operating fishing grounds for *O. bartramii* across time (month) (Chen et al. 2003):

$$FG_{Long} = 141.535 + 1.8435 \\ \times T \left(R_{Long} = 0.8736 \right)$$

$$FG_{Lat} = 39.75 + 0.41628 \times T \left(R_{Lat} = 0.5589 \right)$$

In the formula, FG_{Long} is the longitude of the gravity center related to the distribution of *O. bartramii*, and the unit is degree (10-point system); FG_{Lat} is the latitude of the gravity center related to the distribution of *O. bartramii*, and the unit is degree (10-point system); T is the month (8 (August), 9 (September), and 10 (October)); and R_{Long} and R_{Lat} are the correlation coefficients. Taking the significance level of $\alpha = 0.01$, $R_{Long} = 0.8736 > R(7,0.01) = 0.798$, a significant result; $R_{Lat} = 0.5589 < R(7,0.01) = 0.798$, a nonsignificant result.

Therefore, utilizing a single variable (time) to forecast the central fishing ground in the longitudinal direction is sufficient, illustrating that there is no significant change in the distribution of the central fishing ground for *O. bartramii* from August to October in the sea area between 150°

and 165°E. In the latitudinal direction, because the strength of the Kuroshio Current directly affects the distribution of the *O. bartramii* fishing grounds, the stronger the force of the branches of the Kuroshio Current, the more northerly is the distribution of the *O. bartramii* fishing grounds; otherwise, they shift to the south. For this reason, in terms of the latitudinal direction, marine environmental conditions that express the strength of the Kuroshio Current should be included.

Because the main fishing grounds for *O. bartramii* from August to October are located in the sea areas of the second branch and third branch of the Kuroshio Current, we utilize the position of the 20 °C isotherm on the 155°E and 160°E longitude lines as an indicator of the strength of the Kuroshio Current branches and then insert that value into the model for fishing ground forecasting. After linear regression, the model for fishing ground forecasting in the latitudinal direction is obtained (Chen et al. 2003):

$$FG_{Lat} = -8.461 + 1.165 \times T \\ + (Lat_{155} + Lat_{160})/2 \, (R_{Lat} = 0.8225)$$

In the formula, Lat_{155} is the latitude value for the 20 °C isotherm on the 155°E latitude line, and the unit is degree (10-point system); Lat_{160} is the latitude value for the 20 °C isotherm on the 160°E latitude line, and the unit is degree (10-point system); $(Lat_{155} + Lat_{160})/2$ is a constant; and R_{Lat} is the correlation coefficient.

Taking the significance level of $\alpha = 0.01$, $R_{Lat} = 0.8225 > R(7,0.01) = 0.798$, a nonsignificant result. Therefore, this model better reflects the relationship between the position of the central fishing ground for *O. bartramii* in the latitudinal direction and timing and strength of the Kuroshio Current.

6.2 Catch Prediction

Catch refers to the weight or quantity of fresh commercial aquatic animals and plants collected in natural waters. Catch prediction is an important component of fisheries forecasting. In addition to being related to resource abundance, fishing effort, and so on, catch is also closely related to marine environmental factors, meteorological factors, and other factors. Scientific prediction of catch is conducive to the realization of the sustainable utilization and scientific management of resources, and it can also provide a scientific basis for reasonable production per fishing vessel. There are many catch prediction methods, and they mainly utilize multivariate linear statistics, grey systems, the habitat suitability index (HSI), time series analysis, and neural networks, among other methods. In this section, catch predictions on important commercial species are used as cases for analysis.

6.2.1 Catch Prediction Based on Multivariate Linear Regression: Case Studying for Largehead Hairtail (*Trichiurus haumela*) in the East China Sea

Trichiurus haumela is the most important fishery in the East China Sea, and the winter fishing season for *Trichiurus haumela* in the inshore area of Zhejiang is the largest-scale fishing season in China, with a yield accounting for approximately more than 60% of the *Trichiurus haumela* yield in the entire East China Sea area. Therefore, predicting the catch of the *Trichiurus haumela* is of important significance for understanding the dynamics of the fish stock and guiding fishery practices. Forecasting of the winter fishing season for the *Trichiurus haumela* started in the late 1950s, and many scholars have carried out prediction research on *Trichiurus haumela* catch.

Variation in catch is comprehensively affected by numerous environmental factors; when establishing a forecast equation, it is necessary to screen and analyze the many influencing factors related to catch. The ribbon fish catch during the winter fishing season serves as the most important resource index of *Trichiurus haumela* in the summer and autumn, with total fishing effort in the winter fishing season a secondary factor. Because the actual value cannot be obtained before the fishing season, an estimated

value can be provided when forecasting. Based on data from previous years, two forecast equations have been established to predict the *Trichiurus haumela* catch during the winter fishing season (Chen 2014, 2016, 2021):

$$Y_{\text{catch}} = 14.48 + 4.997 X_{\text{SH}} + 0.133 X_{\text{effort}} \text{ (1954 to 1983)}$$

$$Y_{\text{catch}} = 103.4 + 6.625 X_{\text{NB}} + 1.820 X_{\text{RE}} \text{ (1970 to 1983)}$$

where Y_{catch} is the *Trichiurus haumela* catch during the winter fishing season; X_{SH} is the *Trichiurus haumela* resource index from May to September from Shanghai Fisheries company; X_{effort} is the total fishing effort during the winter fishing season; X_{NB} is the *Trichiurus haumela* resource index from May to August from the Ningbo Fishery Company; and X_{RE} is the corrected number for the relative resource of *Trichiurus haumela* in September. The total fishing effort during the winter fishing season refers to the sum of the fishing vessels multiplied by the fishing days (unit: 100 fishing vessel.day). Based on previous analyses, the forecast accuracy rate is above 80% for most years, with the forecast accuracy rate reaching 96% in the early 1980s.

Shen and Fang (1982) proposed that the amount and strength of the runoff volume of the Yangtze River directly affected the coastal current in China and thus indirectly affected the fishing grounds and fishery development of *Trichiurus haumela*; therefore, the runoff volume of the Yangtze River was added to fisheries forecasting. The *Trichiurus haumela* resource index, the total fishing effort in each fishing season, the runoff volume of the Yangtze River, and so on were used to establish forecast equations:

$$Y_{\text{ZJ}} = 58.10 + 6.780 X_{\text{SH}} + 0.062 X_{\text{ZJ}} - 0.156 X_{\text{YR}}$$

$$Y_{\text{SS}} = 138.34 + 5.39 X_{\text{SH}} + 0.007 X_{\text{SS}}{'} - 0.313 X_{\text{YR}}$$

where Y_{ZJ} and Y_{SS} are, respectively, the total *Trichiurus haumela* yield in all fishing seasons in the inshore area of Zhejiang and the total *Trichiurus haumela* yield in all fishing seasons in the Shengshan fishing ground; X_{SH} is the *Trichiurus haumela* resource index from May to September from Shanghai Fisheries company; X_{ZJ} and X_{SS}' are the total fishing effort invested in the inshore area of Zhejiang and in the Shengshan fishing ground, respectively, in all fishing seasons that year; and X_{YR} is the average runoff volume of the Yangtze River (September).

6.2.2 Catch Prediction Based on the Grey System

Prediction entails drawing support from past investigations to speculate and understand future development trends. Grey prediction involves scientific quantitative prediction on the future state of a system through the discovery and mastery of system development patterns by processing the original data and establishing a grey model (GM). For catch prediction, the application of grey system theory into fisheries has been very successful (Chen 2003).

Principles of Grey Dynamic Modeling

Grey prediction modeling uses the concept of grey modules as a basis. Grey system theory deems that all random quantities are grey quantities and grey processes that change within a certain scope and in terms of a certain time period. The processing of grey quantities does not involve determining statistical patterns and probability distributions but, rather, to establish patterns from irregular original data; that is, processing the data through a chosen mode generates more regular time series data, after which a new model is established. We refer to the module composed of known data as the white module, and the module extrapolated to the future, using the white module, is referred to as the grey module.

In general, in regard to the given original data sequence,

$$X_{(0)} = \left\{ x_{(0)}^{(1)}, x_{(0)}^{(2)}, x_{(0)}^{(3)}, \dots, x_{(0)}^{(N)} \right\}$$

The data cannot be directly used for modeling because these data are mostly random and irregular. If the original data sequence undergoes one cumulative generation, then a new data sequence can be obtained:

$$X_{(1)} = \left\{ x_{(1)}^{(1)}, x_{(1)}^{(2)}, x_{(1)}^{(3)}, \dots, x_{(1)}^{(N)} \right\}$$

among which $x_{(1)}^{(i)} = \sum_{k=1}^{i} x_{(0)}^{(i)}$.

The newly generated data sequence is a monotonous growth curve; obviously, it enhances the regularity of the original data sequence, and the randomness is weakened. For a nonnegative data sequence, the greater the accumulation, the more obvious is the weakening of randomness, and the stronger is the regularity; therefore, it is easier to approximate using an exponential function. The processed data weakens the randomness of the original data sequence; thus, the regularity of the change is found and provides intermediate information for establishing a dynamic model.

The reason that the grey system theory can be used to establish a differential equation model is because the grey system theory treats random quantities as grey quantities that change within a certain scope and treats random processes as grey processes that change within a certain amplitude range and within a certain time zone. Second, after the grey system theory modifies the irregular original data, making it a more regular generated data sequence, modeling is carried out again. Therefore, GM modeling actually generates a data model, but the original data model is used in general modeling. In addition, the grey system theory adjusts, corrects, and improves the precision of the model through the different generation of grey numbers, the different choice of data, and the generation of residual models for different levels.

The GM (n, h) model refers to the differential equation of n-th order and h variables, and GM models with different n and h have different meanings and uses. GM models can be roughly classified into the following two categories.

The GM (*n*, 1) Model

The commonly used GM (*n*, 1) model is a GM model with only one variable. The requirement for the data sequence is a "comprehensive effect" time series. As *n* becomes larger, the calculation becomes more complicated, but the precision may not necessarily be higher; therefore, in general, the *n* value is below the third order. The most commonly used *n* = 1 order model is simple to calculate and has broad applicability; denoted as GM (1, 1), it is referred to as a single-sequence first-order linear dynamic model.

The differential equation for the GM (1, 1) model is $\frac{dx^{(1)}}{dt} + ax^{(1)} = u$.

The coefficient vector is $\widehat{a} = [a, \mu]^T$.

The corresponding time function is $\widehat{x}^{(1)}(t+1) = \left(x^{(0)}(1) - \frac{u}{a}\right)e^{-at} + \frac{u}{a}$.

After taking the derivative and integration, one can obtain: $\widehat{x}^{(0)}(t+1) = -a\left(x^{(0)}(1) - \frac{u}{a}\right)e^{-at}$.

The two aforementioned equations are the basic calculation formulas for grey prediction with the GM (1, 1) model.

GM (2, 1) is a second-order model with two characteristic roots, and its dynamic process can reflect different situations, which may be a monotonic, nonmonotonic, or oscillating (vibrating) situation.

The differential equation for the GM (2, 1) model is:

$$\frac{d^2x^{(1)}}{dt^2} + a_1\frac{dx^{(1)}}{dt} + a_2x^{(1)} = u$$

Its coefficient vector is $\widehat{a} = (a_1, a_2, u)^T$.
Its time response function is:

$$x^{(1)}(t) = C_1e^{\lambda_1 t} + C_2e^{\lambda_2 t} + \frac{u}{a^2}$$

In the formula, λ_1 and λ_2 are the two characteristic roots, and the main dynamic features of the system can be analyzed according to the following different situations.

1. If $\lambda_1 = \lambda_2$, then the dynamic process is monotonic.
2. If $\lambda_1 \neq \lambda_2$ and both are real numbers, the dynamic process may be nonmonotonic.
3. If λ_1 and λ_2 are conjugate complex roots, then the dynamic process is periodically oscillating.

The GM (1, *h*) Model

The GM (1, 1) and GM (2, 1) models introduced above are generally used for prediction. However, as a state analysis model, the GM (1, *h*) model is commonly used, which can reflect the effect of *h*-1 variables on the first derivative of the dependent variable. Since *h* > 1, it is referred to as a first-order linear dynamic model of *h* sequences. The steps for modeling are as follows:

Let there be *h* variables $X_1, X_2, \ldots, X_h$ that compose the original data sequence $x_i^{(0)} = \{x_i^{(0)}(1), x_i^{(0)}(2), \ldots, x_i^{(0)}(n)\}$ ($i = 1, 2, \ldots, h$). A new data sequence is obtained by performing cumulative generation one time on $X_i^{(0)}$:

$$X_i^{(1)} = \left\{x_i^{(1)}(1), x_i^{(1)}(2), \ldots x_i^{(1)}(n)\right\}$$
$$\times (i = 1, 2, \ldots, h)$$

Establish the differential equation:

$$\frac{dx_1^{(1)}}{dt} + ax_1^{(1)} = b_1x_2^{(1)} + b_2x_3^{(1)} + \cdots + b_{h-1}x_h^{(1)}$$

Its coefficient vector, $\widehat{a} = (b_1, b_2, \ldots, b_{h-1})^T$, is solved by using the least-square method, that is,

$$\widehat{a} = \left(B^T B\right)^{-1} B^T Y_N$$

where B is the cumulative matrix and Y_N is the constant term vector, which are, respectively:

$$\mathbf{B}=\begin{bmatrix} -\frac{1}{2}\left(x^{(1)}(1)+x^{(1)}(2)\right) & x_2^{(1)}(2) & \cdots & x_h^{(1)}(2) \\ -\frac{1}{2}\left(x^{(1)}(2)+x^{(1)}(3)\right) & x_2^{(1)}(3) & \cdots & x_h^{(1)}(3) \\ \cdots & & \cdots & \cdots & \cdots \\ -\frac{1}{2}\left(x^{(1)}(n-1)+x^{(1)}(n)\right) & x_2^{(1)}(n) & \cdots & x_h^{(1)}(n) \end{bmatrix}$$

$$Y_N = \left[x_1^{(0)}(2), x_1^{(0)}(3), \ldots, x_1^{(0)}(n)\right]^T$$

Then, the solution of the differential equation can be obtained:

$$x_1^{(1)}(t+!) = \left(x_1^{(0)}(1) - \sum_{i=2}^{h}\frac{b_{i-1}}{a}x_i^{(1)}(t+!)\right)e^{-at}$$
$$+ \sum_{i=2}^{h}\frac{b_{i-1}}{a}x_i^{(1)}(t+!)$$

The GM (1, 1) Model

The GM (1, 1) model is actually grey data sequence prediction, which entails the prediction of the quantitative size of time series data, i.e., population prediction, labor prediction, yield prediction, output value prediction, and various trend predictions, and it generally utilizes statistical data in previous years to predict future developments. Such prediction is not only widely applied, but the steps of the method also have common significance. The basic steps for establishing the GM (1, 1) model are as follows:

Step 1: Perform cumulative generation one time on the data sequence $X^{(0)} = \{x^{(0)}(1), x^{(0)}(2), \ldots, x^{(0)}(N)\}$ to obtain

$$X^{(1)} = \left\{x^{(1)}(1), x^{(1)}(2), \ldots, x^{(1)}(N)\right\}$$

among which $x^{(1)}(t) = \sum_{k=1}^{t} x^{(0)}(k)$.

Step 2: Construct the cumulative matrix B and the constant term vector Y_N, which are:

$$B = \begin{bmatrix} -\frac{1}{2}\left(x^{(1)}(1)+x^{(1)}(2)\right) & 1 \\ -\frac{1}{2}\left(x^{(1)}(2)+x^{(1)}(3)\right) & 1 \\ \vdots & \vdots \\ -\frac{1}{2}\left(x^{(1)}(N-1)+x^{(1)}(N)\right) & 1 \end{bmatrix}$$

$$Y_N = \left[x_1^{(0)}(2), x_1^{(0)}(3), \ldots, x_1^{(0)}(N)\right]^T$$

Step 3: Use the least-square method to solve the grey parameter $\widehat{a}$:

$$\widehat{a} = \begin{bmatrix} a \\ u \end{bmatrix} = \left(B^T B\right)^{-1} B^T Y_N$$

Step 4: Substitute the grey parameter into the time function:

$$\widehat{x}^{(1)}(t+1) = \left(x^{(0)}(1) - \frac{u}{a}\right)e^{-at} + \frac{u}{a}$$

Step 5: Take the derivative and integration of $\widehat{X}^{(1)}$ to obtain:

$$\widehat{x}^{(0)}(t+1) = -a\left(x^{(0)}(1) - \frac{u}{a}\right)e^{-at}$$

or $\widehat{x}^{(0)}(t+1) = \widehat{x}^{(1)}(t+1) - \widehat{x}^{(1)}(t)$

Step 6: Calculate the difference between $x^{(0)}(t)$ and $\widehat{x}^{(0)}(t)$, $\varepsilon^{(0)}(t)$, and the relative error $e(t)$:

$$\varepsilon^{(1)}(t) = x^{(0)}(t) - \widehat{x}^{(0)}(t)$$

$$E(t) = \varepsilon^{(0)}(t)/x^{(0)}(t)$$

Step 7: Test the model precision and apply the model for forecasting.

To analyze the reliability of the model, one must perform a precision test on the model. At present, the more commonly used diagnostic method is a posteriori error test. That is, first calculate the deviation of the observed data s_1:

$$s_1^2 = \sum_{t=1}^{m} \left(x^{(0)}(t) - \bar{x}^{(0)}(t) \right)^2$$

and the deviation of the residual error s_2:

$$s_2^2 = \frac{1}{m-1} \sum_{t=1}^{m-1} \left(q^{(0)}(t) - \bar{q}^{(0)}(t) \right)^2$$

Recalculate the posteriori error ratio $c = \frac{s_1}{s_2}$ and the small error probability: $p = \left\{ \left| q^{(0)}(t) - \bar{q}^{(0)} \right| < 0.6745 s_1 \right\}$.

Diagnose the model based on the posteriori error ratio c and the small error probability p. When $p > 0.95$ and $c < 0.35$, then the model can be deemed as reliable, and it can be used for prediction. At this time, system behavior can be predicted using the model.

The aforementioned seven steps are the entire analysis process for modeling and prediction. When the residual error of the established model is comparatively large and the precision is not ideal, to improve the precision, one should generally use residual error GM (1, 1) modeling on the residual error to correct the forecast model.

Grey Disaster Prediction

Grey disaster prediction is essentially the prediction of outliers. Researchers often determine what values are regarded as outliers based on experience and historical values. The task of grey disaster prediction is to estimate the moment that the next or next several outlier(s) will appear, so as to facilitate preparing in advance and adopting preventive countermeasures.

Now, let X be the original sequence:

$$X_\xi = (x[q(1), x[q(2)], \cdots, x[q(m)])$$

and the disaster sequence; then

$$Q^{(0)} = (q(1), q(2), \cdots, q(m))$$

is referred to as the disaster date sequence.

Disaster prediction is the search for the regularity in disaster date series and the prediction of

the dates on which the next several disasters will occur. Disaster prediction with the grey system is realized by establishing the GM (1, 1) model for the disaster date sequence.

Let $Q^{(0)} = (q(1), q(2), \cdots, q(m))$ be the disaster date series; its one-time cumulative sequence is:

$$Q^{(1)} = (q(1), q(2), \cdots, q(m))$$

The sequence generated by the adjacent mean value $Q^{(1)}$ is $Z^{(1)}$; then, $q(k) + az^{(1)}(k) = b$ is referred to as the disaster GM (1, 1) model.

Now, let $X = (x(1), x(2), \ldots, x(n))$ be the original sequence, in which n is the date. Given a certain outlier ξ, the corresponding disaster date series is:

$$Q^{(0)} = (q(1), q(2), \cdots, q(m))$$

among which $q(m)(\leq n)$ is the date of occurrence for the most recent disaster; then, $\hat{q}(m+1)$ is referred to as the predicted date of the next disaster. For any $k > 0$, $\hat{q}(m+k)$ is referred to as the predicted date of the k-th disaster in the future.

Testing Methods for the Grey Prediction Model

Testing Method with Absolute Correlation

Now, let there be the original sequence:

$$X^{(0)} = \left(x^{(0)}(1), x^{(0)}(2), \cdots, x^{(0)}(n) \right)$$

Its corresponding relative error sequence is:

$$\Delta = \left(\left| \frac{\varepsilon(1)}{x^{(0)}(1)} \right|, \left| \frac{\varepsilon(2)}{x^{(0)}(2)} \right|, \cdots, \left| \frac{\varepsilon(n)}{x^{(0)}(n)} \right| \right) = \{\Delta_k\}_1^n$$

In the above formula, $X^{(0)}$ is the original sequence, $\hat{X}^{(0)}$ is the corresponding analog sequence, and ε is the absolute correlation between $X^{(0)}$ and $\hat{X}^{(0)}$. If for the given $\varepsilon_0 > 0$ there is $\varepsilon > \varepsilon_0$, then the model is referred to as a qualified correlation model.

Testing Method with the MSE Ratio and Small Error Probability

Table 6.6 Different grade for precision testing (Chen 2003)

Critical value for index Grade	Relative error a_0	Correlation rate ε_0	MSE ratio value C_0	Small error probability p_0
Grade I	0.01	0.90	0.35	0.95
Grade II	0.05	0.80	0.5	0.80
Grade III	0.10	0.70	0.65	0.70
Grade IV	0.20	0.60	0.80	0.60

Let $X^{(0)}$ be the original sequence, $\widehat{X}^{(0)}$ be the corresponding analog sequence, and $\varepsilon^{(0)}$ be the residual error sequence; then, the mean value of and variance in $X^{(0)}$ are, respectively:

$$\bar{x} = \frac{1}{n}\sum_{k=1}^{n} x^{(0)}(k) \text{ and } S_1^2 = \frac{1}{n}\sum_{k=1}^{n}\left(x^{(0)}(k) - \bar{x}\right)^2$$

The mean value of and variance in the residual error are, respectively:

$$\bar{\varepsilon} = \frac{1}{n}\sum_{k=1}^{n}\varepsilon(k) \text{ and } S_2^2 = \frac{1}{n}\sum_{k=1}^{n}(\varepsilon(k) - \bar{\varepsilon})^2$$

$C = \frac{S_2}{S_1}$ is referred to as the variance ratio value. For the given $C_0>0$, when $C<C_0$, the model is referred to as a qualified MSE model.

If $p = P(|\varepsilon(k) - \bar{\varepsilon}| < 0.6745 S_1)$ is among the values, it is referred to as the small error probability. For the given $p_0>0$, when $p>p_0$, the model is referred to as a qualified small error probability model.

Through the aforementioned analysis, three methods for testing the model are given. These three methods all judge the precision of the model by inspecting the residual error; among them, the requirement is that the mean relative error $\overline{\Delta}$ and the simulation error are better if they are smaller, while the absolute correlation ε is better if it is larger, the MSE ratio value C is better if it is smaller (because a small C indicates that S_2 is small and S_1 is large. That is, the residual error variance is small, and the variance in the original data is large, which illustrates that the residual errors are more concentrated and the oscillation amplitude is small and that the original data are more scattered and the oscillation amplitude is large; therefore, a good simulation effect requires S_2 to be as small as possible compared with S_1), and the small error probability p is better if it is

larger. If the given set α, ε_0, C_0, and p_0 take on value, the simulation precision has been determined for the tested model.

Commonly used precision grades are provided in Table 6.6. In general, the most commonly used is the relative error test index.

Cases of Catch Prediction with the Grey System

Catch Prediction Based on Grey Theory: Case Studying for *Penaeus chinensis* Yield in the Bohai Sea

Guo (1992) utilized general multivariate linear regression and grey system prediction models GM (0, h) and GM (1, h) to carry out modeling of the shrimp *Penaeus chinensis* yield in the Bohai Sea, ultimately comparing their precision. Xs$_1$ represented the relative shrimp yield in the Bohai Sea; Xs$_2$, Xs$_3$, and Xs$_4$ represented the relative quantity of juvenile shrimp in Bohai Bay, Laizhou Bay, and Liaodong Bay, respectively; and the shrimp yield in the Bohai Sea in 1969 was set as 100%. The relative shrimp yield and the relative quantity of juvenile shrimp in the subsequent 14 years are provided in Table 6.7.

We refer to the data (Xs$_1$) in the Table 6.7 as the original sequence, denoted as follows: $\left\{X_1^{(0)}(i)\right\}$; $k = 1, 2, 3, 4$; and $i = 1, 2, \ldots 15$.

Utilizing the data in Table 6.7, the ordinary multivariate regression equation is established as: $\widehat{X}_1^{(0)}(i) = 0.933 X_2^{(0)}(i) + 0.459 X_3^{(0)}(i) + 0.366 X_4^{(0)}(i) - 1.694$;

second, the grey static multivariate model GM (0, 4) is established as:

$$\widehat{X}_1^{(1)}(i) = 0.987 X_2^{(1)}(i) + 0.493 X_3^{(1)}(i)$$
$$+ 0.297 X_4^{(1)}(i) - 32.509$$

Table 6.7 Prediction results for the shrimp yield in the Bohai Sea (Guo 1992)

Year	Relative quantity of juvenile shrimp			Relative yield in shrimp (Xs_1)	Forecasted values		
	Bohai Bay (Xs_2)	Laizhou Bay (Xs_3)	Liaodong Bay (Xs_4)		Regression	GM (0, h)	GM (1, h)
1969	119	48	1	100.0	133.5	108.9	100.0
1972	20	157	16	100.5	100.6	101.9	114
1973	66	243	100	236.8	216.9	214.6	221.4
1974	13.9	165	251	313.4	301.7	293.0	305
1975	100	314	114	254.0	288.9	287.3	281.1
1976	64	37	39	87.4	90.6	93.0	95.6
1977	158	123	44	212.7	222.3	229.6	231.1
1978	163	223	42	320.0	276.3	283.3	279.5
1979	305	191	133	404.9	426.3	434.6	441.1
1980	176	276	2	313.2	300.0	310.3	303.1
1981	119	117	46	205.6	184.2	188.8	188.9
1982	20	61	23	58.2	55.6	55.6	55.0
1983	91	72	25	147.1	128.1	132.7	133.4
1984	48	21	27	53.5	63.4	65.7	67.4
1985	24	323	33	174.3	192.8	192.7	177.8

Additionally, the grey dynamic multivariate model GM (1, 4) is:

$$\frac{dX_1^{(1)}}{dt} + 1.731X_1^{(1)} = 1.778X_2^{(1)} + 0.763X_3^{(1)} + 0.568X_4^{(1)}$$

Its consequent time-corresponding equation is:

$$\widehat{X}_1^{(1)}(i) = \begin{aligned} &\left(100 - 1.027X_2^{(1)}(i) - 0.441X_3^{(1)}(i)\right.\\ &\left. - 0.325X_4^{(1)}(i)\right)e^{-1.72t(i-1)}\\ &+ 1.027X_2^{(1)}(i) + 0.441X_3^{(1)}(i)\\ &+ 0.328X_4^{(1)}(i) \end{aligned}$$

We used the above three models to predict the relative shrimp yield (Table 6.9). The absolute deviations and relative deviations of the predicted values from the three models relative to the original sequence values are provided in Table 6.8.

The correlation between the GM predicted sequence and the measured sequence was compared through a correlation between the regression model predicted sequence and the measured sequence (Table 6.9). A comparison of the average absolute error, average relative error, and maximum deviation of the predictions is provided in Table 6.10.

The average errors and maximum deviations of the predictions generated by the grey multivariate model are reduced relative to those generated by the regression model, indicating that the grey multivariate model is the optimal prediction model (Table 6.11).

Prediction of the Abundance of the Eel Fry Catch

Xie et al. (1998) used the 25 years of data from 1972 to 1996 to conduct disaster prediction for the lack of abundance in eel fry (Table 6.12). The specific calculation process is as follows:

Step 1: Set the original data sequence

$$X^{(0)} = \left\{X^{(0)}(k)|k = 1, 2, \cdots, n\right\}$$

$$= \{11, 5.4, 11.2, \ldots, 12\}$$

Step 2: Determine the disaster threshold ξ. Stipulate that a year with an annual eel yield greater than 12 t is an abundant year; that is, $\xi = 12$.

Step 3: Perform disaster mapping based on ξ.

Table 6.8 Prediction results generated by the three models (Guo 1992)

Year	Regression model		GM (0, h)		GM (1, h)	
	Absolute error	Relative error (%)	Absolute error	Relative error (%)	Absolute error	Relative error (%)
1969	33.5	33.5	8.9	8.9	0	0
1972	0.1	0.1	1.4	1.4	13.6	13.6
1973	19.9	8.4	22.2	9.4	13.4	6.5
1974	11.7	3.7	20.4	6.5	8.3	2.7
1975	34.9	13.7	33.3	13.1	27.1	10.7
1976	3.2	3.7	5.6	6.4	8.2	9.3
1977	10.1	4.7	16.9	8.0	18.4	8.6
1978	43.7	13.7	36.7	11.5	40.5	12.7
1979	21.4	5.3	29.7	7.3	36.2	8.9
1980	13.2	4.2	2.9	0.9	10.1	3.2
1981	21.4	10.4	16.8	8.2	16.7	8.1
1982	2.6	4.5	1.6	2.7	3.2	5.6
1983	19.0	12.9	14.4	9.8	13.7	9.3
1984	9.9	18.5	12.2	22.9	13.9	26.0
1985	18.5	10.6	18.3	10.5	3.4	2.0

Table 6.9 Results generated by the three prediction models (Guo 1992)

Model	Regression	GM (0, h)	GM (1, h)
Correlation rate	0.6116	0.6281	0.6399
Improvement rate	0.0%	2.7%	4.7%

Table 6.10 Comparison of the errors in the three prediction models (Guo 1992)

Model	Average absolute error	Average relative error (%)	Maximum absolute deviation
Regression model	17.544	9.86	43.7
GM (0, h)	16.087	8.5	36.7
GM (1, h)	15.2	8.48	40.5

Table 6.11 Comparison of the errors in the three catch prediction models (Guo 1992)

Model	Average absolute error	Average relative error	Maximum deviation
GM (0, h)	8.3%	13.8%	16%
GM (1, h)	13.4%	14%	7.3%

Table 6.12 Eel fry catch from 1972 to 1996 (Unit: t) (Xie et al. 1998)

No.	1	2	3	4	5	6	7	8	9	10	11	12	13
Year	1972	1973	1974	1975	1976	1977	1978	1979	1980	1981	1982	1983	1984
Yield	11	5.4	11.2	2.3	11.3	5	9	22	3	6	7	5	22
No.	14	15	16	17	18	19	20	21	22	23	24	25	
Year	1985	1986	1987	1988	1989	1990	1991	1992	1993	1994	1995	1996	
Yield	7	2	13	3	8	40	12	12	10	6	15	12	

$$\xi : X^{(0)} \to X^{(0)}\xi$$

$$X^{(0)}\xi = \left\{ X^{(0)}\xi(k') \mid X^{(0)}(k^{\cdot}) \geq \xi, \right.$$

k' is equivalent to $K = \{22, 22, 13, 40, 15\}$ in $X^{(0)}\xi(k^{\cdot}) = X^{(0)}(k)$.

The disaster date set is obtained:

$$PD = k' \big| k' X^{(0)}\xi(k')$$

$$= |8, 13, 16, 19, 24|$$

$$= Y[8, 13, 16, 19, 24]$$

Step 4: Determine the parameters:

$$B = \begin{bmatrix} -14.5 & 1 \\ -29 & 1 \\ -46.5 & 1 \\ -68 & 1 \end{bmatrix}$$

$$a = [a, \mu]^T$$

$$a = \left[B^T B \right]^{-1} B^T Y_N$$

$$= (-0.20294024, 9.837248)$$

Step 5: Establish a GM (1, 1) model:

The GM (1, 1) prediction model is $\frac{dPD}{dk'} - 0.20294024PD = 9.9837248$.

Thus, the time-corresponding function of GM (1, 1) is:

$$PD = 11.60724672e^{0.20294024k'}$$

Step 6: Precision testing.

Precision testing of the model occurs as follows:

Calculated values:

$$PD^{(0)}(2') = 12.9, PD^{(0)}(3') = 12.9, PD^{(0)}(4') = 12.9, \text{ and}$$

$$PD^{(0)}(5') = 12.9$$

Original values:

$$PD^{(0)}(2) = 13, PD^{(0)}(3) = 16, PD^{(0)}(4) = 19, \text{ and}$$

$$PD^{(0)}(5) = 24$$

Residual errors:

$$q(2') = 0.131, q(3') = 0.235, q(4') = -0.312, \text{ and}$$

$$q(5') = 0.343$$

Relative errors:

$$e(2') = 1.0079\%, e(3') = 1.472\%, e(4') = -1.640\%, \text{ and}$$

$$e(5') = 1.429\%$$

Its precision test is passed.
$c = 0.019$
$p = 1$
Step 7: Model prediction:

$$p^{(0)}(6') = 28.9, p^{(0)}(7')$$

$$= 35.5, p^{(0)}(6') - p^{(0)}(5') = 28.9 - 23.7 = 5.2, \text{ and}$$

$$p^{(0)}(7') - p^{(0)}(6') = 35.5 - 28.9 = 6.6$$

The last abundant year in the table is 1995; therefore, the next abundant year will be 1995 + 5 = 2000, and the next abundant year will be 2000 + 7 = 2007.

If a year with an annual eel fry yield less than 6 t is set as a lack of abundance, the same method can be used to find the years in which a year with a lack of abundance will appear. Such a prediction is only for the year, not the yield. The yield can be predicted through other grey system models.

6.3 Prediction of Abundance or Abundance Index

The amount of abundance or abundance index is an important research content in fisheries resource science, and it is also one of the main components of fisheries forecasting. The scientific prediction of resource amounts and the assessment of resource abundance are conducive to the sustainable use and scientific management of resources. Variations in resource amounts, or resource abundance, are due to factors such as fishing and the marine environment, which can cause year-to-year changes. In the scientific prediction of the amount of resources or resource abundance, one must first understand the life history process, habitat, and migration of the prediction target. In this chapter, important commercial species are used as cases for analysis.

6.3.1 Prediction of the Abundance of *O. bartramii* in the North Pacific Ocean

Chen et al. (2005) utilized statistical production data from squid-jigging vessels in China from 1995 to 2002 in combination with sea temperature data (SST and its SSTA) for *O. bartramii* spawning grounds and feeding grounds to explore the relationship between changes in the resource abundance of *O. bartramii* in the fishing grounds of the Northwest Pacific Ocean and the year-to-year variation in the surface temperature and established a relationship between resource abundance and surface temperature.

The feeding grounds (operating fishing grounds) were divided into two sea areas based on the differences in the marine environmental conditions of the Northwest Pacific Ocean and the

distribution of the operating squid-jigging fishing grounds: west of 150°E at 39° and 45°N and within 150°–165°E and 39°–45°N.

The average daily yields for each year in the aforementioned two sea areas and the entire sea area were obtained. Because *O. bartramii* was the target fish species of squid-jigging operations, there was no bycatch, and the operating vessel types were basically consistent; therefore, the average daily yield (CPUE) could be roughly used as an indicator of resource abundance.

According to the previous research results, the spawning grounds for *O. bartramii* in the sea area of the Northwest Pacific Ocean (west of 170°E) are located at 20°–30°N and 140°–170°E, and spawning occurs from January to April; therefore, the surface temperatures and their mean anomaly values at the spawning grounds and the two feeding grounds between January and April from 1995 to 2002 were calculated.

Due to the small number of samples, grey relational analysis was used to analyze the relationship between SST and mean anomaly values at the spawning grounds and the feeding grounds and resource abundance. The variables were surface temperature, mean anomaly value for surface temperature, and CPUE. Among them, CPUE was the parent sequence, and surface temperature and its mean anomaly value were the sub-sequences. Averaging was used to transform the original data. The resolution coefficient was 0.5. The method for calculating the grey correlation degree is as follows:

Suppose an evaluation is conducted at m sample points and the evaluation index system is composed of n indicators. The measured values of all indicators at each sample point constitute a data sequence, denoted as:

$$x_i(k) = \{x_i(1), x_i(2), \cdots, x_i(n)\} \quad (i = 1, 2, \cdots, m)$$

Among the m sample points participating in the evaluation, the optimal values are selected, and the optimal values of all individual indicators compose the reference data sequence, which is denoted as:

$$x_0(k) = \{x_0(1), x_0(2), \cdots, x_0(n)\}$$

The numerical values of the reference data sequence are at the optimal level for each sample point in each index system. In fact, the reference data sequence is the "ideal mode" at each sample point. In addition, this is used as the standard for grey system evaluation, and other sample points are used in conjunction for a comparative analysis to conduct a quantitative evaluation.

For the aforementioned $m + 1$ sequences, $\{x_0\}$, $\{x_1\}$, $\{x_2\}$, ..., $\{x_m\}$, if the dimension or order of magnitude or index type differs, then initialization or normalization has to be carried out for the evaluation results to be comparable and to reduce interference from random factors.

Calculate the correlation coefficient $\theta_i(k)$ between each sample point and the evaluation standard $\{x_0\}$; the formula is as follows:

$$\theta_i(k) = \frac{\Delta \min + Rc \cdot \Delta \max}{\Delta_i(k) + Rc \cdot \Delta \max} \quad (i = 1, 2, \ldots m; \\ k = 1, 2, \ldots n)$$

In the formula, $\Delta_i(k) = |x_i(k) - x_0(k)|$;

$$\Delta_{\min} = \min_i \left[\min_k \Delta_i(k) \right];$$

$$\Delta_{\max} = \max_i \left[\max_k \Delta_i(k) \right]; \text{ and}$$

Rc is the resolution coefficient $(0 < Rc < 1)$.

The grey correlation degree is defined as $r_i = \frac{1}{n} \sum_{k=1}^{n} \theta_i(k)$.

The main factors affecting resource abundance were obtained through grey relational analysis, and then, a multivariate statistical method was utilized to establish the model for resource abundance and multiple variables.

Relationship Between the Year-to-Year Variation in the SST of Spawning Grounds and Resource Abundance

In the spawning grounds for *O. bartramii*, the SST for each year varied between 22.6 and 24.4 °C, and the mean SST anomaly value was 0.19–1.46 °C. From 1995 to 1997, the SST was at a comparatively normal level, with average SSTs of 22.6–22.8 °C, mean SST anomaly values of 0.19–0.27 °C, and CPUEs for squid-jigging operations each year of 1.90–2.60 t/day. In 1998 and from 2000 to 2002, the SST was slightly higher, with average SSTs of 23.4 to 23.7 °C, mean SST anomaly values of 0.61–0.81 °C, and CPUEs for squid-jigging operations each year of 2.56 t/day and 1.43–1.86 t/day, respectively. In 1999, the surface temperature was high, with an average surface temperature of 24.34 °C, mean SST anomaly value of 1.46 °C, and CPUE of 2.046 t/day (Fig. 6.8).

After grey relational analysis, the correlations between SST and mean SST anomaly value and resource abundance were 0.644 and 0.626, respectively (Table 6.18), indicating that the average SST and mean SST anomaly values at spawning grounds for *O. bartramii* have a comparatively close relationship with resource abundance.

Relationship Between the Year-to-Year Variation in the SST of Feeding Grounds and Resource Abundance

In the feeding grounds for *O. bartramii* (39°–45°N and west of 150°E), the SSTs for each year were between 15.3 and 17.6 °C, and the mean SST anomaly values were between 0.03 and 1.69 °C. In 2002, the SST was at a comparatively low level, with an average SST of 15.3 °C, mean SST anomaly value of 0.03 °C, and CPUE of 0.87 t/day. From 1995 to 1997 and in 1998 and 2001, the SSTs were at a medium level, with average SSTs of 15.6–16.8 °C, mean SST anomaly values of 0.27–1.07 °C, and CPUEs of 1.84 to 2.78 t/day, 3.49 t/day, and 2.08 t/day, respectively. From 1999 to 2000, SSTs were at a

Fig. 6.8 Relationship between SST variations at spawning grounds and average daily catch (CPUE) from 1995 to 2002 (Chen et al. 2005)

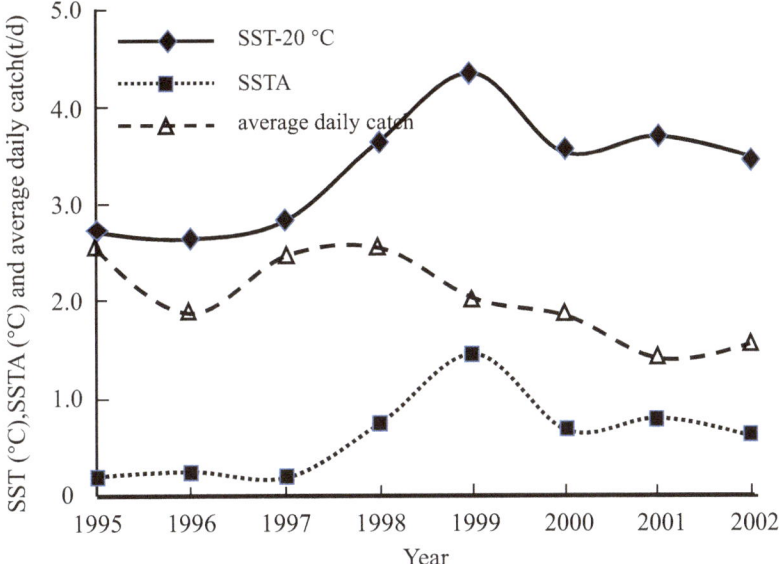

comparatively high level, 17.4–17.5 °C, with mean SST anomaly values of 1.60–1.69 °C and CPUEs of 1.15–2.46 t/day (Fig. 6.9).

After grey relational analysis, the correlations between SST and the mean SST anomaly value and resource abundance were 0.606 and 0.761, respectively (Table 6.18). Between 1995 and 2002, except for 2001, in which a small increase appeared in resource abundance, the changes in

the average SSTs and mean SST anomaly values were basically consistent with changes in resource abundance (Fig. 6.9).

In the sea area containing the feeding ground for *O. bartramii* (39°–45°N and 150–165°E), the SSTs for each year were between 14.5 and 17.0 °C, and the mean SST anomaly values were between −0.56 and 1.52 °C. In 1995, 1997, and 2002, the SSTs were at comparatively

Fig. 6.9 Relationship between SST variations at feeding grounds in the sea area west of 150°E and average daily catch (CPUE) from 1995 to 2002 (Chen et al. 2005)

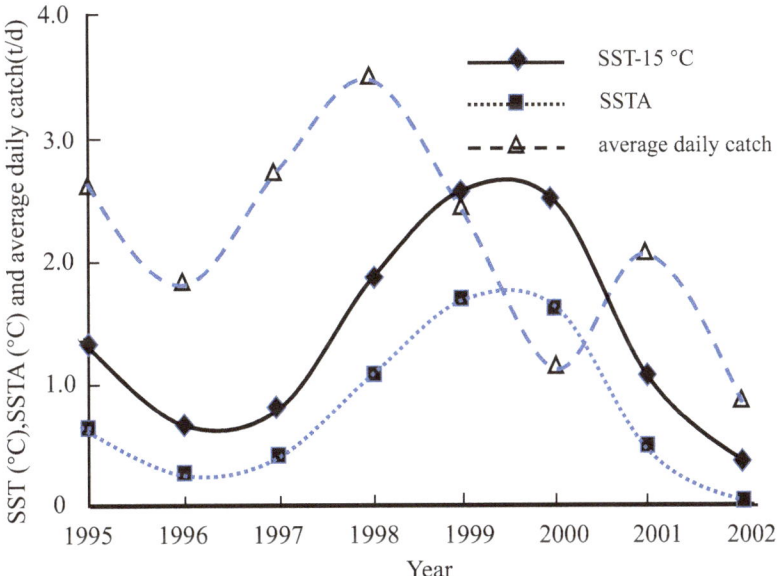

Fig. 6.10 Relationship between SST variation at feeding grounds in the sea area from 150 to 165°E and average daily catch (CPUE) (Chen et al. 2005)

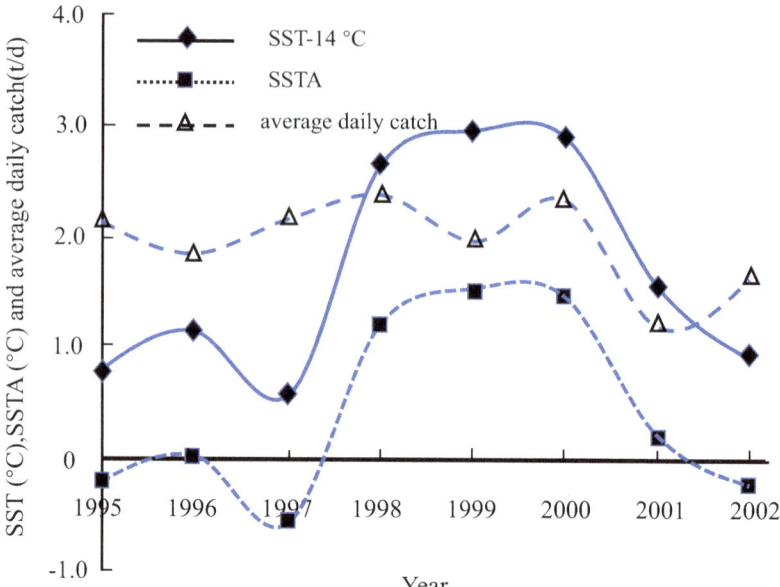

low levels, lower than the normal level in previous years, with average SSTs of 14.5–14.9 °C; negative mean SST anomaly values, that is, −0.56 to −0.20 °C; and CPUEs of 1.68–2.19 t/day. In 1996 and 2001, SSTs were at a medium level, with average SSTs of 15.1–15.6 °C, mean SST anomaly value of 0.02–0.27 °C, and CPUEs of 1.87 and 1.25 t/day, respectively. From 1998 to 2000, the SSTs were at comparatively high levels, 16.6–17.0 °C, the mean SST anomaly values ranged from 1.20 to 1.53 °C, and the CPUEs were 1.99–2.41 t/day (Fig. 6.10).

After grey relational analysis, the correlations between SST and the mean SST anomaly value and resource abundance were 0.683 and 0.534, respectively (Table 6.13). Between 1995 and 2002, except for 1999 and 2002, in which minor anomalies appeared in resource abundance, the

changes in the average SST and mean SST anomaly values were basically consistent with changes in resource abundance (Fig. 6.10).

Relationship Between Resource Abundance and Mean SST Anomaly Value

Based on grey relational analysis, resource abundance had a comparatively close relationship with SST and mean SST anomaly value at spawning grounds and feeding grounds (the grey correlation degrees were all above 0.5). Because the mean SST anomaly value better represents variations in SST across years, the mean SST anomaly values for spawning grounds and feeding grounds as well as their reciprocal sympathetic factors were used to establish multivariate linear equations.

Table 6.13 Correlations between the average SST and mean SST anomaly values in each sea area and average daily catch (CPUE) (Chen et al. 2005)

Sea area	Average sea temperature and resource abundance	Mean SST anomaly value and resource abundance
140–170°E, 20–30°N	0.644	0.626
West of 150°E, 39–45°N	0.606	0.761
150°–165°E, 39–45°N	0.683	0.534

Table 6.14 Relationship between the resource abundance of *O. bartramii* in the sea area west of 150°E in the North Pacific Ocean and SST (Chen et al. 2005)

Model 1	Model 2
CPUE = 1.0700 + 1.6840 T_1 + 2.7596 T_2–2.3177T_3 CPUE: Average daily yield (ton/day) T_1: Mean SST anomaly value for spawning grounds (20–30°N, 140–170°E) T_2: Mean SST anomaly value for feeding grounds (39–45°N, west of 150°E) T_3: $T_1 \times T_2$ is the sympathetic effect between the spawning grounds and the feeding grounds Residual standard deviation (SD) = 0.5877 Multiple correlation coefficient R = 0.8687 $F_{0.05} = 1.026 > p = 0.6037$ Average relative error 8.58%	CPUE = 1.9938–2.3044T_1 + 2.4257 T_2 CPUE: Average daily yield (ton/day) T_1: Mean SST anomaly value for spawning grounds (20–30°N, 140–170°E) T_2: Mean SST anomaly value for feeding grounds (39–45°N, west of 150°E) Residual SD = 0.6371 Multiple correlation coefficient R = 0.6508 $F_{0.05} = 0.7349 > p = 0.5764$ Average relative error 13.77%

The fisheries forecasting model for the sea area west of 150°E is shown in Table 6.14. The test for both models was significant under the 0.05 confidence limit level. However, after adding T_3, the multiple correlation coefficient for the model was 0.8687, while the average relative error was only 8.58%. For the model in which T_3 had not been added, the multiple correlation coefficient was 0.6508, and the average relative error was 13.77%.

The fisheries forecasting model for the sea area between 150° and 160°E is shown in Table 6.15. The test for both models was significant under the 0.05 confidence limit level. However, after T_3 was added, the multiple correlation coefficient for the model was 0.8771, while the average relative error was only 0.49%. For the model in which T_3 had not been added, the multiple correlation coefficient was 0.7367, and the average relative error was 1.00%.

6.3.2 Prediction of the Abundance of Argentine Shortfin Squid (*Illex argentinus*) in the Southwest Atlantic Ocean

Illex argentinus is an important commercial cephalopod that is widely distributed throughout the Southwest Atlantic Ocean. Existing research findings and production practices indicate that the amount of *Illex argentinus* resources in the Southwest Atlantic has changed drastically year-to-year and is significantly affected by the marine environment. Wang (2015) used the standardized CPUE for *Illex argentinus* as an indicator of resource abundance to find the key sea areas that affected resource abundance by analyzing the correlation between the surface temperature and mean SST anomaly value at spawning grounds and the CPUE and, in an attempt to establish a forecast model between different environmental

Table 6.15 Relationship between the resource abundance of *O. bartramii* in the sea area between 150° and 160°E in the North Pacific Ocean and SST (Chen et al. 2005)

Model 1	Model 2
CPUE = 2.5311–1.5226T_1 + 0.1519 T_2 + 0.6720 T_3 CPUE: Average daily yield (ton/day) T_1: Mean SST anomaly value at spawning grounds (20–30°N, 140–170°E) T_2: Mean SST anomaly for feeding grounds (39–45°N, west of 150°E) T_3: $T_1 \times T_2$ is the sympathetic effect between the spawning grounds and the feeding grounds Residual SD = 0.2434 Multiple correlation coefficient R = 0.8771 $F_{0.05} = 4.4476 > p = 0.09174$ Average relative error 0.49%	CPUE = 2.3339–0.9152T_1 + 0.5158 T_2 CPUE: Average daily yield (ton/day) T_1: Mean SST anomaly value for spawning grounds (20–30°N, 140–170°E) T_2: Mean SST anomaly value for feeding grounds (39–45°N, west of 150°E) Residual SD = 0.3065 Multiple correlation coefficient R = 0.7367 $F_{0.05} = 2.966 > p = 0.1414$ Average relative error 1.00%

influencing factors and resource abundance, utilized factors such as SST and the mean SST anomaly value in key sea areas as environmental indicators that affected resource abundance, providing a reference for the scientific management and production of *Illex argentinus* in sea areas of the Southwest Atlantic Ocean.

Previous research indicates that the sea areas of 30°S to 45°S and 40°W to 65°W are the spawning grounds of *Illex argentinus* in the Southwest Atlantic Ocean. During the spawning months (June to August), the correlations between the time sequence values composed of the SST and SSTA at each point and the time sequence values composed of the CPUE in the coming year were calculated and analyzed, and the SST and SSTA of sea areas with high correlations were selected as the influencing factors for the recruitment of *Illex argentinus*.

Calculating the ratio of the range occupied by the optimal SST of spawning grounds to the total area of spawning grounds is one of the important methods for measuring whether the habitat of the spawning grounds is good or bad. An SST of 16–18 °C was defined as the optimal SST for the spawning grounds, the ratio of the range occupied by the optimal SST to the total area of the spawning grounds (P_S) was calculated, and P_S was used to express the degree of suitability of the habitat at spawning grounds. Therefore, Ps was selected as the influencing factor for the recruitment of *Illex argentinus*, and the correlations between the temporal values for Ps and the temporal values for CPUE in the coming year were calculated and analyzed.

A multivariate linear model was established for the correlated factors that significantly affect the recruitment of *Illex argentinus* resources and CPUE. Error backpropagation (EBP) neural network is also built to forecast the CPUE of *Illex argentinus*.

An EBP neural network is a multilayer forward neural network that uses an EBP supervised algorithm, which can learn and store a large number of schema mapping relationships, and has been widely applied to various fields.

The EBP model used MSE as the standard for judging the optimal model. The fitting residual

error was obtained by comparing the forecasted values with the actual values. The function definition formula is $MSE = \frac{1}{N} \sum_{k=1}^{N} (y_k - \widehat{y}_k)^2$, where y_k is the actual CPUE value and $\widehat{y}_k$ is the forecasted CPUE value.

Correlation Analysis and the Selection of Key Areas

For spawning months (June to August) in spawning grounds (30°S to 45°S and 40°W to 65°W) (Fig. 6.11), the analysis was conducted to assess the correlation between the SST in each $0.1° \times 0.1°$ every month from June to August and the CPUE of the following year. The SST of three continuous areas in June presented a significant correlation with the CPUE of the following year (Table 6.16). The three areas were Area 1, with a distribution range of 38°–39°S and 54°–55°W; Area 2, with a distribution range of 40.5°–41.5°S and 51°–52°W; and Area 3, with a range of 39.9°–40.4°S and 42.6°–43.1°W. The SSTA in each $0.5° \times 0.5°$ every month from June to August did not present a continuous area that significantly correlated with the CPUE of the following year. The correlation analysis between the Ps time sequence and the CPUE time sequence indicated (Table 6.17) that there was no significant correlation between the ratio of the range occupied by the optimal SST to the total area of the spawning grounds from June to August and the CPUE of the following year.

Forecast Model Realization and Comparison of the Results

Multivariate Linear Model

A multivariate linear model was established by utilizing a sample composed of the selected surface temperatures of three consecutive areas in June and the CPUE (t/day) of the following year: $CPUE = 0.152 SST_{Area1} + 0.17 SST_{Area2} + 0.58 SST_{Area3} - 5.8$. The correlation coefficient R is 0.943 ($p = 0.007 < 0.05$) (Wang 2015).

Fig. 6.11 *Illex argentinus.* Major oceanographic features of the southwest Atlantic including the inferred hatching area of winter-spawned squid 32°–39°S, 49°–61°W; and the fishery area of the Falklands Interim Conservation and Management Zone and the Falklands Outer Conservation Zone (Waluda et al. 2001)

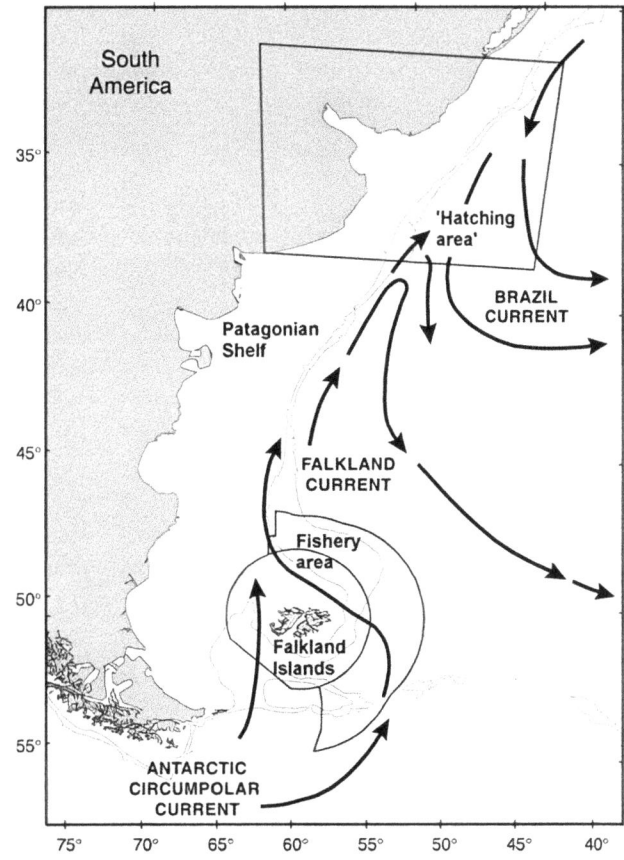

Table 6.16 Parameters for the correlation analysis of SST in key areas in June and the CPUE of the following year (Wang 2015)

Parameter	Area 1 in June	Area 2 in June	Area 3 in June
Longitude and latitude ranges	38°–39°S 54°–55°W	40.5°–41.5°S 51°–52°W	39.9°–40.4°S 42.6°–43.1°W
R value	0.8754	0.78	0.8655
p value	0.002	0.0132	0.0026

Table 6.17 Parameters for the correlation analysis of the optimal surface temperature of the spawning grounds divided into Ps and the CPUE of the following year (Wang 2015)

Statistical parameter	P_S in June	P_S in July	Ps in August
R value	0.266	0.42	0.36
p value	0.499	0.254	0.312

EBP Forecast Model

Utilizing the selected SST of three consecutive areas in June and different combinations of Ps in July as the input factors for the EBP forecast model, a variety of EBP forecast models were constructed (Wang 2015):

Scheme 1: Select a total of three factors—SST of Area 1, SST of Area 3, and Ps—as the input

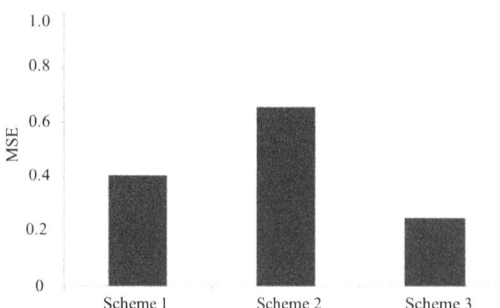

Fig. 6.12 Simulation results of different neural network models (Wang 2015)

layer to construct a 3:4:1 EBP network structure.

Scheme 2: Select a total of three factors—SST of Area 2, SST of Area 3, and Ps—as the input layer to construct a 3:4:1 EBP network structure.

Scheme 3: Select a total of four factors—SST of Area 1, SST of Area 2, SST of Area 3, and Ps—as the input layer to construct a 4:5:1 EBP network structure.

The MSE under the three schemes were obtained by utilizing MATLAB for the calculation (Fig. 6.12). As seen in Fig. 6.12, the MSE of Scheme 3 was the smallest, and its accuracy rate was 96.4%.

6.3.3 Prediction of the Recruitment of *Dosidicus gigas* in the Southeast Pacific Ocean

Dosidicus gigas is an oceanic cephalopod that is widely distributed in the sea area of the East Pacific Ocean. At present, many countries and regions in the world have developed and utilized *Dosidicus gigas* resources, and it has become an important commercial fishing species in the southeast Pacific. *Dosidicus gigas* is easily affected by marine environmental factors, and the year-to-year changes in resources are drastic. Wang (2015) tried to find more suitable marine environmental indicators that affected the recruitment of *Dosidicus gigas* resources and utilized an

EBP neural network to establish a more accurate forecast model for resource recruitment, providing a reference for the scientific management and production of *Dosidicus gigas* in the sea area of the Southeast Pacific Ocean.

The production data for *Dosidicus gigas* in the Southeast Pacific Ocean were obtained from the squid-jigging Science and Technology Group of Shanghai Ocean University, and the time period was from January 2003 to December 2012. The data included date, longitude, latitude, daily yield, and number of fishing vessels.

The SST, sea surface height (SSH), and chlorophyll-a (Chl-a) data were obtained from http://oceanwatch.pifsc.noaa.gov/las/servlets/dataset; the temporal resolution was monthly, the spatial resolution for the SST was $0.1° \times 0.1°$, the spatial resolution for the SSH and Chl-a was $0.25° \times 0.25°$, and the longitude and latitude ranges were 20°S–20°N and 110°W–70°W.

The calculation for the CPUE of *Dosidicus gigas* is as follows:

$$\text{CPUE}_Y = \frac{Catch_Y}{Ves_Y}$$

In the formula, $Catch_Y$ and Ves_Y, respectively, represent the fishing yield and the number of operating fishing vessels in year Y. The average daily CPUE ($t \cdot day^{-1}$) of a single vessel per year was calculated as the resource abundance index for *Dosidicus gigas*.

The recruitment of *Dosidicus gigas* resources is closely related to the environment at the spawning grounds and feeding grounds. Therefore, the correlations between the temporal values for SST, SSH, and Chl-a at each point (longitude and latitude $1° \times 1°$) from January to December and the temporal values for the CPUE in the current year and the subsequent year were calculated and analyzed. The SST, SSH, and Chl-a values for sea areas with high correlations were selected as the influencing factors for the recruitment of *Dosidicus gigas* resources. Among them, the sea areas with SST, SSH, and Chl-a values highly correlated with the current year's CPUE represented the effect of the feeding habitat environment on resource recruitment, and the sea areas with SST, SSH, and Chl-a values highly correlated with the subsequent year's CPUE represented the effect of the spawning habitat environment on resource recruitment.

The ratio of the optimal SST range for the spawning grounds and the feeding grounds to the total area indicates whether the habitat environment is good or bad. Previous research indicates that the suitable SST for *Dosidicus gigas* spawning in September is 24–28 °C and that the suitable SST for *Dosidicus gigas* feeding in July is 17–22 °C. Therefore, the ratios of the optimal SST during range for spawning in September and feeding in July to the total area (represented by P_S and P_F, respectively) were calculated, and P_S and P_F were used to express the degree of suitability of the habitat environment in spawning grounds and feeding grounds.

A multilayer forward neural network was used to establish a forecast model for resource abundance. Samples from 2003 to 2011 were used as the training samples, and samples from 2012 were used as the verification samples. The parameters of the network design, i.e., the number of neurons in the input layer, were determined based on the significantly correlated factors selected previously and the combination of P_F as well as P_S. The one neuron in the output layer was the CPUE, and the number of neurons in the hidden layer was obtained based on an empirical formula. The learning rate was 0.1, and the momentum parameter was 0.5. The parameters for the termination of network training, i.e., the maximum training batches, were 100 times, and the maximum error given was 0.001. The model was trained 10 times, and the optimal results were taken, preventing the appearance of a state of overfitting.

The EBP model used MSE as the standard for judging the optimal model. The fitting residual error was obtained by comparing the forecasted value with the actual value. The function definition formula is $MSE = \frac{1}{N} \sum_{k=1}^{N} (y_k - \widehat{y}_k)^2$, where y_k is the actual value of the CPUE and $\widehat{y}_k$ is the forecasted value of the CPUE.

Analysis of the Characteristic Environmental Factors

The correlation analysis indicated that the maximum value for the correlation between SST and the current year's CPUE occurred at 13°N and 102°W (Point 1) in July (Table 6.18, Fig. 6.13a) and that the maximum value for the correlation between SST and the subsequent year's CPUE occurred at 8°N and 103.5°W (Point 2) in June (Table 6.18, Fig. 6.13c).

The correlation analysis indicated that the maximum value for the correlation between SSH and the current year's CPUE occurred at 11°N and 102°W (Point 3) in September (Table 6.18, Fig. 6.13b) and that the maximum value for the correlation between SSH and the subsequent year's CPUE occurred at 12°N and 97.5°W (Point 4) in February (Table 6.18, Fig. 6.13d).

The correlation analysis indicated that the maximum value for the correlation between Chl-a and the current year's CPUE occurred at 8°S and 107°W (Point 5) in March (Table 6.18, Fig. 6.13e) and that the maximum value for the correlation between Chl-a and the subsequent year's CPUE occurred at 10°S and 93.5°W (Point 6) in October (Table 6.18, Fig. 6.13f).

Table 6.18 Parameters for the correlation analysis between the environmental factors in key sea areas and CPUE (Wang 2015)

Parameter	SST at Point 1 in July	SST at Point 2 in June	SSH at Point 3 in September	SSH at Point 4 in February	Chl-a at Point 5 in March	Chl-a at Point 6 in October
Position	13°N, 102°W	8°N, 103.5°W	11°N, 102°W	12°N, 97.5°W	8°S, 107°W	10°S, 93.5°W
R value	0.86	0.91	0.91	0.92	0.94	0.92
p value	0.001	0.0002	0.0002	0.0002	0.00003	0.0001

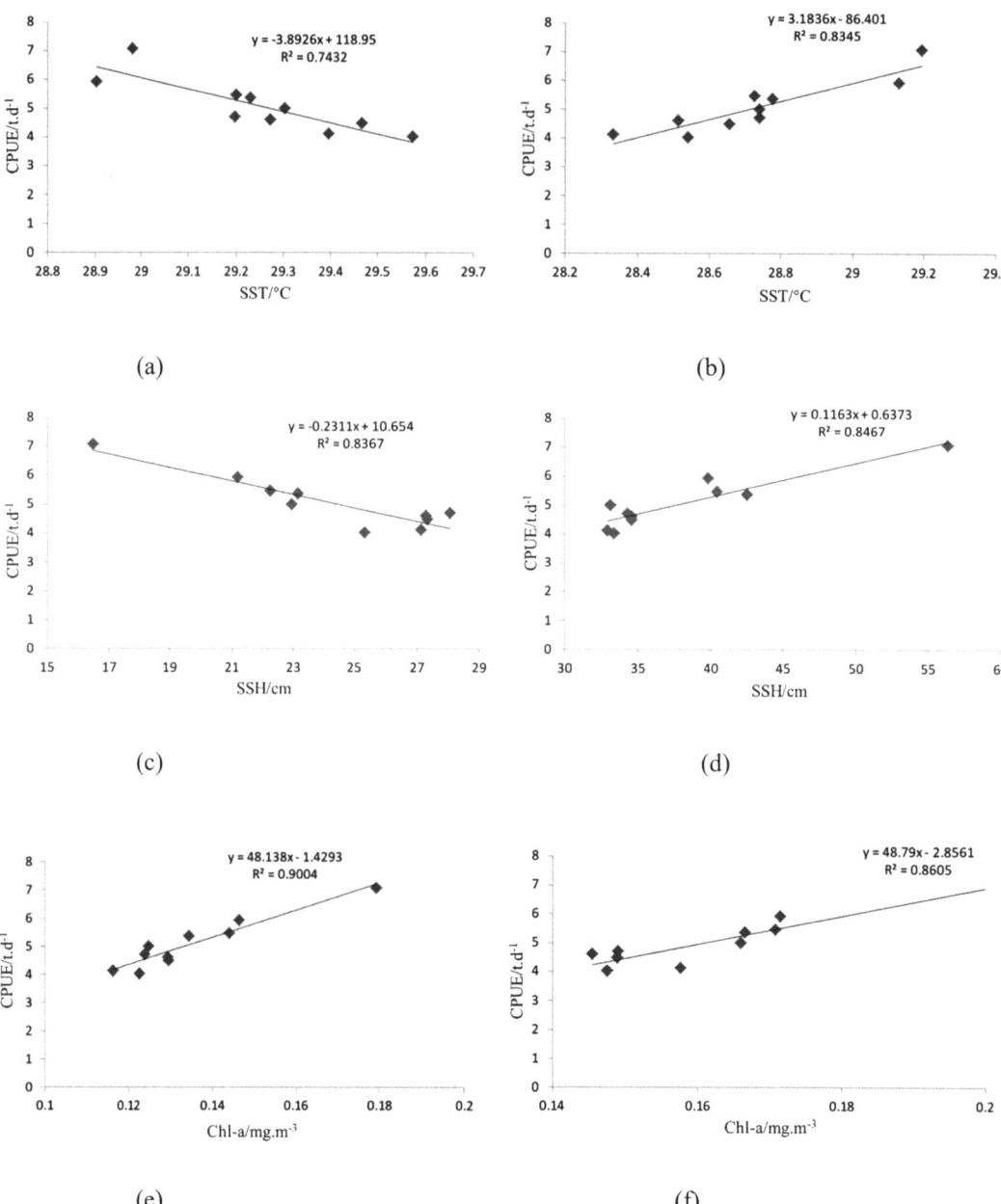

Fig. 6.13 Linear relational expressions for characteristic environmental factors and the recruitment of *Dosidicus gigas* (Wang 2015). (**a**) the correlation between SST and the current year's CPUE occurred at 13°N and 102°W in July; (**b**) the correlation between SST and the subsequent year's CPUE occurred at 8°N and 103.5°W in June; (**c**) the correlation between SSH and the current year's CPUE occurred at 11°N and 102°W in September; (**d**) the correlation between SSH and the subsequent year's CPUE occurred at 12°N and 97.5°W in February; (**e**) the correlation between Chl-a and the current year's CPUE occurred at 8°S and 107°W in March; (**f**) the correlation between Chl-a and the subsequent year's CPUE occurred at 10°S and 93.5°W in October

Forecast Model Realization and Comparison of the Results

Utilizing the selected environmental factors in the key sea areas and different combinations of P_S and P_F as the input factors for the EBP forecast model, a variety of EBP forecast models were constructed:

Scheme 1: Select a total of four factors—SST of Point 1, SSH of Point 3, Chl-a of Point 5, and P_F—as the input layer to construct a 4:5:1 EBP network structure, which represents a forecast model established by utilizing the key influencing factors of the feeding environment.

Scheme 2: Select a total of four factors—SST of Point 2, SSH of Point 4, Chl-a of Point 6, and Ps—as the input factors to construct a 4:5:1 network structure, which represents a forecast model established by utilizing the key influencing factors of the spawning environment.

Scheme 3: Select a total of eight factors—SST of Point 1 and Point 2, SSH of Point 3 and Point 4, Chl-a of Point 5 and Point 6, P_S, and P_F—as the input factors to construct an 8:9:1 EBP network structure, which represents a forecast model established by utilizing the key factors of the comprehensive environment.

MATLAB was utilized for modeling and to calculate the MSE under the three schemes (Fig. 6.14). The MSEs of Scheme 2 and Scheme 3 were close and were better than that of Scheme 1, and their accuracy rate was approximately 90% (Wang 2015).

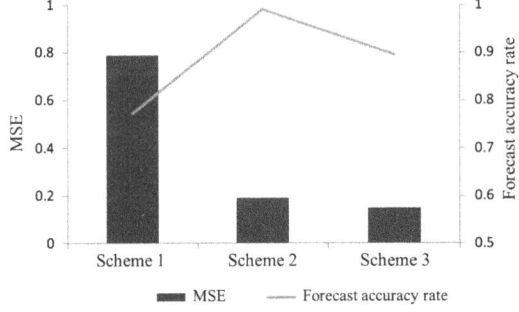

Fig. 6.14 Simulation results and accuracy rates of different neural network models (Wang 2015)

6.3.4 Prediction of the Abundance of Antarctic Krill (*Euphausia superba*)

Euphausia superba is an important part of the marine ecosystem of the Southern Ocean, and it is also the main target species for commercial fishing. In the early 1960s, the former Soviet Union pioneered in sending vessels to Antarctica to catch krill. Subsequently, Japan, Poland, Germany, Chile, and other countries successively launched research on the development and utilization of *Euphausia superba*. By the early 1970s, small-scale commercial fishing had developed, with the highest catch in history, 528,000 tons, recorded in 1982. The marine environment of the Southern Ocean is comparatively complicated, and the environment, especially sea ice, plays a very important role in the abundance and distribution of *Euphausia superba*. Research indicates that the summer abundance of *Euphausia superba* resources in Sea Area 48 is proportional to the area of sea ice in the winter of the previous year. The changing marine environment of the Southern Ocean, for example, the scope and concentration of sea ice, sea temperature, and circulation mode, is vital to the life history of *Euphausia superba*. The comprehensive effect of these factors makes the monitoring and assessment of the status of *Euphausia superba* resources more difficult and directly impacts the recruitment of *Euphausia superba* and the distribution of operating fishing grounds. Chen et al. (2011) analyzed the effect of the variation in the scope of sea ice in the winter and spring on the resource abundance of *Euphausia superba*, providing a basis for the scientific development and utilization of *Euphausia superba* resources.

The production data for *Euphausia superba* in previous years were obtained from the Commission for the Conservation of Antarctic Marine Living Resources (CCAMLR) at www.ccamlr. org; the data fields included year and month of operation, catch (unit: ton), fishing effort (unit: hour), and fishing sea area. The time span was from 1997 to 2008, and the resolution was monthly.

The Antarctic sea ice data were obtained from the National Ice and Snow Data Center at the University of Colorado at Boulder (http://nsidc. org/data/seaice_index/); the data fields included year, month, and area of sea ice. The time span was from 1996 to 2008, and the resolution was monthly.

Relevant studies have shown that as the habitat for *Euphausia superba*, sea ice provides a very good feeding environment for adults and larvae that are overwintering. When spring comes, the plankton growing under the ice reproduce in large numbers in the surface layer after the ice floes have melted. Thus, the larger the scope of the ice floes, the larger is the scope of the plankton distribution. On the other hand, when there are suitable environment conditions in the winter and spring, more plankton biomass inhabits the surface layer of the sea area. Therefore, in this study, the winter and spring sea ice data (July to November) was extracted, and one-way ANOVA and correlations were utilized to analyze the relationship between the sea ice status from 1996 to 2008 and the resource abundance of *Euphausia superba* in the summer.

Based on the results of the ANOVA and correlation analysis, linear regression was utilized to establish the relationship model for CPUE in the

summer of the current year and the area of sea ice in the winter and spring of the prior year.

$$CPUE = a_0 + a_1 x + \varepsilon$$

where CPUE (t/h) is the resource abundance index for *Euphausia superba* in the summer and x is the area of sea ice in the winter and spring (unit: 10^6 km^2).

Variations in *Euphausia superba* Catch

Taking a wide view of the variations in *Euphausia superba* catch from 1997 to 2008 (Fig. 6.15), the operating fishing grounds for krill were mainly concentrated in Area 48, with the total catch stabilizing between 100,000 and 160,000 tons and the annual average catch holding at 112,000 tons. However, there were differences in the distribution of the catch in each sub-area. In Area 48.1 (50°–65°W and 50°–65°S), the annual catch from 1997 to 2001 was comparatively stable, holding between 40,000 and 70,000 tons and accounting for 38–64% of the catch in Area 48. However, the fluctuations in annual catch from 2002 to 2008 were comparatively large, with the catch in 2008 being the lowest, only 2884 tons and accounting for 1.84% of the total annual catch in Area

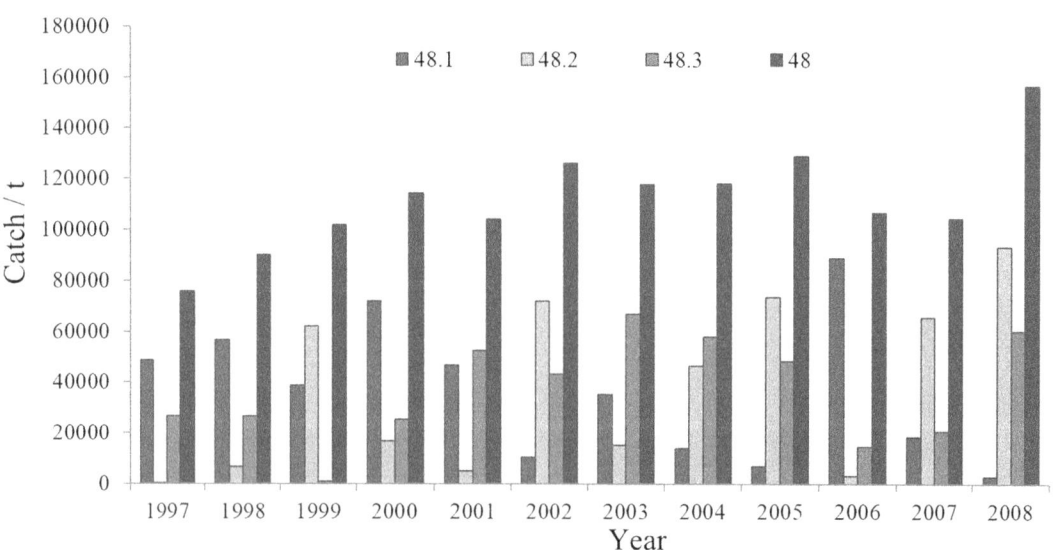

Fig. 6.15 Catch distribution of *Euphausia superba* in each fishing area (Chen et al. 2011)

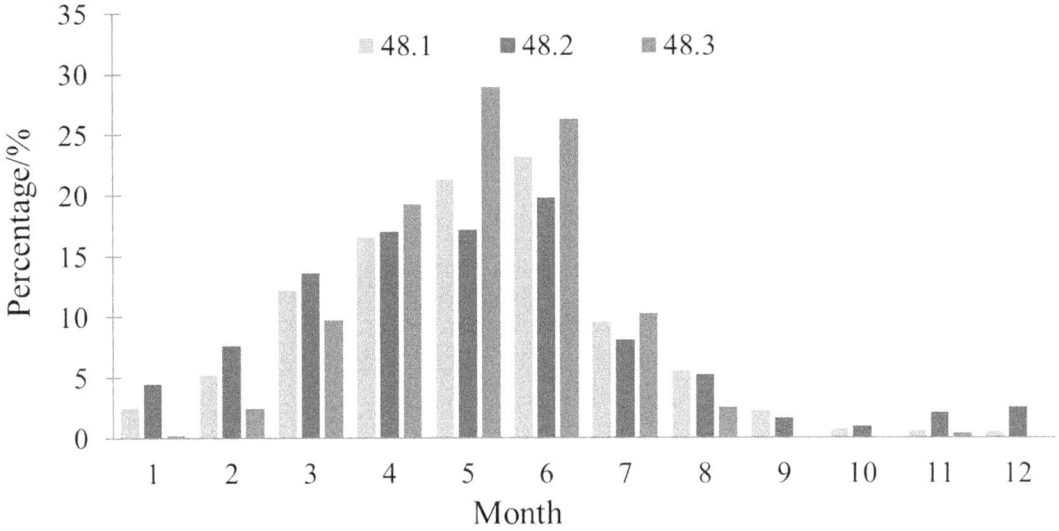

Fig. 6.16 Monthly distribution of *Euphausia superba* catch (percent) in each fishing area (Chen et al. 2011)

48, and with the catch in 2006 being the highest, reaching 8.89×10^4 tons, and accounting for 83.4% of the annual catch in Area 48 (Fig. 6.16). The overall average catch was $(3.7 \pm 2.7) \times 10^4$ tons, and the peak fishing season was from March to July (Fig. 6.17).

In Area 48.2 (30°–50°W and 55°–65°S), the variations in the interannual catch were comparatively large, with an average catch of $(3.8 \pm 3.4) \times 10^4$ tons (Fig. 6.17). The peak fishing season was from March to July (Fig. 6.16), with the cumulative catch of the peak fishing season accounting for 75.5% of the

catch in Area 48.2. The highest catch was 9.3×10^4 tons in 2008, accounting for 59.7% of the total catch in Area 48. The lowest catch appeared in 1997 at 98 tons, accounting for only 0.1% of the catch in Area 48 (Fig. 6.16).

In Area 48.3 (30°–50°W and 50°–55°S), the variation in its annual average catch were relatively small at $(3.7 \pm 2.1) \times 10^4$ tons (Fig. 6.17). The annual catch from 2000 to 2005 was comparatively stable, holding between 40,000 and 60,000 tons. The peak fishing season occurred from April to June, and the cumulative catch accounted for approximately 75% of that in

Fig. 6.17 Variations in the annual average *Euphausia superba* catch in each fishing area from 1997 to 2008 (Chen et al. 2011)

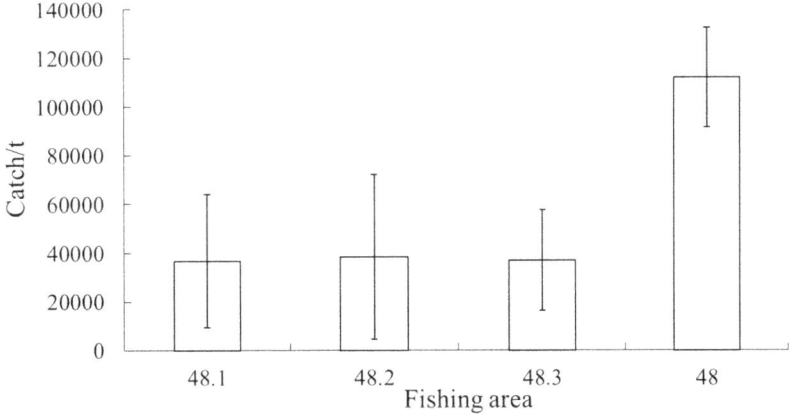

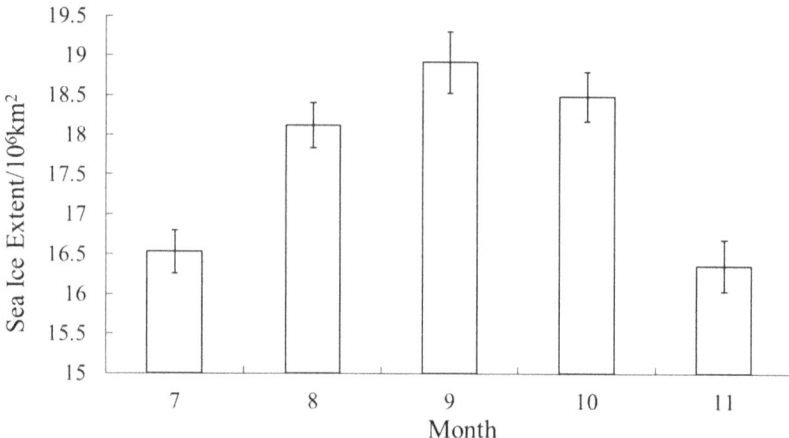

Fig. 6.18 Variations in the area of Antarctic sea ice between July and November from 1996 to 2008 (Chen et al. 2011)

Area 48.1 (Fig. 6.16). The highest catch was 6.7×10^4 tons (in 2003), accounting for approximately 56.8% of the total catch in Area 48 for that year, and the lowest catch was only 985 tons (in 1999), accounting for only 1% of the total catch in Area 48 that year.

Year-to-Year and Seasonal Changes in Sea Ice in the Spring and Winter (July to November)

Taking a wide view of the variations in the area of sea ice between July and November from 1996 to 2008 (Fig. 6.18), the sea ice presented significant changes between each month within the year (ANOVA, $F_{5,72} = 389.22$, $p < 0.0001$). The average area of sea ice increased from the minimum value of $(16.36 \pm 0.31) \times 10^6$ km^2 (mean $\pm$ SD) in early spring in November to the maximum value of $(18.91 \pm 0.37) \times 10^6$ km^2 in September (Fig. 6.18). However, the change in the area of sea ice each month was substantial; in particular, the variation in sea ice was greatest in September and least in July ($s^2_7 = 0.26$), and the monthly variance (s_i^2) in sea ice was $s^2_9 > s^2_{10} > s^2_{11} > s^2_8 > s^2_7$. The area of sea ice was the largest in September 2006, reaching 19.4×10^6 km^2, and the smallest area was in November 2001, at only 15.8×10^6 km^2. The year-to-year change in sea ice was not significant

(ANOVA, $F_{12,65} = 0.12, p > 0.05$). The statistics indicated that the seasonal variation in sea ice was greater than the year-to-year variation.

Establishment of the Relationship Between the CPUE for Krill and Sea Ice in the Winter and Spring (July to November) and the Regression Model

The correlation analysis results indicated that the resource abundance of krill in the summer was significantly correlated with the prior year's area of sea ice in September ($r = -0.756$, $p < 0.05$), the area of sea ice in October ($r = -0.674$, $p < 0.05$), and the average area of sea ice from July to November ($r = -0.721$, $p < 0.05$). This result indicated that the scope of sea ice in the winter and spring (July to November) had a significantly negative effect on the resource abundance of *Euphausia superba* in the subsequent year; it also indicated that the size of the area of sea ice in the 2 months of September and October in the winter and spring had an important influence on the resource abundance of krill. Therefore, the average area of sea ice in the winter and spring (July to November) was used as the independent variable to establish a regression model for the resource abundance CPUE of the subsequent year.

Table 6.19 Regression analysis results (Chen et al. 2011)

	Coefficient	p value	Lower limit 95%	Upper limit 95%
a_0	177.705	0.006249	62.86408	292.546
a_1	−9.59428	0.008126	−16.0886	−3.1

ANOVA $F = 10.83$, Significance $F = 0.008$

Regression statistics	
Correlation coefficient R	0.7556
Coefficient of determination R^2	0.5710
Adjusted coefficient of determination R^2	0.5281
Standard error	2.3644

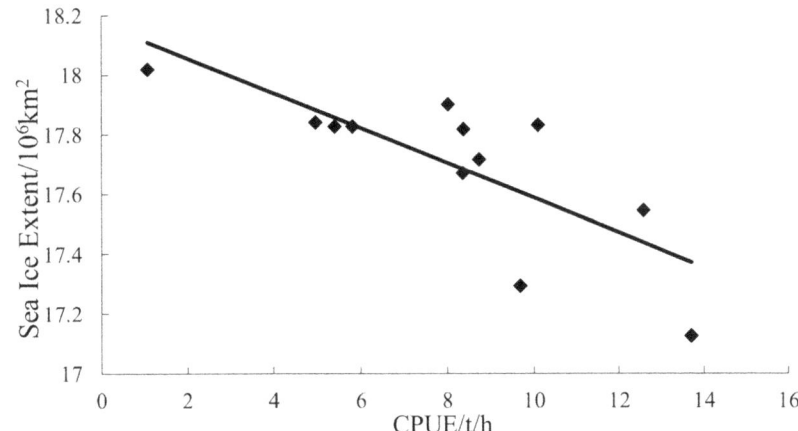

Fig. 6.19 Relationship between the average area of sea ice in the winter and spring from July to November and the CPUE for *Euphausia superba* in the summer of the subsequent year (Chen et al. 2011)

$$CPUE = a_0 + a_1 x + \varepsilon$$

In the formula, CPUE (t/h) is the resource abundance index for *Euphausia superba* in the summer, x is the average area of sea ice in the winter and spring (July to November) (unit: 10^6 km^2), and ε is the random error.

The regression analysis indicated that (Table 6.19) the scope of sea ice in the winter and spring presented a significantly negative correlation ($a_1 = -9.59$, $p < 0.05$) with the CPUE for krill in the summer of the subsequent year. This model explained 57.1% of the variation in the CPUE for krill in the summer ($R^2 = 0.571$). In the winter and spring of 2006, the area of sea ice was comparatively large, and the CPUE for krill in the summer of the following year was the lowest. The CPUE in the summer from 1997 to 1998 was at a moderate level, and the scope of sea ice in the winter and spring was reduced

compared to that in 2006. When the CPUE was comparatively high in the summer of 2008, the sea ice in the winter and spring was reduced substantially (Fig. 6.19).

References

Chen F, Chen XJ, Liu BL et al (2011) Effect of sea ice on the abundance index of Antarctic Krill *Euphausia superba*. Oceanologia Et Limnologia Sinica 42(04):495–499. (in Chinese)
Chen XJ (1995) Relationship between fishing ground of *Ommastrephes bartramii* and water temperature factors in Northwest Pacific Ocean. J Shanghai Ocean Univ 03:181–185. (in Chinese)
Chen XJ (1997) An analysis of marine environmental factors for the formation of fishing ground for *Ommastrephes bartramii* in the Northwest Pacific Ocean. J Shanghai Ocean Univ 04:263–267. (in Chinese)
Chen XJ (2003) Grey system theory in fisheries science. China Agriculture Press. (in Chinese)
Chen XJ (2014) Fisheries resources and fisheries oceanography. Ocean Press. (in Chinese)

Chen XJ (2016) Theory and method of fisheries forecasting. Ocean Press. (in Chinese)

Chen XJ (2021) Fisheries oceanography. Science Press. (in Chinese)

Chen XJ, Qian WG, Xu LX et al (2003) Study on *Ommastrephes bartramii* fishing ground and forecasting model from 150°E to 165°E in the North Pacific Ocean. Mar Fish Res 04:1–6. (in Chinese)

Chen XJ, Tian SQ, Xu LX (2005) Analysis on changes of surface sea temperature in the spawning and feeding ground of *O. bartramii* and its relationship with abundance index in the Northwestern Pacific Ocean. J Shanghai Ocean Univ 02:168–175. (in Chinese)

Guo M (1992) Grey forecasting of white shrimp yield in the Bohai Sea. Fish Sci 03:10–14. (in Chinese)

Ichii T, Mahapatra K, Sakai M et al (2009) Life history of the neon flying squid: effect of the oceanographic regime in the North Pacific Ocean. Mar Ecol Prog Ser 378:1–11

Japan Fisheries Agency (2021) Skipjack- Western and Central Pacific Ocean. Current status of international marine fisheries resources in 2021. https://kokushi.fra.go.jp/genkyo-R02.html (in Japanese)

Lehodey P, Bertignac M, Hampton J et al (1997) El Nino Southern Oscillation and tuna in the western Pacific. Nature 389:715–718

Shen JN, Fang RS (1982) Discussion on forecasting methods for *Trichiurus lepturus* yield during the winter fishing season in the inshore area of Zhejiang. Fish Sci Technol Inf 05:1–5. (in Chinese)

Waluda C, Rodhouse P, Podestá G et al (2001) Surface oceanography of the inferred hatching grounds of *Illex argentines* (Cephalopoda: Ommastrephidae) and influences on recruitment variability. Mar Biol 139 (4):671–679

Wang JT (2015) Fisheries forecasting and stock assessment for commercial oceanic ommastrephid squid. Shanghai Ocean Univ. (in Chinese)

Wang JT, Chen XJ (2013) Changes and prediction of the fishing ground gravity of skipjack (*Katsuwonus pelamis*) in the Western Central Pacific. Period Ocean Univ China (Nat Sci) 43(8):44–48. (in Chinese)

Wang YG, Chen X J (2005) Economic oceanic squid resources and fisheries in the world. Ocean press. (in Chinese)

Wei C, Zhou BB (1988) Study on the short-term *S. niphonius* fisheries forecasting in the Yellow Sea and the Bohai Sea. Acta Oceanol Sin 02:216–221. (in Chinese)

Xie J, Huang ZH, Xiao XZ et al (1998) Disaster prediction for grey years with lack of abundance in eel fry in Taiwan province. Reservoir Fish 01:3–5. (in Chinese)

Effects of Global Climate Changes on Marine Fishery Resources

7

Xinjun Chen

Abstract

Since the 1990s of the last century, the world's marine fisheries have entered a turning point. In 1990, the global marine catch declined for the first time, and most of the high-value traditional fishery resources were fully exploited or overexploited. Since 2016, there has been a small increase in the production of the marine fishing industry, reaching 84.4 million tons in 2018, but still below the peak catch of 86.4 million tons in 1996. The catch statistics from FAO show that the global marine catch has shown a relative change among different years, in which major changes in the production belong to oceanic species such as Peruvian anchovy or anchoveta (*Engraulis ringens*) and Pacific sardine (*Sardinops sagax*) and Ommastrephidae squid such as Argentinean flying squid (*Illex argentinus*) and jumbo flying squid (*Dosidicus gigas*); one of the main reasons for these fluctuations in these species is the changes of global climate and marine environmental factors. For this reason, this chapter briefly describes the current state of development of the world's marine fisheries and the characteristics of the marine environment and the main economic species resources in each sea area, such as the Northwest Pacific Ocean (statistical area 61) and Southeast Pacific Ocean (statistical area 87). The global climate events, such as water temperature rise, ocean acidification, and El Niño-Southern Oscillation (ENSO), and their impacts on marine fisheries are summarized. In this chapter, the impacts of global climate change and marine environmental change on resources and fishing ground for tuna, Pacific saury (*Cololabis saira*), Chilean jack mackerel (*Trachurus murphyi*), Peruvian anchovy, cod, salmon, and cephalopod are analyzed in detail, which will provide some cases for the scientific research of fisheries forecasting and also provide a scientific basis for the sustainable use and scientific management of global marine fishery resources.

Keywords

Global climate change · Marine fisheries · Resources and fishing ground

7.1 Current Status of Marine Fishery Development in the World

7.1.1 Development of Marine Fishery

Since the 1990s, people have recognized a turning point in the world's fisheries; for the

X. Chen (✉)
College of Marine Sciences, Shanghai Ocean University, Lingang Newcity, Shanghai, China
e-mail: xjchen@shou.edu.cn

traditionally largest fishery producers, the marine fishing yield has been unstable, showing a continued slump. As determined by a global marine catch analysis by the Food and Agriculture Organization (FAO) of the United Nations, the annual growth rate for the marine catch decreased in the 1980s, with the first decrease in the global marine catch appearing in 1990, i.e., a decrease of 3% compared to 1989; this trend continued to exist within the next few years, with an average annual decrease of 1.5% between 1990 and 1992. Among these fishing yields, most of the high-value resources were fully exploited or overexploited. In 2000, the total yield of the marine fishing industry reached 84.97 million tons (Fig. 7.1). From 2009 to 2016, the yield of global capture fisheries continued to stabilize at approximately 80 million tons (Fig. 7.1); the fishing yield reached 84.61 million tons in 2018, which is lower than the peak of 86.4 million tons in 1996. In 2019 and 2020, the annual marine landings fall to 80.41 million tons and 78.78 million tons (not including aquatic plants, whale, seal, and other animals, hereinafter), respectively, as the COVID-19 in the world affects marine fishing industry since 2020 (Fig. 7.1).

The FAO assessment results indicated that the percentage of fish populations at a biologically sustainable level had reduced from 90% in 1974 to 65.8% in 2017 (Fig. 7.2). In contrast, the proportion of populations in which the catch was at a biologically unsustainable level had increased continuously, from 10% in 1974 to 34.2% in 2017, and the growth rate was especially steep in the late 1970s and in the 1980s. These calculations treated all fish populations equally, without considering their respective biomass and catch. Biologically sustainable populations accounted for 78.7% of the landing.

In 2017, the proportion of populations that were sustainably fished at maximum yield accounted for 59.6% of the total assessed populations, and the underfished populations accounted for 6.2%. The proportion of the underfished populations continued to trend downward from 1974 to 2017, and the proportion of populations that were sustainably fished at maximum yield decreased continuously from 1974 to 1989, after which it started to rebound, increasing to 59.6% by 2017.

In 2017, among the 16 main statistical areas recognized by the FAO, the statistical area 37 (Mediterranean Sea and the Black Sea) had the highest proportion of populations fished at an unsustainable level (62.5%); the next is the statistical area 87 (Southeast Pacific Ocean; 54.5%) and the statistical area 41 (Southwest Atlantic

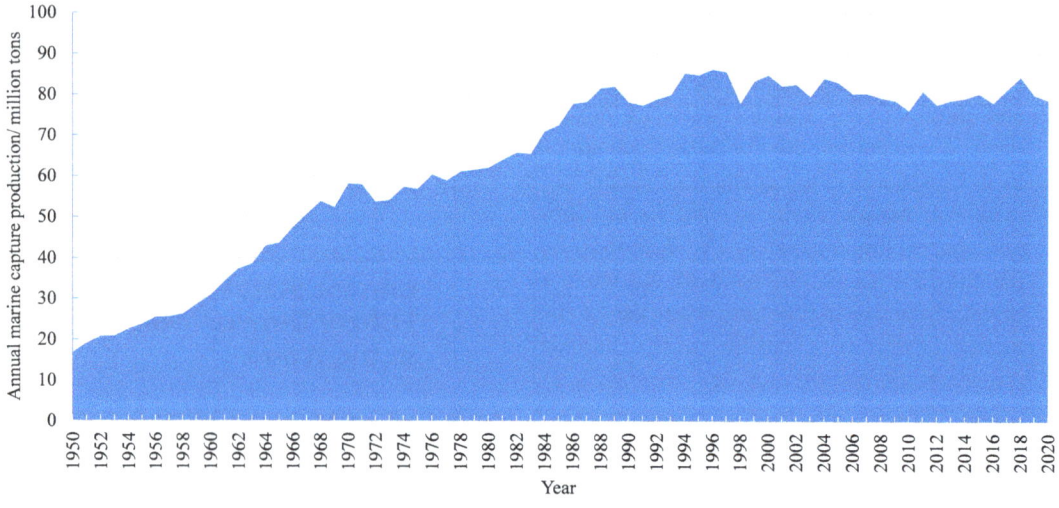

Fig. 7.1 World capture production from 1950 to 2020 (not including aquatic plants, whale, seal, and other animals) (FAO 2022)

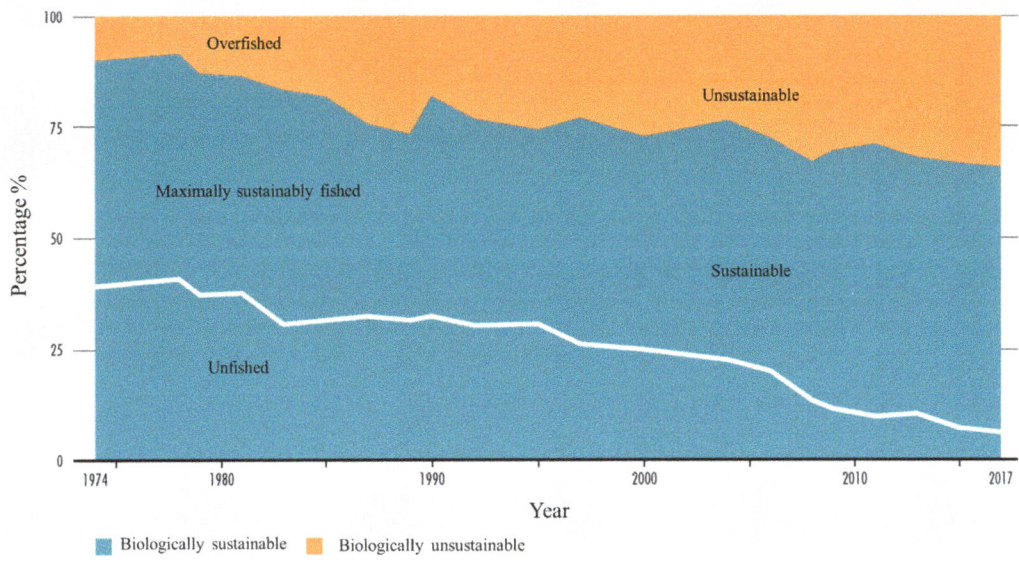

Fig. 7.2 Changes in the state of marine fish stocks during 1974 to 2017 in the world (FAO 2020)

Ocean; 53.3%) (Fig. 7.3). On the contrary, the statistical area 77 (Eastern Central Pacific Ocean), the statistical area 81 (Southwest Pacific Ocean), the statistical area 67 (Northeast Pacific Ocean), and the statistical area 71 (Western Central Pacific Ocean) had the lowest proportions of populations being fished at a biologically unsustainable level (13–22%). In 2017, the proportions in other areas accounted for 21–44% (Fig. 7.3).

The time patterns of the landing in the various areas were also not the same and were affected by many factors, such as ecosystem productivity, fishing intensity, management, and the status of fish populations. Excluding the Antarctic and Arctic areas, which had very little landings, three patterns were observed in the other areas (Fig. 7.4):

1. The first pattern: continuous upward trend starting from 1950
2. The second pattern: fluctuation in yield around the global stability value since 1990 and landings dominated by pelagic species with shorter life spans
3. The third pattern: overall downward trend after reaching a historical peak value

The proportion of biologically sustainable populations in the first group (71.5%) was higher than those in the second pattern (64.2%) and the third pattern (64.5%). There was no direct association between the fishing mode and the population status. Generally, a continuous increase in catch usually indicates improvements in population status or an increase in fishing intensity, and it is highly probable that a continuous decrease is due to a decrease in abundance. However, the decrease in catch may also be caused by other factors, such as environmental changes as well as fishery measures adopted to decrease fishing intensity and restore overfished populations.

7.1.2 Status of Marine Fishery Development in Each Statistical Area by FAO

Northwest Pacific Ocean

The statistical area 61 by FAO is the Northwest Pacific Ocean, which is one of the most fully utilized fishing areas in the world. There are numerous species in this area; pelagic fish are especially abundant. The cold Oyashio Current

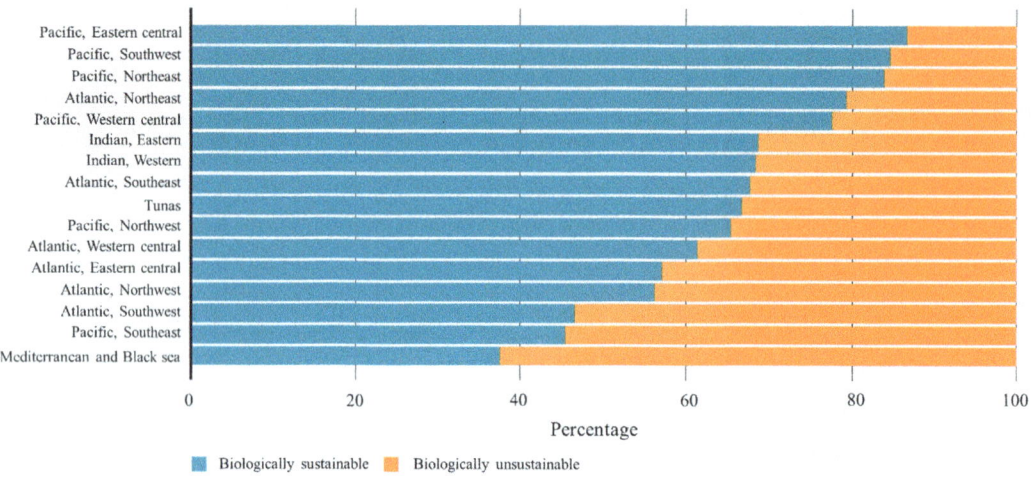

Fig. 7.3 Proportions of stocks fished at biologically unsustainable and sustainable levels by different statistical areas in 2017 (FAO 2020). Note: Tuna stocks are assessed separately as they are largely migratory and straddling across different statistical areas

and the warm Kuroshio Current converge and mix in the northeastern area of Japan, generating many eddies in the current boundary area, where these marine environmental conditions are helpful to form good fishing grounds. The main fishing targets in this area are sardine (*Sardina pilchardus*), jack mackerel (*Trachurus trachurus*), chub mackerel (*Scomber japonicus*), anchovy, herring (*Clupea pallasi*), sauries (*Cololabis saira*), salmon and trout, skipjack tuna (*Katsuwonus pelamis*), tuna, squid, Alaska pollock (*Gadus chalcogrammus*), flatfish (Pleuronectiformes), whale (Cetacea), etc. (Chen 2014, 2016).

Among the FAO defined statistical areas, the yield in this area is the highest in the all statistical areas and attained 19.15 million tons (not including aquatic plants, whale, seal, and other animals, hereinafter), with the landing accounting for 24.31% of the global marine capture catch in 2020. The total catch in this area fluctuated between 17 million and 24 million tons during 1980 to 2000, yielding approximately 21.25 million tons with highest landing in 2014. Based on catch data by FAO, the Japanese pilchard (*Sardinops melanostictus*) and Alaskan pollock (*Theragra chalcogramma*) were the species with

the highest yields, with peak yields of 5.4 million tons and 5.1 million tons, respectively, but their yields have decreased greatly in the past 25 years. In contrast, squid, cuttlefish (Sepiida), octopus, and shrimp landings have increased substantially since the 1990s. Based on the report by FAO (2020), two populations of Japanese long-tailed anchovies (*Engraulis japonicus*) were overfished, and two populations of *Theragra chalcogramma* were sustainably fished, while another population was overfished. Generally, among the fish populations monitored by the FAO (2020) in the Northwest Pacific Ocean, approximately 65.4% of the populations were captured within biologically sustainable limits (FAO 2020; Chen 2021).

Northeast Pacific Ocean

The Northeast Pacific Ocean is statistical area 67 by FAO and also includes the eastern part of the Bering Sea and the Gulf of Alaska. In this area, the main currents are the Alaska Current and California Current. The important economic species are soles, salmon and trout, herring (*Clupea pallasi*), cod (Gadiformes), king crab, and shrimp, among others (Chen 2014, 2016).

In 2017, the highest landing in the statistical area 67 remained approximately 3.39 million

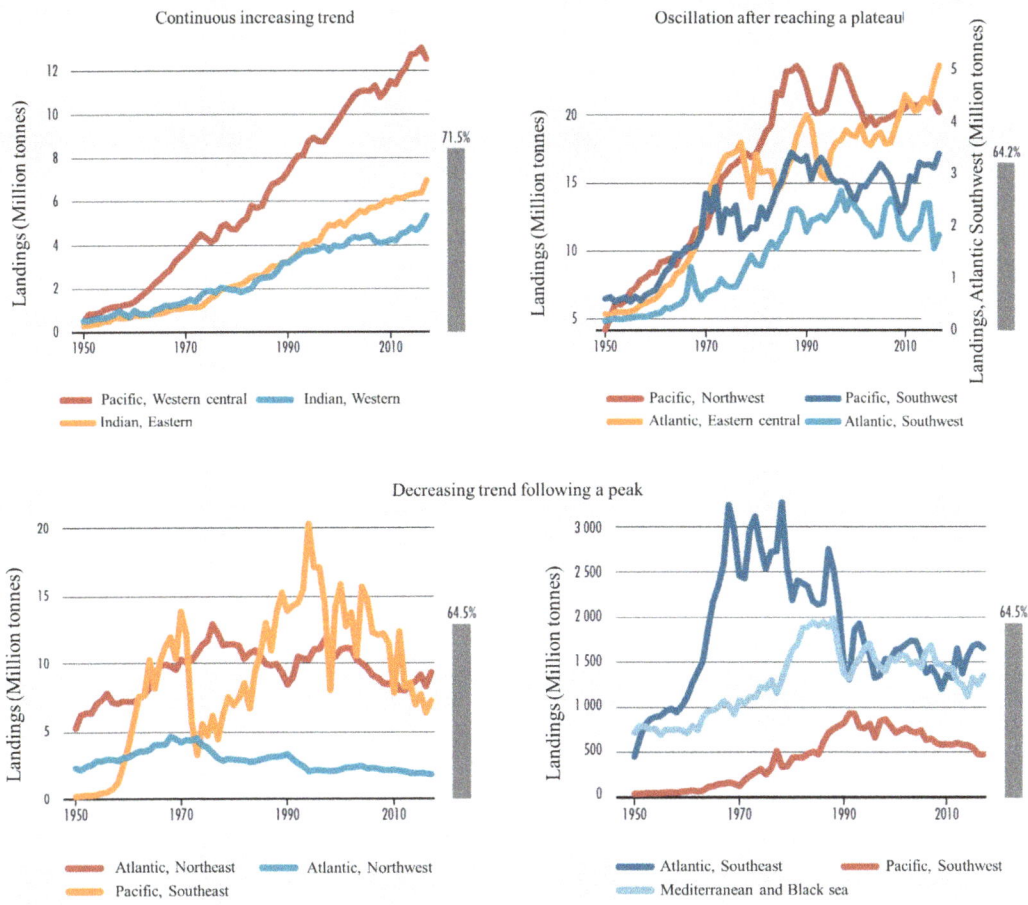

Fig. 7.4 Three patterns for marine fishing landings from 1950 to 2017 in the world (FAO 2020). Note: The bar indicates the percentage of stocks at a biologically sustainable level in each pattern in 2017

tons, which is no obvious change in the species structure of the catch since 2010. But the annual landings in 2019 and 2020 fall to 3.16 million tons and 2.86 million tons, respectively. *Theragra chalcogramma* was the species with the most abundant resources, accounting for nearly 50% of the total landing. Pacific cod (*Gadus microcephalus*), hake (*Merluccius*), and sole also accounted for high percentages of the catch. In the past 10 years, the average annual yield in salmon, trout, and smelt (*Osmerus mordax*) has changed significantly, fluctuating between 230,000 and 520,000 tons, with a highest catch of 516,700 tons in 2013 and a catch of 235,594 tons in 2020. With the exception of the salmon

populations, all assessed populations in this area seem to have been sustainably managed. Generally, 83.9% of the assessed populations in this area were fished within biologically sustainable limits in 2017 (FAO 2020; Chen 2021).

Western Central Pacific Ocean

The Western Central Pacific Ocean is statistical area 71 by FAO. The main fishing grounds are distributed in the continental shelf along the western coast of the Pacific Ocean, of which there are tuna fishing grounds. This area is mainly controlled by the North Equatorial Current system. Affected by the Kuroshio Current in the north waters, the current is comparatively stable, while the surface current in the south part is affected by the prevailing monsoon, and the direction of flow changes as the monsoon changes (Chen 2014, 2016).

This area is one of the important marine fisheries in the world. The landing in this area ranks second, with the total yield presenting a continuously increasing trend starting in the 1950s, reaching the highest landings of 13.33 million tons in 2018 and 2019 (accounting for 15.77% and 16.64% of the global total, respectively). The landing in 2020 slightly falls to 13.26 million tons. The main species are tuna and tuna-like species, accounting for approximately 21% of the total landing. Sardines and long-tailed anchovies are also the main species in this area. A few populations are considered underfished, especially in the west part of the South China Sea. The reported catch has always been maintained at a high level, maybe because of the expansion of the fishing range to new areas or the changes that have occurred in the nutritional level of the target species. The tropical and subtropical features and limited available data of this area generate considerable uncertainty regarding population assessments. Generally, 77.6% of the assessed populations in this area were captured within biologically sustainable limits in 2017 (FAO 2020; Chen 2021).

Eastern Central Pacific Ocean

The Eastern Central Pacific Ocean is statistical area 77 by FAO. This area has the California Current distributed in the northern part and the Peru Current distributed in the southern part. The subsurface Equatorial Counter Current, which is also an important current, is also present. These currents create environmental conditions for the formation of excellent fishing grounds (Chen 2014, 2016).

In the last several decades, the landings in this area fluctuated between 1.63 million and 2.06 million tons. The total landing in 2020 was 1.69 million tons, accounting for 2.15% of the global total. Most of the catches in this area are small-to-medium-size pelagic fish, including California sardine, long-tailed anchovy, and Pacific jack mackerel (*Trachurus symmetricus*), squid, and

white shrimp. These species populations with shorter life spans are more easily affected by changes in marine conditions; even if the fishing effort is maintained at a reasonable level, the yield will still fluctuate. At present, overfishing has affected some of the high-value inshore resources, such as grouper and shrimp. The proportion of the assessed populations fished within biologically sustainable limits in this area has been maintained at 86.7% since 2015 (FAO 2020; Chen 2021).

Southwest Pacific Ocean

The Southwest Pacific Ocean is statistical area 81 by FAO. The fishing grounds are mainly distributed in the waters around New Zealand and Australia. The main currents in this area are the South Equatorial Current, the East Australian Current, and the West Wind Drift. The current system around New Zealand is complex and volatile (Chen 2014, 2016).

The potential allowable catch in the Southwest Pacific Ocean is not high, approximately 2.1 million tons. However, this fishing area currently produces the lowest yield, peaking at 857,000 tons in 1998, i.e., only 40.7% of the potential catch. From 1994 to 2000, the fishing yield in this area was unstable, with a continuous decrease in two consecutive years, i.e., 1999 and 2000, at 807,000 and 753,000 tons, respectively, less than 1/2 of the potential catch. In 2020, the fishing yield was 431,888 tons, which is the lowest recorded level in the last decade. This area is also an fishing ground for distant-water fishing nations and regions (FAO 2020; Chen 2021).

Southeast Pacific Ocean

The Southeast Pacific Ocean is statistical area 87 by FAO. There is extensive upwelling in this area. The fishing grounds are mainly located along the western coast of South America. The most important marine fishery in this area is anchovy capture fisheries. This fishery started to develop in the second half of the 1950s; increased through the first half of the 1960s, reaching a historical peak of 13.06 million tons in 1970; but decreased sharply afterward. Other main fisheries include tuna and hake; the Peruvian and Chilean shrimp fisheries and Chile's shellfish and mollusk fisheries are also valuable (Chen 2014, 2016).

The highest landing in the Southeast Pacific Ocean in 2018 was 10.3 million tons in the last decade, accounting for approximately 12.25% of the global landing. The landing in 2020 is 8.40 million tons. The two species with the highest yield were Peruvian anchovy (*Engraulis ringens*) and jumbo flying squid (*Dosidicus gigas*), which are fished within biologically sustainable limits, with the landings of nearly seven million tons and 890,000 tons in 2018, respectively. The FAO reports considered that two pelagic species, Chilean jack mackerel (*Trachurus murphyi*) and Pacific mackerel (*Scomber japonicus*), are also fished within biologically sustainable limits (FAO 2020). But Pacific sardine (*Sardinops sagax*) is being seriously overfished, and Patagonian toothfish (*Dissostichus eleginoides*) is being fished at an unsustainable level. Generally, the percentage of assessed populations in this area fished within sustainable limits accounted for 45% (FAO 2020; Chen 2021).

Northwest Atlantic Ocean
The Northwest Atlantic Ocean is statistical area 21 by FAO. The main oceanographic features in this area are closely related to the Gulf Stream and the Labrador Current; these two currents converged and mixed at the Grand Banks south of Newfoundland, forming the world-famous excellent fishing ground (Chen 2014, 2016).

The landing in the Northwest Atlantic Ocean was 1.75 million tons in 2017, presenting a continuously decreasing trend since the early 1970s with peaking landing of 4.5 million tons. The landing in 2020 further falls to 1.54 million tons. The resources status of Atlantic cod (*Gadus morhua*), bilinear hake (*Merluccius bilinearis*), white hake (*Urophycis tenuis*), and haddock (*Melanogrammus aeglefinus*) is poor. Since the late 1990s, the landing of the above cods and hakes has been maintained at 100,000 tons, which is only 5% of the highest catch of 2.2 million tons for these populations. Although the fisheries sector has significantly reduced fishing operations, these populations still have not yet recovered. By contrast, the American lobster (*Homarus americanus*) catch has increased rapidly, reaching 160,000 tons in 2017. Based on the FAO report in 2020, the percentage of the assessed populations in this area fished within biologically sustainable limits was 56.2% in 2017 (FAO 2020).

Northeast Atlantic Ocean
The Northeast Atlantic Ocean is statistical area 27 by FAO, and it is one of the main fishing areas in the world. The fishing grounds in this area include the North Sea fishing ground, the Iceland fishing ground, the fishing ground in the northern waters of Norway, the fishing ground in the southeastern part of the Barents Sea, and the fishing ground in the continental shelf from Bear Island to Spitsbergen. The main hydrological features of this area are the North Atlantic Current and its branch currents (Chen 2014, 2016).

The landing in the Northeast Atlantic Ocean ranked 9.35 million tons in 2017, which occupied 11.47% of the global total. The landing began to decrease year by year after peaking at 13 million tons in 1976 and stabilized at the range of 8.1 million to 9.4 million tons during 2010 to 2020. The species in the Northeast Atlantic Ocean faced extremely large fishing pressure in the late 1970s and early 1980s. After the 1990s, various countries eased fishing pressure in hopes of restoring the overfished stocks. Most of the stocks have been maintained at the same status since 2015, and notably, some of the stocks are no longer listed as overfished stocks. Based on the FAO report, the percentage of the assessed populations fished within biologically sustainable limits in this area was 79.3% in 2017 (FAO 2020; Chen 2021).

Western Central Atlantic Ocean
The Western Central Atlantic Ocean is statistical area 31 by FAO. The fishing grounds are mainly distributed in the Caribbean Sea and the Gulf of Mexico. The main current in this area is the extension of the equatorial current, including the Caribbean Current. These currents create suitable conditions forming excellent fishing grounds (Chen 2014, 2016).

The landing in the Western Central Atlantic Ocean peaked at 2.3 million tons in 1984 and then gradually decreased to 1.18 million tons in 2014, rebounding slightly in 2016 to 1.56 million tons and falling to 1.24 million tons in 2020 again. The catches of important species such as large-scale menhaden (*Brevoortia patronus*), skipjack (*Katsuwonus pelamis*), and round sardinella (*Sardinella aurita*) have decreased, but they are still in a biologically sustainable state based on the estimates. Snapper and grouper have been heavily fished since the 1960s, but some of the stocks in the Gulf of Mexico have started to recover since the stricter management regulations were implemented. Some invertebrate species including the spiny lobster (*Panulirus argus*) and the queen conch (*Lobatus gigas*) are being fully fished in the Gulf of Mexico. Although the fishing effort has been reduced in recent years, the white shrimp resources on the continental shelves of the Caribbean Sea and Guyana have not shown signs of recovery. In addition, the oyster (*Crassostrea virginica*) population in the Gulf of Mexico is currently being overfished. In 2017, the percentage of the assessed populations fished within biologically sustainable limits in this area was 61.4% (FAO 2020; Chen 2021).

Eastern Central Atlantic Ocean

The Eastern Central Atlantic Ocean is statistical area 34 by FAO. The main fishing ground is the continental shelf along the coast of western Africa. The main current systems in this area are the Canary Current, the Benguela Current, and the Equatorial Counter Current. The cold Canary Current, which descends southward along the waters of northern West Africa, converges with the Equatorial Counter Current, which ascends northward along the southern coast of West Africa and forms seasonal upwelling in the waters of northern West Africa, where the continental shelf is wider, thus forming a good fishing ground (Chen 2014, 2016).

The landing in this area attained the peak of 5.49 million tons in 2018, accounting for 6.50% of the global total. But the landings in 2019 and 2020 fall to 5.37 million and 4.95 million tons, respectively. Sardine is the most important fishing

target, with an annual reported landing of approximately one million tons since 2004. Another important small pelagic species is *Sardinella aurita*, the landing has presented an overall decreasing trend since 2001, and the landing in 2017 was approximately 220,000 tons, approximately 50% of the peak level. The demersal species in this area are being intensively fished, and the status of each population is different; some populations are being fished at a sustainable level, but some are being fished at an unsustainable level. In 2017, the percentage of the assessed populations fished within biologically sustainable limits in this area was only 57.2% (FAO 2020; Chen 2021).

Southwest Atlantic Ocean

The Southwest Atlantic Ocean is statistical area 41 by FAO. The main fishing ground is distributed in the continental shelf along the eastern coast of South America. This area is affected by the Brazil Current and the Falkland Current. The latter ascends northward along the coast to Rio de Janeiro and converges with the Brazil Current, where the water masses mix; the water quality is highly fertile, eddy currents are generated, and the sea water is exchanged vertically, forming an extensive fishing ground (Chen 2014, 2016).

In this area, the annual landing has fluctuated between 1.70 and 2.42 million tons during 2010 and 2020, with a highest landing of 2.42 million tons and only accounting for 3% of the global total in 2014. Regarding the landings, the most important species is Argentine flying squid (*Illex argentinus*), which accounts for 10–40% of the total catch in this area. However, in the last 2 years, the landing of this species has decreased sharply, decreasing from one million tons in 2015 to 360,000 tons in 2017. The catches in southern blue whiting (*Micromesistius australis*) and hake (*Macruronus magellanicus*) have also shown a continuously decreasing trend in the past 20 years. Based on landing, the second most important species in this area is Hubbsi hake (*Merluccius hubbsi*). In the past 10 years, the landing for *Merluccius hubbsi* had stabilized at approximately 350,000 tons; although signs of slow recovery had emerged, this species is still at an unsustainable level. Based on the FAO report published in 2020, the percentage of the assessed populations fished within biologically sustainable limits in this area was only 46.7% in 2017 (FAO 2020; Chen 2021).

Southeast Atlantic Ocean

The Southeast Atlantic Ocean is statistical area 47 by FAO. The fishing ground is mainly distributed in the waters above the continental shelf along southern part of western Africa. The main current in this area is the Benguela Current and West Wind Drift. When the Benguela Current moves northward along the western coast of Africa, strong upwelling is generated due to the action of offshore winds, which leads to the formation of excellent fishing ground (Chen 2014, 2016).

The landing in this area presents a decreasing trend during the past 50 years; the total yield decreased from 3.3 million tons in the early 1970s to 1.36 million tons in 2020, a slight rebound compared to 1.34 million tons in 2011. The most important fishing targets in this area are jack mackerel and hake (*Merluccius*). These populations have been actively replenished since 2006, and strict management measures have been implemented; therefore, these populations along the coast of South Africa and Namibia have recovered. The status of the South African sardine (*Sardinops ocellatus*) population degenerated, requiring the Namibian and South African fishery management departments to take special conservation measures. The very important sardine (*Sardinella aurita* and *S. maderensis*) populations along some of the coastal areas of Angola and Namibia are maintained at biologically sustainable levels. Based on the FAO report, the percentage of the assessed populations fished within biologically sustainable limits in this area was 67.6% in 2017 (FAO 2020; Chen 2021).

The Mediterranean Sea and the Black Sea

The Mediterranean Sea and the Black Sea are statistical area 37 by FAO. The Mediterranean Sea is a large nearly closed body of water. The Atlantic water system enters the Mediterranean

Sea through the Strait of Gibraltar, mainly flowing along the coast of Africa, and can reach the eastern part of the Mediterranean Sea. The low-salinity water of the Black Sea is brought into the Mediterranean Sea through currents. The Nile River is the main source of fresh water in the Mediterranean Sea. The Suez Canal brings high-temperature surface water from the Red Sea into the Mediterranean Sea (Chen 2014, 2016). All the above currents and water systems affect the distribution of productivity and fishing ground in this area.

The total landing in this area peaked at approximately two million tons in the mid-1980s and then dipped to a low point of 1.12 million tons in 2014; the annual landing ranged from 1.39 million tons 1.18 million tons since 2015. Compared with that for small pelagic populations, the fishing mortality rate for demersal populations in this area is higher. The fishing pressure on important targets such as European hake (*Merluccius merluccius*) and turbot (*Scophthalmus maximus*) is especially prominent, and the biomass levels of many long-tailed anchovy (*Engraulis encrasicolus*) and sardine populations are lower than biologically sustainable levels. In recent years, the fishing rates for some of the populations have continued to decrease (such as Black Sea turbot, *Scophthalmus maeoticus*) because this area is still facing an overfishing problem. Based on the FAO report in 2020, the percentage of the assessed populations fished within biologically sustainable limits in this area was only 37.5% (FAO 2020; Chen 2021).

East Indian Ocean

The East Indian Ocean is statistical area 57 by FAO. In the eastern area of the Indian Ocean, the fishing grounds are mainly distributed in the continental shelf of different countries. The countries surrounding this area include Bangladesh, India, Thailand, Indonesia, Myanmar, and Australia, with the catches mainly obtained by these countries. Distant-water fishery countries and regions mainly fish for tunas in this area (Chen 2014, 2016).

The landing in the East Indian Ocean presents a steadily increasing trend, peaking at 7.09

million tons in 2017. The annual landings fall to 6.50–6.80 million tons during 2018 and 2020. Due to limited catch data, the status of most species in this area has not been properly assessed, and some assessment results should be interpreted with caution. Some results indicate from FAO that largehead hairtail (*Trichiurus*), croaker (Sciaenidae), toli shad (*Tenualosa toli*), sea catfish (Ariidae), oil sardine (*Sardinella longiceps*), and sardinella (*Sardinella* spp.) may be overfished, but anchovy (Engraulidae), Indian mackerel (*Rastrelliger kanagurta*), scad (*Decapterus* spp.), shad (*Tenualosa ilisha*), giant tiger prawn (*Penaeus monodon*), banana prawn (*Penaeus merguiensis*), bobtail squid (Sepiolidae), and cuttlefish (Sepiidae) are at sustainable fishing levels. The stock assessment in 2017 indicates that the percentage of the assessed populations fished within biologically sustainable limits in this area attained 68.6% (FAO 2020; Chen 2021).

West Indian Ocean

The West Indian Ocean is statistical area 51 by FAO. The surrounding countries include India, Iran, Kenya, Madagascar, Maldives, Mozambique, Oman, Pakistan, Somalia, South Africa, Sri Lanka, Tanzania, and Yemen. This area produces sardines, croaker and drums, skipjack, yellowfin tuna (*Thunnus albacares*), Bombay duck (*Harpadon nehereus*), Japanese Spanish mackerel (*Scomberomorus niphonius*), largehead hairtail, and shrimp, among other species (Chen 2014, 2016).

In this area, the fishing grounds are mainly distributed in the continental shelf of western Africa. The coastal countries account for 90.6% of the total landing in this area. Currently, the distant-water fishery countries and regions engaged in marine fishing in this area are Japan, France, Spain, South Korea, and Taiwan Province of China, among others, and they mainly fish for tuna and demersal fish, accounting for less than 10% of the total catch in this area.

The total landing in the West Indian Ocean continues to increase, reaching 5.62 million tons in 2020. The assessed results indicate that the

white shrimp (main export products that earn foreign exchange) in the Southwest Indian Ocean are being overfished, requiring the relevant countries to implement stricter management. The International Fisheries Commission continues to update the assessment results on main fishing species. The stock assessment in 2017 indicated that the percentage of the assessed populations in this area fished within biologically sustainable limits was 66.7% (FAO 2020; Chen 2021).

Antarctic Sea

The Antarctic Sea includes statistical areas 48, 58, and 88 by FAO, and it is connected with the three oceans, with the northern boundary being the Antarctic convergence line and the southern boundary being the Antarctic continent; half of the area of the Antarctic Sea is covered by ice in the winter. This area abounds in krill, but there are not many fish species, and only a few species, such as Antarctic cod and ice fish, have fishery value (Chen 2014, 2016).

The main surface currents in the Antarctic Sea have been analyzed and studied. Upwelling appears in the convergence zone of the low-pressure band at approximately 65°S, and upwelling also appears in the waters near the Antarctic continent. The marine environment of the Antarctic Sea is a deep-sea system with significant circulation. The upwelling brings rich nutritious substances to the surface, the biological productivity in the summer is very high, and the biomass in the winter is reduced.

There are about 60 species of pelagic fish and about 90 species of demersal fish in the Antarctic area (including Subantarctic waters), but the quantities of these fishes are still unclear. In the fertile waters (convergence zone) of the Antarctic-Pacific area, the average dry weight of lanternfish (Myctophidae) is 0.5 g/m^2; the lanternfish resources in the southern central convergence zone are abundant, and the yield via middle-layer trawling by the former Soviet Union was 5–10 tons every 2 h.

The largest biomass in the Antarctic area is krill. The scientists in the world have made different estimates for krill. The Russian scholar estimated 0.15–5 billion tons of krill based on whales feeding on krill; using the primary productivity of the Antarctic Sea, United Nations expert Gulland calculated 500 million tons, with an annual catchable amount being 100–200 million tons; and French scholar proposed that the biomass of krill in total was 210–290 million tons, the consumption by cetaceans and other animals every year was 130–140 million tons, and krill that reached the catchable specifications did not exceed 40–50% of the total biomass. It was estimated in surveys in recent years that the annual catchable amount of krill is 50 million tons (FAO 2020; Chen 2021).

7.2 Overview of Changes in the Global Marine Environment

Oceans, which account for approximately 71% of the area of Earth, contain abundant renewable resources—marine biological resources. Approximately 90% of the animal protein on Earth exists in oceans. They are not only an important resource foundation for the future development of humans but also important components of the life system on Earth. The effects of the global climate changes (including ocean acidification) on the habitat distribution and variations in marine fishery resources have become increasingly significant, which has been reported in the biennial State of World Fisheries and Aquaculture published by FAO as important topics for discussion. A deep understanding and mastery of the mechanism by which global climate change affects the marine ecological environment and biological resources, as well as the consequences that may be generated, and having knowledge regarding international frontier research trends in this field are conducive to the sustainable exploitation of marine fishery resources and conducive to the conservation and recovery of marine biological resources as well as the stability of the marine ecosystem.

7.2.1 Global Climate Warming and Its Effects on the Marine Ecosystem

Global climate warming is a "natural phenomenon." When fossil fuels are burned and forests are cut and burned, carbon dioxide and many other greenhouse gases are produced. Because the greenhouse gases are transmitted to visible light via solar radiation and absorb longwave radiation reflected from Earth, they can intensely absorb infrared waves from ground radiation, which lead to global climate warming, often referred to as the "greenhouse effect." The consequences of global warming will redistribute global precipitation, melt glaciers and permafrost, raise sea levels, and so on, which will directly or indirectly affect marine ecosystems as well as marine fisheries. Therefore, the global warming has always been a hot topic of concern for scientists in the world.

The Concept of Global Warming and Its Causes

Global warming means the phenomenon of climate change in which the temperature of Earth's atmosphere and oceans increases due to the greenhouse effect in a given period, which is one of the tragedies of the commons. In the last more than 100 years, the global average air temperature has undergone four changes, cold → warm → cold → warm, with an increasing trend for air temperature on the whole. After the 1980s, the average air temperature in the world has increased noticeably.

It is generally believed that there are two main causes of the generation of global warming:

(1) Human factors—dramatic population increase; pollution of the atmospheric environment; degradation of the marine ecosystem; damage to land, such as erosion and salinization; and sharp reduction in forest resources, among others

(2) Natural factors—volcanic activity and variation in the trajectory of Earth's periodic rotation, among others

Trends in Global Warming and the Consequences

According to the prediction by the Intergovernmental Panel on Climate Change (IPCC), the earth will fully enter a warming world in the next 50 to 100 years. Due to the effect of human activities, the concentration of greenhouse gases has increased very rapidly in the twenty-first century, which will make the temperature rise rapidly globally, in the East Asia region, and in China in the next 100 years; the average surface temperature in the world will increase by 1.4–5.8 °C. The realities of global warming are a cause for alarm in all countries in the world, and climate warming has already seriously affected the sustainable development of economy, society, and ecosystem.

The consequences of global climate warming are extremely serious, with the following manifestations: (1) As the climate becomes warmer, glaciers will melt, and sea levels will rise, causing the loss of ecologically function groups, such as coastal tidal flat wetlands, mangroves, coral reefs, and so on. (2) The water area will increase. Water evaporation has increased, rainy seasons are becoming prolonged, and the floods are becoming increasingly more frequent. The chances of being inundated with flood water have increased, and the extent and severity of the effects of windstorms have increased. (3) The increase in air temperature may cause the ice and snow on the Antarctic and the Arctic Ocean to melt, resulting in the gradual extinction of endangered marine animal, such as polar bears. (4) There will be changes to native ecosystems and impacts on the spheres of production, such as aquaculture and fisheries. (5) Many small islands will be submerged, affecting local socioeconomic development.

Effects of Global Warming on the Marine Ecosystem

With the increase in global air temperature, the amount of water vapor that evaporates from the oceans has increased substantially, exacerbating the phenomenon of ocean warming, but the ocean warming is geographically uneven. Many evidence indicates that seasonal upwelling may be affected by ocean warming, thereby affecting the entire food web. The consequences of ocean warming may affect the productivity, plankton, and fish. In a general case, it is concluded that ocean warming changes will drive the distribution for most marine species to shift toward the two poles, expanding the distribution of species habited in the warm water and shrinking the distribution of species habited in the cold water. Changes in pelagic fish communities will also occur, with an expected shift toward deep waters to offset the increase in sea water temperature. Moreover, ocean warming may change the predator-prey matching relationship, thereby affecting the marine ecosystem.

There have been many reports of the effects of ocean warming on marine living species. According to a report from the National Oceanic and Atmospheric Administration (NOAA), there has been an increase in the incidents of death in stranded jumbo flying squid (*Dosidicus gigas*) on the west coast of the United States in the past 20 years; this important squid generally lives in the south of the Gulf of California and along the coast of Peru. However, with the ocean warming, they swam northward, and the northern limit for their range of distribution has expanded from the area at 40°N in the 1980s to 60°N in the 2010s.

Based on the statistics data, in the past 100 years, the water temperature of the three oceans has been increasing, i.e., 0.51 °C overall (Fig. 7.5). The increasing trend for water temperature in each ocean is different, ranging from 0.43 to 0.71 °C/100 years (Table 7.1). In the inshore areas of China, the increasing trend for water temperature is even greater; the water temperature in the area of the Yellow Sea is increasing 1.21 °C/100 years, and those in the northern and southern parts of the East China Sea are increasing 1.21 °C/100 years and 1.14 °C/100 years, respectively (Fig. 7.6).

7.2.2 Ocean Acidification and Its Harm to Marine Biological Resources

In 2003, the terminology "ocean acidification" first appeared in the well-renowned UK-based

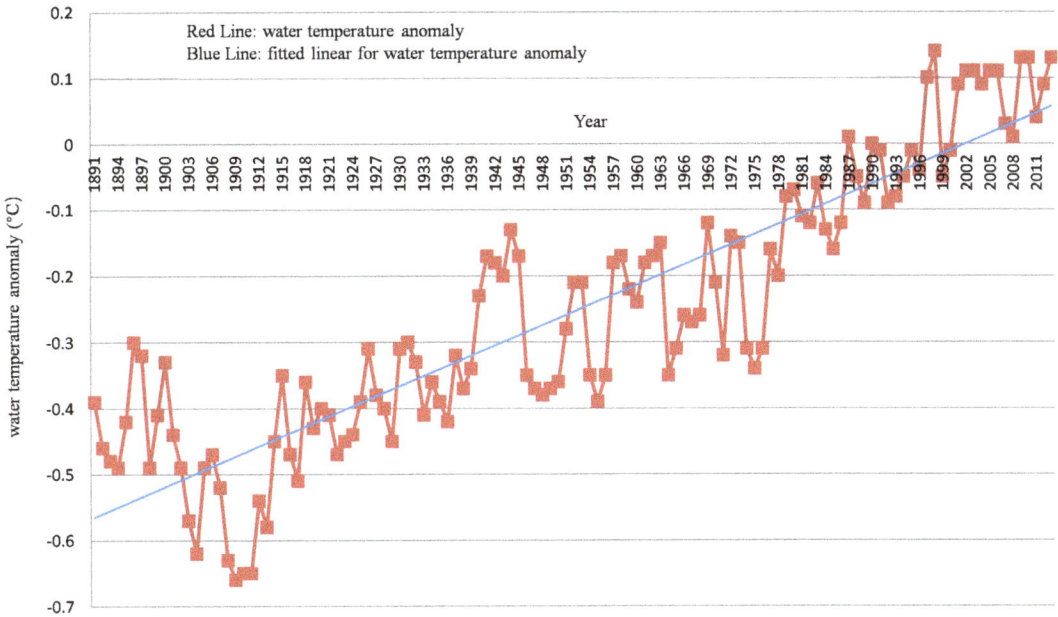

Fig. 7.5 Schematic diagram of the increasing trend for global water temperature from 1981 to 2013 (Japan Meteorological Agency 2018)

scientific journal *Nature* (Caldeira and Wickett 2003). Orr et al. (2005) sketched out the potential threats of "ocean acidification" over the twenty-first century and its impact on calcifying organisms. Fifty-five million years ago, a biological extinction event occurred in the oceans, and the culprit was carbon dioxide dissolved in sea water, the total volume of which was estimated to have reached 4.5 trillion tons; afterward, it took a period of at least 100,000 years to return to normal for the oceans. On August 13, 2009, more than 150 top global marine researchers gathered in Monaco and signed the *Monaco Declaration*. The signing of this declaration further meant that global scientists had expressed serious concerns about the impact of ocean acidification on the global marine ecosystem. This declaration noted that abrupt changes in the acid-base values (pH levels) of sea water were occurring 100 times faster than the natural changes in the past. These rapid changes in recent decades have already severely affected marine species, food web, ecological diversity, and fisheries. This declaration called on policymakers to stabilize carbon dioxide emissions within a safe range to avoid dangerous problems. If carbon dioxide emissions in the atmosphere continue to increase, coral reefs will not be able to survive in most areas of ocean by 2050, thus leading to seriously threaten commercial fishing target and food security of the humans.

Table 7.1 Water temperature changes in the three oceans (Japan Meteorological Agency 2018)

Sea area	Long-term trend
1: North Pacific Ocean	0.45 °C/100 years
2: South Pacific Ocean	0.43 °C/100 years
3: North Atlantic Ocean	0.61 °C/100 years
4: South Atlantic Ocean	0.71 °C/100 years
5: Indian Ocean	0.57 °C/100 years

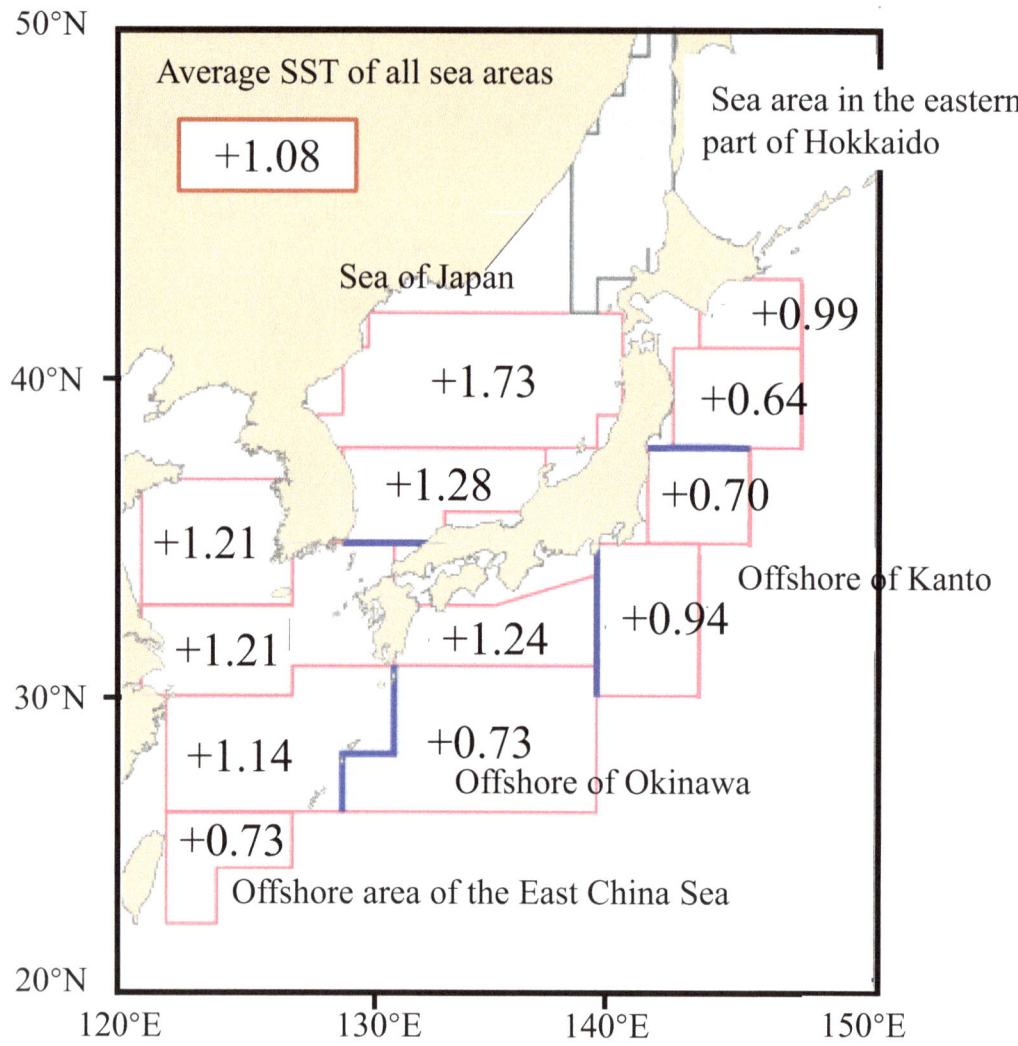

Fig. 7.6 Water temperature increase in the Northwest Pacific Ocean over the last 100 years (Japan Meteorological Agency 2018)

The Concept of Ocean Acidification and Its Causes

The Concept of Ocean Acidification

Ocean acidification refers to the phenomenon in which, due to the absorption of excess carbon dioxide in the air by sea water, there is a decrease in the degree of acidity and alkalinity. The degree of acidity and alkalinity is generally expressed as the pH value, which ranges from 0 to 14; a pH of 0 represents the strongest acidity, and a pH of 14 represents the strongest alkalinity. Sea water is weakly alkaline, and its pH value of the surface water is equal to nearly 8.2. When excess carbon dioxide in the air enters the oceans, the oceans acidify. Studies have shown that due to the effect of human activities, excess carbon dioxide emissions had already reduced the pH value of the sea water surface layer by 0.1 by 2012, indicating that the acidity of sea water had already increased by 30%.

Causes of Ocean Acidification

The oceans continuously carry out gas exchange with the atmosphere, and any component emitted into the atmosphere ultimately dissolves in the oceans. Before the arrival of the Industrial Age, changes of carbon in the atmosphere were caused by natural factors, and such natural changes resulted in fluctuations of global climate. However, since the Industrial Revolution, humans have mined and used fossil fuels, such as coal and petroleum, and cut down large areas of forests; by the beginning of the twenty-first century, more than 500 billion tons of carbon dioxide had been emitted. This has caused the carbon content in the atmosphere to increase year to year.

Affected by the sea breeze, atmospheric components first dissolve into the surface layer (several hundred meters deep) of the oceans, and in the subsequent centuries, these components gradually spread to the sea floor. Studies have shown that in the nineteenth and twentieth centuries, the oceans had already absorbed 30% of the carbon dioxide emitted by humans; currently, they are absorbing carbon dioxide at approximately one million tons per hour. The results from US and European scientists in 2012 indicated that the oceans are undergoing the fastest acidification in 300 million years and the rate of acidification exceeds that at the time of the biological extinction 55 million years ago. Human activities cause the constant acidification of sea water; it is estimated that the acidity of the surface layer of sea water will increase to pH 7.8 by 2100, 150% higher than in 1800.

The Harm of Ocean Acidification

Effects on Phytoplankton

Because phytoplankton is considered as the basis of the marine food web and primary productivity, their damage likely causes numerous marine animals to face shocks, including fish, shrimp, sharks, and whales. In sea water with a comparatively low pH, the feeding value of the nutrient salts will be reduced, and the changes on the ability of phytoplankton to absorb various nutri-

ent salts will also happen. Meanwhile, sea water with increasingly acidification will also corrode the bodies of marine living organisms. The studies have shown that calcified algae, coral insects, shellfish, crustaceans, and echinoderms are less efficient at forming calcium carbonate skeletons and shells in the acidified sea water. Due to global warming, the upper surface layer of oceans that absorb CO_2 from the atmosphere has also become less dense due to the increase in temperature, leading to weaken the water exchange between the surface sea water and the middle and deep water layers, which is not beneficial to the growth of phytoplankton.

Effects on Coral Reefs

The tropical coral reefs are very important in the marine ecosystem, which provide habitat, feeding ground, and breeding grounds of fish, and their production accounts for 12% of global catches. The research team found that the state of growth of corals is well when the average value of pH in sea water is 8.1. When the value of pH is 7.8, alcyonarian coral can survive. If the value of pH drops below 7.6, the corals cannot survive. The weak alkalinity in sea water is conducive to the utilization of calcium carbonate by marine organisms to form shells. The studies have shown that the value of pH in sea water is expected to reach approximately 7.8 by the end of this century, the acidity will be substantially higher than the normal state, and thus corals may be reduced greatly or even disappear.

Effects on Mollusks

Some studies have deemed that the sea waters in the southern part of equator will generate a corroding result on snail shells by 2030. These mollusks are one of the important food sources for salmon in the Pacific Ocean; a decrease in numbers or their disappearance in some areas will affect the salmon fishing industry. In addition, in acidified oceans, the internal shell of the cuttlefish will thicken and become more dense, causing cuttlefish to swim slowly, thereby affecting its feeding and growth.

Effects on Fish

Experiments have shown that when fish are in sea water in which the acidity is normal and where all other conditions are the same, only 10% will be caught in 30 h; however, when the same fish are placed in acidified experimental waters, such as the conditions in the vicinity of the Great Barrier Reef, they will be consumed by predators in the vicinity within 30 h. Based on a recent report from the *Proceedings of the National Academy of Sciences of the United States of America*, after simulating the acidity of sea water in the next 50–100 years, in sea water with the highest acidity, fish larvae instinctively avoid predators at first but then are very quickly attracted by the scent of the predators because of the damage of their olfactory system.

Effects on Marine Fisheries

Ocean acidification directly affects the growth and death of marine species, leading to permanent damage in commercial fish species, which would ultimately affect the yield of marine fisheries, threatening the food supply of millions of people. Although there is still no convincing prediction on how great an effect changes in the PH value will bring to marine catch, it is certain that ocean acidification will lead to a decrease in catch and an increase in fishing costs.

Ocean acidification reduces the suitable habitats of fish. In the Pacific region, the coral reefs are the important habitats for marine animals including fish and shrimp, which provide nearly 90% of the animal protein for the island countries. According to estimates, the value created for humans each year by corals and coral ecosystems exceeds USD 375 billion. A substantial reduction in coral reefs would generate a major effect on the marine ecosystem, social, and economic development.

The acidification of the oceans has reduced the availability of food for fish. Ocean acidification impedes the ability of certain large plankton at the bottom of the food chain to form calcium carbonate, making it difficult for these organisms to grow and leading to lower fish production at the top of the food chain.

The FAO estimates that more than 500 million people worldwide depend on fishing and aquaculture as a source of protein and income. For the poorest 400 million, fish provide about half of their daily requirements for animal protein and micronutrient. The impact of ocean acidification on marine living resources is bound to jeopardize the livelihoods of these poor people.

7.2.3   Phenomena of Global Climate Change with Greater Effects on Fisheries

El Niño-Southern Oscillation (ENSO)

In the world, a climatic phenomenon with the most obvious effects on fishery resources is El Niño and its corresponding La Niña phenomenon.

El Nino is a climatic phenomenon generated by the loss of equilibrium after the interaction of oceans with the atmosphere within an extensive range, and its significant feature is wide-ranging, continuous, and abnormal warming of the sea surface temperature (SST) in the areas of the eastern and central parts of the equatorial Pacific Ocean. La Niña refers to wide-ranging, continuous, and abnormal cooling of the SST in the eastern and central parts of the equatorial Pacific Ocean. Southern oscillation refers to the large-scale pressure fluctuation oscillation between the Indian Ocean and the Pacific Ocean; because El Niño is closely related to southern oscillation, they are collectively referred to as ENSO.

Oceanic Warm Pool and Cold Pool

The oceanic warm pool is also referred to as the thermal reservoir or warm pile, generally referring to the warm area where the average SST of the tropical West Pacific Ocean and the eastern part of the Indian Ocean remains above 28 °C for many years. Due to the combined action of solar radiation, heat exchange, and trade winds that blow from east to west, a large amount of warm water gradually accumulates in the warm pool area, causing the SST in this area to be 3–9 °C higher than that in the East Pacific Ocean. Its total area occupies 26.2% of the tropical marine area and accounts for 11.7% of the global marine area, with an east-west span of 150 longitudinal degrees and a north-south extension of approximately 35 latitudinal degrees; the depth of the West Pacific Warm Pool is between approximately 60 and 100 m. Changes in its scope can be used as one of the bases for predicting the El Niño event.

The "cold pool" refers to the low-temperature area that appears underwater in the area of the northern part of the Bering Sea in the summer; the emergence of the cold pool is caused by various factors, such as the formation of sea ice in the winter and the heating of the sea water surface layer in the spring and summer. With climate change, the scope of the cold pool shrinks or enlarges.

Aleutian Low
The Aleutian Low refers to a wide-ranging subpolar depression (cyclone) belt located in the vicinity of the Aleutian Islands near 60°N. The Aleutian Low is located in the region of the Aleutian Islands in the winter, moving northward when summer arrives and almost disappears. It attracts the surrounding air, causing it to rotate counterclockwise and thus stirring the body of water at the surface layer of the surrounding ocean to form a counterclockwise circulation system. A cyclonic circulation system centered on the Aleutian Low and composed of the warm Alaska Current, the Oyashio Current, and the warm North Pacific Current is generated north of 45°N in the North Pacific Ocean.

North Atlantic Oscillation (NAO)
The NAO refers to the north-south alternative variations between the Azores High and the Iceland Low, regulating the strength of the West Wind between 40° and 60°N in the North Atlantic Ocean and primarily affecting climate change in North America and Europe; when the NAO is in a positive state, the West Wind strengthens and moves north, and when the NAO is in a negative state, the opposite effect occurs.

Pacific Decadal Oscillation (PDO)
In addition to the interannual variation in the ENSO existing in the Pacific Ocean, variations on an interdecadal scale also exist in the North Pacific and North American regions in climate models of sea-atmosphere interactions, primarily manifesting as the PDO. The PDO is a type of long-period climatic fluctuation phenomenon that generally appears once every 20–30 years. Similar to El Nino variations, the PDO is also divided into warm and cold stages based on SST. During the PDO warm (cold) period, the SST in the Eastern Tropical Pacific Ocean is on the high (low) side, but the SST in the North Pacific Ocean is significantly reduced (increased). Existing studies have also shown that the long-term ENSO and PDO climate changes are closely related in time and space.

7.3 Effects of Climate Change on the Fishing Grounds for the World's Main Species

Currently, the main fishing targets in the world are tuna, Peruvian anchovy (*Engraulis ringens*), saury (*Cololabis saira*), Chilean jack mackerel (*Trachurus murphyi*), cod, salmonids, and cephalopods. In this section, we provide a general analysis of the research progress on the relationship between resource variations and fishing ground changes for these fish and climate change.

7.3.1 Tuna

Worldwide, tuna is an important commercial fish and is broadly distributed in the temperate and tropical areas of each ocean; rich in resources, tuna is the main fishing target of the world's distant-water fisheries and coastal countries. Tuna species include yellowfin tuna (*Thunnus albacares*), bigeye tuna (*T. obesus*), albacore tuna (*T. alalunga*), southern bluefin tuna (*T. maccoyii*), and skipjack (*Katsuwonus pelamis*), among others.

Among tuna, skipjack accounts for the highest annual total fishing yield; it is mainly distributed in the Western Central Pacific Ocean, East Pacific Ocean, Indian Ocean, and East Atlantic Ocean, and it occupies an important position in the world's tuna fisheries. The skipjack yield in the area of the Western Central Pacific Ocean is the highest. The West Pacific Ocean has the oceanic warm pool (West Pacific Warm Pool) with the highest global SST, and this warm pool area has a high temperature, low salinity, and comparatively low primary productivity. The data show that the skipjack yield in the West Pacific Warm Pool area is the highest in the entire West Pacific area. The reason for the high yield is the very large upwelling in the eastern waters of this warm pool, forming a band with low temperature, high salinity, and high primary productivity—a cold-water tongue; the convergence zone between the warm pool and the cold-water tongue, an area where phytoplankton and microzooplankton gather, is rich in bait organisms and has more nutritious substances, forming a good feeding ground for skipjack and thus becoming a good purse seine fishing ground for skipjack. The tuna tagging program implemented by the Secretariat of the Pacific Community (formerly the South Pacific Commission, SPC) from 1990 to 1992 and the calculation and analysis by Lehodey et al. (1997) of the catch per unit effort (CPUE) data for skipjack caught by US purse seiners (Fig. 7.7), the Southern Oscillation Index (SOI), and the 29 °C isotherm of the SST in the warm pool area showed that the distribution variations in the skipjack fishery are closely related to the occurrence of ENSO and the movement of West Pacific Warm Pool (Fig. 7.8).

How does the occurrence of the ENSO affect changes in the distribution of skipjack fishing grounds? Many studies have shown that the occurrence of the ENSO has affected changes in the scope of the West Pacific Warm Pool area, in turn resulting in changes in the skipjack fishing grounds. When El Nino occurs, the Eastern Equatorial Pacific Ocean warms, the upwelling weakens, the West Pacific Warm Pool expands eastward, warm water occupies the central and eastern Pacific regions of the equator, and the convergence zone between the warm pool and the cold-water tongue also moves eastward with the eastward expansion of the warm pool; therefore, the high-yield skipjack fishing ground formed in this convergence zone also moves eastward. The process is opposite during the occurrence of La Niña; i.e., the West Pacific Warm Pool shrinks westward, the convergence zone also moves westward as the warm pool moves westward, and the skipjack fishery also moves westward along with them.

Yellowfin tuna and bigeye tuna are important capture species; between them, the annual total fishing yield of yellowfin tuna in the world is second only to that of the skipjack. The large-scale ocean current system in the area where the yellowfin tuna and bigeye tuna fishing grounds are distributed in the Pacific Ocean includes the east-to-south equatorial current and the west-to-north equatorial current and Kuroshio Current, which are closely related to El Niño, and the westward equatorial current and the east-to-north Equatorial Counter Current; changes in its development have an important effect on changes in the distribution of the yellowfin tuna and bigeye tuna fishing grounds. Research has found (Lu et al. 2001) that during the periods in which El Niño occurred, the STT in the area of the tropical Pacific Ocean increased compared to that in a normal year (the equatorial Pacific area east of 155°W), the angling rate for yellowfin tuna was higher, but the area with a high angling rate for bigeye tuna was located at the western edge of the Eastern Equatorial Pacific Ocean. In a La Niña year, with the decrease in the SST in the Eastern Equatorial Pacific Ocean, the angling rate for yellowfin tuna in this area was also lower, and the area with a higher angling rate for yellowfin tuna moved to the area of the North Equatorial Pacific Ocean where the SST was higher in a La Niña year; with the decrease in SST in the Eastern Equatorial Pacific Ocean, the angling rate for bigeye tuna in this area also decreased obviously.

Albacore tuna is an important capture species. The fishing grounds for the Pacific albacore tuna are mainly distributed in the area of the Northwest Pacific Ocean in the vicinity of 30°N and in the

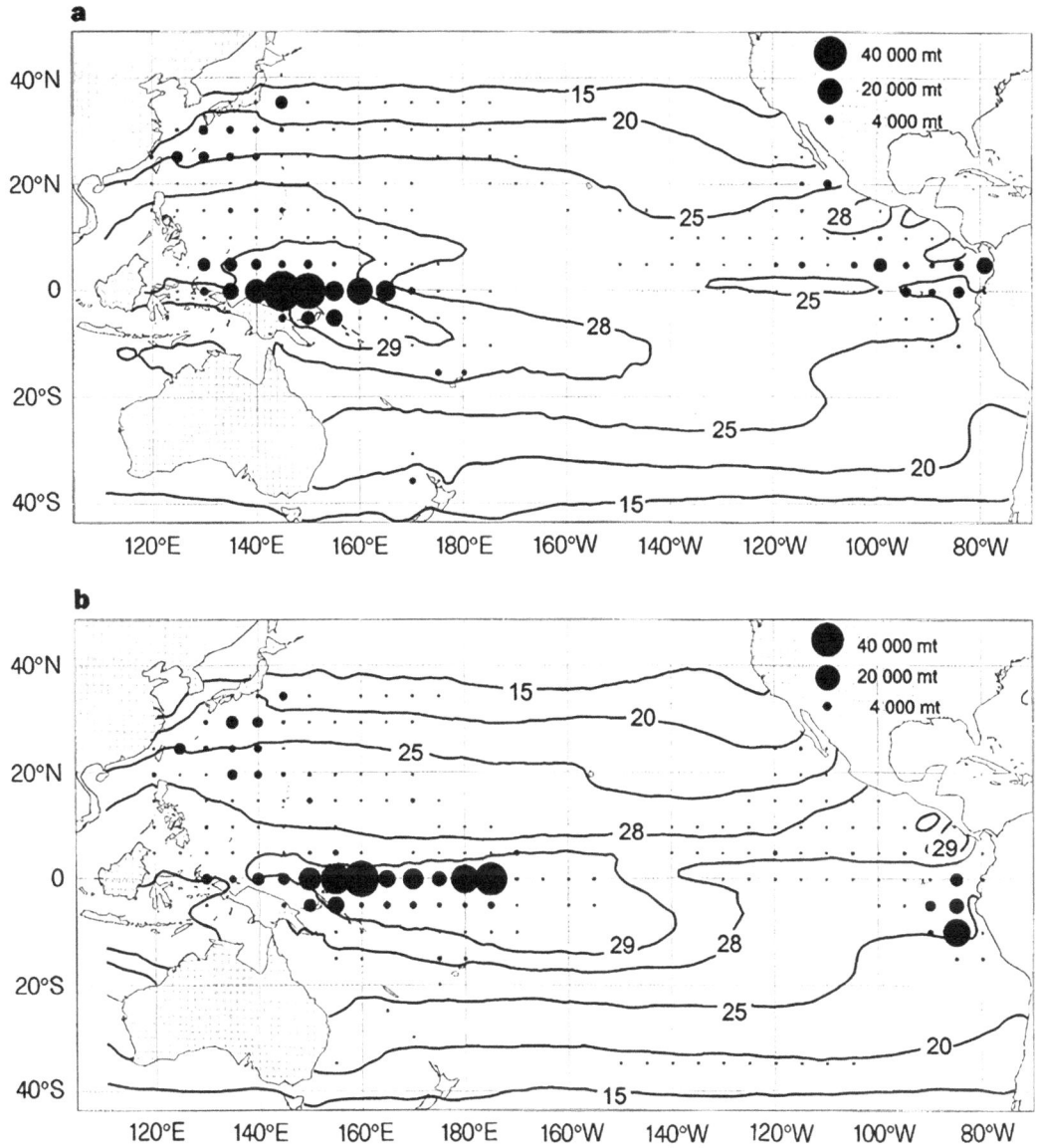

Fig. 7.7 Distribution of skipjack tuna catch (tons) and mean sea surface temperature (SST, in 8C) in the Pacific Ocean. (**a**), In the first half of 1989 (La Niña period). (**b**) In the first half of 1992 (El Niño period). The effect of ENSO on the location of the warm pool (SST > 28–29 °C) and the distribution of skipjack catch is clearly evident (Lehodey et al. 1997)

area of the Southwest Pacific Ocean between 0° and 40°S. The area in the Southwest Pacific Ocean extends eastward to the vicinity of 120°W, and the waters with high catch are mainly distributed in the vicinity of 10°S and the area east of Australia. Studies have shown (Kimura and Sugimoto 1997; IATTC 2001) that the distribution of albacore tuna fishing grounds is affected by large-scale marine phenomena. After the ENSO occurs (Lu et al. 1998), the CPUE of albacore tuna in the Southwest Pacific Ocean between 0° and 10°S increased, which might

Fig. 7.8 Skipjack tuna CPUE of the US purse seine fleet in the western equatorial Pacific. (**a**) Average CPUE for 1988 to 1995. The rectangle delimits the time-longitude section (**b**) of monthly mean CPUE; the pink line is the longitudinal gravity center of CPUE (G), the blue line is the 29 8C sea surface temperature (SST) isotherm, and the broken red line is the southern oscillation index (SOI); each variable was smoothed with a 5-month moving average. Correlation coefficients between G and the SOI (**c**), and between the 29 8C SST isotherm and the SOI (**d**), are estimated by the cross-correlation function (lags in months; dotted lines represent coefficients significant at $P = 0.05$) (Lehodey et al. 1997)

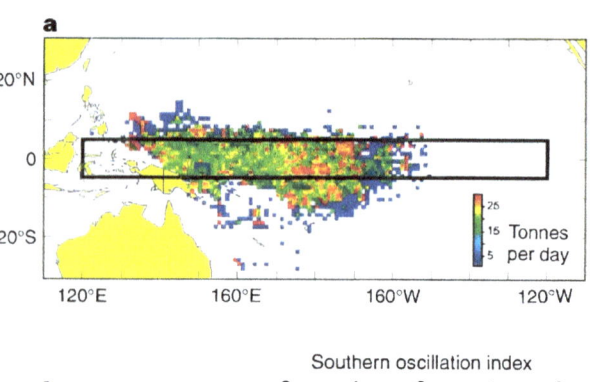

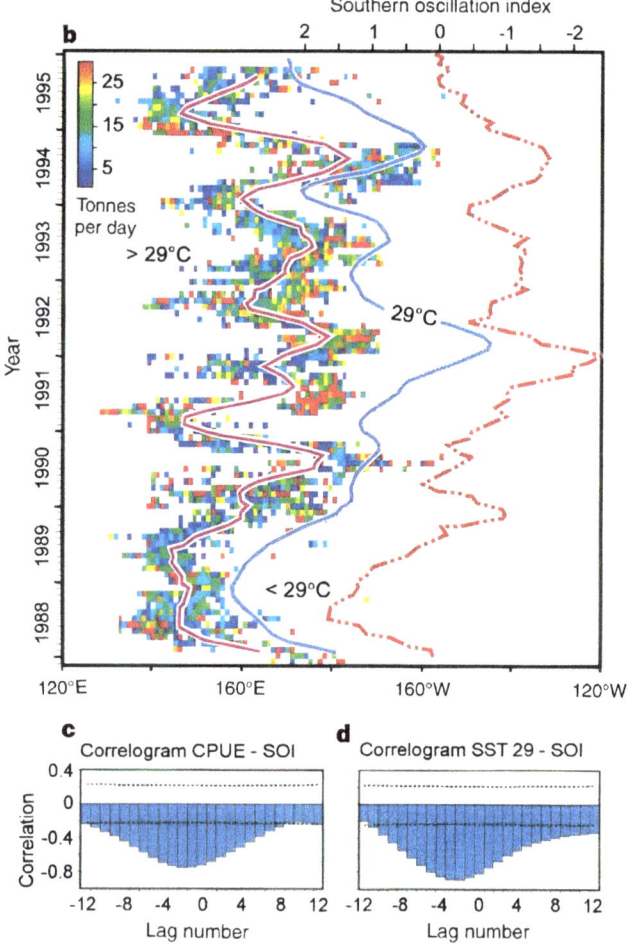

have been due to the northward shrinkage of the area with a suitable water temperature for albacore tuna in the area of the Southwest Pacific Ocean between 0° and 40°S brought by the ENSO, which had led to the northward shrinkage of the albacore tuna fishing ground to between 0° and 10°S. In the area south of 30°S, a decrease in the CPUE of albacore tuna appeared 4 years after the ENSO occurred, and in the area of 10°–30°S, a decrease in the CPUE of albacore tuna appeared 8 years after the ENSO occurred; such a lag phenomenon may be caused by decreases in the spawning and recruitment of albacore tuna during the ENSO period. A Japanese study has deemed that a counterclockwise migration of albacore tuna exists in the North Pacific Ocean, and when El Niño occurs, cold-water areas appear in the central and southwestern parts of the North Pacific Ocean, potentially causing the migration path of the albacore tuna to be wider than in non-El Niño years.

7.3.2 Saury

Saury is a cold-water pelagic fish with abundant resources, and with this species, simple fishing methods result in a high fishing efficiency; it is a fishery species with higher economic benefits. Saury inhabits the subtropical and temperate waters of 18°–66°N in the Pacific Ocean along the coast of Asia and the Americas, and the fishing grounds are mainly distributed in the Northwest Pacific Ocean, with mainly two fishing grounds: the northeastern area of Japan and an offshore fishing ground extending south of the Kuril Islands to the high seas.

The formation and distribution of saury fishing grounds in the Northwest Pacific Ocean are mainly affected by the Kuroshio Current and the Oyashio Current. In spring each year, as the water temperature gradually increases, sauries in the Northwest Pacific Ocean begin a northward feeding migration, reaching the Oyashio Current area along the coast of the Kuril Islands in the summer and forming the summer feeding ground for Pacific saury. After reaching maturity, the schools of fish begin to migrate southward, forming an autumn fishing ground, and are caught along the northeast coast of Japan and in the high seas south of the Kuril Islands. The spawning period of saury in the Northwest Pacific Ocean is comparatively long, lasting from autumn to spring of the following year, and the main spawning ground in autumn is located in the convergence zone of the Kuroshio Current-Oyashio Current in the northern part of the Kuroshio Front, but the spawning ground in winter and spring is located in the waters of the Kuroshio Current. Studies have noted (Shen et al. 2004) that the abundance and distribution of cold-current system zooplankton (such as crustaceans and chaetognaths) and fish

eggs used as feed for saury are affected by the marine environment; in addition, the feeding process for saury also requires suitable environmental conditions; therefore, the formation and distribution of saury fishing grounds are greatly affected by the marine environment. Because the water temperature of the Oyashio Current is comparatively low, it is not suitable for saury survival; therefore, a saury fishing ground is formed in the convergence zone of the Kuroshio Current-Oyashio Current, where the water temperature is more suitable, and the fishing ground moves north or south with the change in seasons and water temperature. In the summer, the water temperature in the vicinity of the Kuril Islands increases, and there is abundant feed; therefore, sauries mainly feed and fatten in the Kuril Islands and the Sea of Okhotsk; as the water temperature decreases in autumn, schools of fish also gradually descend southward to obtain food. Therefore, the strength of the Kuroshio Current will affect the abundance of the saury population, while the force of the Oyashio Current will affect the degree of plumpness of the schools of fish and the formation and location of fishing grounds.

Research has also found (Tian et al. 2003) that the abundance of large saury (body length 28.9–32.4 cm, average body length 30.7 cm) is closely related to the SST of the Kuroshio Current in the winter, while the abundance of medium saury (body length 24.0–28.5 cm, average body length 26.8 cm) is closely related to the SST of the convergence zone of the Kuroshio Current-Oyashio Current as well as the SST of Oyashio Current, indicating that the abundance of large-sized and medium-sized saury are affected by the different marine environments of the subtropical zone and the subfrigid zone, respectively.

Studies have deemed that the ENSO has an obvious effect on large saury resources. Through an analysis of the abundance of Pacific saury and ENSO data from 1950 to 2000, El Niño has a positive effect on large saury, while La Niña has a negative effect on large saury; in El Niño years, the abundance of large saury is higher than that during ordinary years, i.e., three times that during La Niña years. The ENSO basically has no obvious effect on the fluctuation of medium Pacific saury resources, but the abundance of medium Pacific saury is significantly correlated with the North Pacific Index (NPI), indicating that the Aleutian Low may have an effect on medium-sized Pacific saury resources (Tian et al. 2002). These studies have shown that medium-sized and large-sized Pacific saury resources are affected by changes in different marine environmental systems.

7.3.3   Chilean Jack Mackerel

Chilean jack mackerel is an oceanic pelagic fish that is mainly distributed along the coast of Peru and Chile in the Southeast Pacific Ocean, and it is also seen in the long and narrow waters along the southern coast of Argentina and east of New Zealand between 35° and 50°S in the Southwest Atlantic Ocean, which is referred to as the "jack mackerel belt." Studies have deemed (Arcos et al. 2001) that Chilean jack mackerel is a warm fish species and that the 15 °C isotherm along the coast of the Southeast Pacific Ocean is important for the distribution of jack mackerel. Chilean jack mackerel juveniles are suited for inhabiting warmer waters, while adults prefer to inhabit cooler waters; the water temperature of the surface layer gradually increases from south to north along the coast of Chile, and juveniles are mainly distributed in the nearshore area where the SST is higher than 15 °C, while the adults are mainly distributed in the area south of the 15 °C isotherm. The range of distribution for jack mackerel in the inshore areas of Chile changes as the 15 °C isotherm varies, especially during the periods when El Niño and La Niña occur. From 1997 to 1998, El Niño occurred, and as the warm water (>15 °C) of the northwestern area intruded into the waters along the central and southern coast of Chile (the main feeding ground for the Chilean jack mackerel), the 15 °C isotherm deviated southward in both 1997 and 1998, more southward than in a normal year, which directly affected the migration of Chilean jack mackerel, causing the mixing of jack mackerels of different body lengths and sizes, which generally do not cluster together. Viewed from the

catch situation, from 1997 to 1998, Chilean jack mackerel caught in that fishing ground were mainly juveniles, with a body length less than 26 cm, some months accounting for 80% of the total catch.

7.3.4   Peruvian Anchovy

Peruvian anchovy is a clustering pelagic fish distributed in the Southeast Pacific Ocean, and this species occupies a very high position in world fishery. The yield is extremely high but unstable, with an annual yield that can reach approximately ten million tons but that is only approximately one million tons in some year. The main producing countries are Peru and Chile. The 1998 FAO research report analyzed the effects of El Niño from 1997 to 1998 on fluctuations in pelagic fish resources such as Peruvian anchovy (FAO 1999). The report indicated that the East Pacific Ocean, especially the western region of South America, was the region that most seriously suffered the adverse effects of the warm El Niño Current; the increase in the coastal water temperature and the weakening of the increase resulted in severe decreases in biomass and schools of small pelagic fish. The main reasons for the decline were the lack of replenishment and poor growth conditions as well as an increase in the natural mortality rate. Prior to this strong El Niño, abnormal climate changes generated effects on the fluctuation of Peruvian anchovy resources. The Peruvian anchovy production of Peru in 1970 was 12.28 million tons, but due to the occurrence of El Niño in 1972, the catch decreased sharply to 4.45 million tons. The effects of El Niño lasted for 2 to 3 years; in 1973 and 1974, the catches of Peruvian anchovy were 1.5 million tons and three million tons, respectively, 1/8 and 1/4 of the catch in 1970 (Zheng and Yang 1993). Although the anchovy caught by Peru has decreased substantially, the distribution of anchovy has extended into the 200-meter deep-water area of northern Chile in an area where no anchovy had appeared originally. Such a phenomenon is also deemed to be caused by changes that occurred in the traditional Peruvian anchovy fishing grounds as a result of the intrusion of the warm equatorial current into the coast of Peru after the 1972 El Niño.

7.3.5 Cod

Cod (Gadiformes) is the main commercial fish species in the world, inhabiting the ocean bottom and the deep sea; it is widely distributed in each ocean in the world, with a variety of species, including Atlantic cod (*Gadus morhua*) and Alaskan pollock (*Theragra chalcogramma*). Atlantic cod is primarily distributed on both coasts of the North Atlantic Ocean; in Europe, it is mainly distributed in inshore areas of countries such as the United Kingdom, Iceland, and Norway and in the area of Spitsbergen in the Barents Sea. These areas are affected by the warm North Atlantic Current from the Gulf Stream, plus the convergence of multiple currents, such as the warm West Spitsbergen Current, the warm Norwegian Current, the warm West Greenland Current, and the cold East Greenland Current, which form the Northeast Atlantic fishing ground. In North America, Atlantic cod is mainly distributed in the areas of Newfoundland to the Gulf of Maine, and these areas are deeply affected by the confluence of the warm Gulf Stream and the cold Labrador Current. Studies have found (Ottersen and Stenseth 2001) that in the area of the Barents Sea, changes in the NAO and the water temperature can explain 55% of the changes in the abundance of Atlantic cod. In years with high NAO, a strong West Wind increases the warm North Atlantic Current and the warm Norwegian Current flowing from the southwest, making the water temperature of the Barents Sea increase, which is suitable for the survival and growth of Atlantic cod larvae, and the increase in water temperature also increases the number of *Calanus finmarchicus*, the main feed for Atlantic cod larvae. Additionally, the warm Norwegian Current that flows into the Barents Sea also carries a large amount of zooplankton feed, and the increase in its flow is also conducive to the growth of Atlantic cod larvae.

In the area of the Northwest Atlantic Ocean in North America, the catches from the Atlantic cod fishing grounds in the Grand Banks of Newfoundland, Canada, and in the Labrador Sea area are also affected by changes in SST and salinity caused by the NAO (Mantua et al. 1997). Climate change also has a relationship with cod resources (O'Brien et al. 2000). Since 1988, the SST in the area of the North Atlantic Ocean has continued to increase, leading to immature cod under 5 years of age or even under 3 years of age to become the main catch of Atlantic cod in that region due to a lack of replenishment of cod resources. It was reported in *Science* that from the last half of the 1980s to the 1990s, the melting of Arctic Ocean ice had increased, the melting Arctic Ocean ice diluted the salinity of the sea water, and this low-salinity sea water flowed down southward from the Arctic Ocean with the cold Labrador Current through the Davis Strait and into the Northwest Atlantic Ocean, creating a new water temperature boundary; changes occurred in the ecosystem, further changing the growth and composition of plankton and ultimately leading to a decline in Atlantic cod resources. NOAA, after conducting a survey of pollock resources in the Bering Sea in the summer of 2007, reported that pollock that originally inhabited the Bering Sea and Aleutian sea area had moved northward off the northwestern shores of the Pribilof Islands to the area near the exclusive zone of Russia; it was initially thought that global warming might have been the reason for the northward movement of the pollock fishing ground in the Bering Sea. The "cold pool" phenomenon in the northern area of the Bering Sea has generated a similar effect on pollock resources.

Studies have also deemed (Wyllie-echeverria and Wooster 1998) that the cold pool in the area of the Bering Sea is related to the distribution of the pollock population. In winter, as the stronger Aleutian Low moves eastward, the Bering Sea becomes warm, and the scope of the cold pool shrinks. As the weaker Aleutian Low moves westward, the Bering Sea becomes cold, the scope of the cold pool becomes large, and such changes directly affect the pollock population in the Bering Sea.

7.3.6   Salmonids

Salmonids (Salmonidae) are important cold-water commercial fishes in the northern hemisphere. Salmons are anadromous fishes that are distributed in the northern part of the Pacific Ocean and the Atlantic Ocean as well as in the area of the Arctic Ocean and various water system basins along the coast. Studies have found (Mantua et al. 1997) that PDO occurred in 1925, 1947, and 1977 (in 1925, the cold period became a warm period; in 1947, the warm period became a cold period; and in 1977, the cold period became a warm period again). During the warm (cold) period, the SST of the Eastern Tropical Pacific Ocean was higher (lower), while the SST of the North Pacific Ocean was significantly reduced (increased); when the latter two instances of PDO occurred, they greatly affected the catch of salmonids in the North Pacific Ocean.

Climate change also has an effect on Arctic freshwater fish and anadromous spawning fish (Reist et al. 2006). Migratory fish will be subjected to the comprehensive effects of climate change on freshwater environments, estuaries, and marine areas. The increase in air temperature caused by climate change may generate three consequences for Arctic freshwater fish and anadromous spawning fish: local population extinction; northward migration of the distribution range; and genetic changes through natural selection. The distribution of many fishes is restricted by the location of isotherms, and climate change, as reflected in temperature changes and in resource changes, such as bait and food, affects the distribution of fish. When the temperature increases, Atlantic salmon (*Salmo salar*) may disappear from the southern part of the original distribution areas in Europe and North America and migrate to colder rivers (these rivers are warmer than originally and are more suitable for the Atlantic salmon to inhabit). In the western part of the Atlantic Ocean, as the rivers inhabited by Atlantic salmon become richer in productivity, the abundance of Atlantic salmon will also increase. For Arctic char (*Salvelinus alpinus*), the increase in temperature has multiple effects.

Due to the increase in the SST in summer, the long-term continuation of the optimum growth temperature (12–16 °C) and the increase in marine productivity will increase the average body length and weight of Arctic char. However, higher temperatures in spring and the acceleration of ice melting will cause adverse effects on the Atlantic salmon that migrate when the ice melts in spring. Although such a situation may also improve the adaptability of Atlantic salmon residing in the sea, it will reduce its salt tolerance as well as shorten the time for successful upstream migration, and the drastic increase in temperature will also hinder the osmotic pressure adjustment ability of migratory fish, cause an increase in energy consumption, and lead to a decrease in the growth rate as well as an increase in mortality rate during the catadromous process.

7.3.7   Cephalopods

The effect of climate change on cephalopod resources is realized through the effect on their life history processes. Their life history processes usually include feeding migration and spawning migration. Before reaching the feeding area, cephalopod larvae usually move with the current; for example, neon flying squid (*Ommastrephes bartramii*) ascend northward with the Kuroshio Current, and Argentine shortfin squid (*Illex argentinus*) descend southward with the Brazil Current, and this process is an extremely important link that affects the amount of cephalopod resources but may be affected by smaller individuals and weaker mobility.

Spawning grounds are important habitats for cephalopods, and a large number of studies have shown that the degree of suitability of the marine environment of spawning grounds is extremely important for recruitment quantity; therefore, many scholars often utilize the effect of environmental changes at spawning grounds to explain the reasons for changes in the amount of resources and have obtained comparatively good results.

For squids (inshore loliginid and oceanic squid), SST, NAO and time series analysis methods are used to study the effects of marine climate change on the *Loligo pealeii* and *Illex illecebrosus* resources in the Northwest Atlantic Ocean. Results have indicated that changes in the water temperature of spawning grounds affect their embryonic development, growth, and recruitment (Dawe et al. 2000, 2007). Studies have also found (Arkhipkin et al. 2004) that changes in the salinity of spawning grounds affect the activities of Patagonian squid and their distribution in feeding grounds. In addition, when the water temperature at spawning grounds is higher than 10.5 °C, Patagonian squid migrate earlier to feeding grounds. Studies have deemed (Waluda et al. 1999) that a change in the suitable surface temperature of spawning grounds has a very important effect on the recruitment of Argentine flying squid and that the change in the suitable surface temperature at spawning grounds results from the mutual collocation between the Brazil Current and the Falkland Current. Studies have found (Waluda and Rodhouse 2006) that in the area of the Southeast Pacific Ocean, the range of suitable temperature (24 to 28 °C) for spawning grounds in September is positively correlated with the recruitment of jumbo flying squid (*Dosidicus gigas*). Additionally, phenomena such as El Niño and La Niña have notable effects on *Dosidicus gigas* resources, and it is thought that the El Niño and La Niña phenomena cause changes in the primary and secondary productivity at spawning grounds and, in turn, affect the early life stages of *Dosidicus gigas* as well as mature individuals. In the area of the Northwest Pacific Ocean, the suitable water temperature range for spawning grounds and feeding grounds for the western population of squid born in winter and spring can very well explain the changes in their resource abundance (Cao et al. 2009), and the El Niño and La Niña phenomena have significant effects on the recruitment of neon flying squid in the Northwest Pacific Ocean (Chen et al. 2007).

Regarding octopus, embryonic development and the growth of juveniles are closely related to water temperature (Hernández-López et al. 2001). Studies have found (Caballero-Alfonso et al. 2010) that temperature is an important environmental indicator that affects the amount of octopus resources, and the NAO also indirectly affects the amount of octopus resources by changing the water temperature at spawning grounds. Studies have also found that environmental factors affect the habitat, resource density, and distribution of octopus in the area near Brazil (Leitea et al. 2009); moreover, in areas near intertidal zones, smaller octopuses can grow faster in a warm water environment. In addition, small- and medium-sized individual octopuses are mostly distributed in waters 1–2 °C higher than the suitable temperature for early stages because it is conducive to their growth. Environmental factors such as temperature have an obvious influencing effect on the resource density and distribution of octopus species.

In addition to generating an effect on spawning grounds, feeding migration, growth in feeding grounds, and reproductive migration are also important components of the life cycle of cephalopods, but currently, there are comparatively few studies that address this topic. In accordance with the biological data for Japanese flying squid (*Todarodes pacificus*), Japanese scholars utilized bioenergy models and nutritional ecosystem models to study resource variations (Kishi et al. 2009). The results showed that because the predation density in the northern part of the Sea of Japan is higher than that in the central part of the Sea of Japan, *Todarodes pacificus* individuals in the northern part of the Sea of Japan are larger than the squid individuals that migrate from the central part of the Sea of Japan. Additionally, an increase in the global air temperature day by day could cause changes in the migration path of *Todarodes pacificus*. The study found (Choi et al. 2008) that changes in the global climate, which has caused changes in the migration path of *Todarodes pacificus*, accompanied by changes in the marine ecosystem environment, which has also affected the distribution of spawning grounds as well as the survival of juveniles, affected recruitment. The study also deemed that the squid distributed in the North Pacific Ocean would undergo seasonal migration in the south-north direction annually and that the force of the Kuroshio Current and a high or low surface

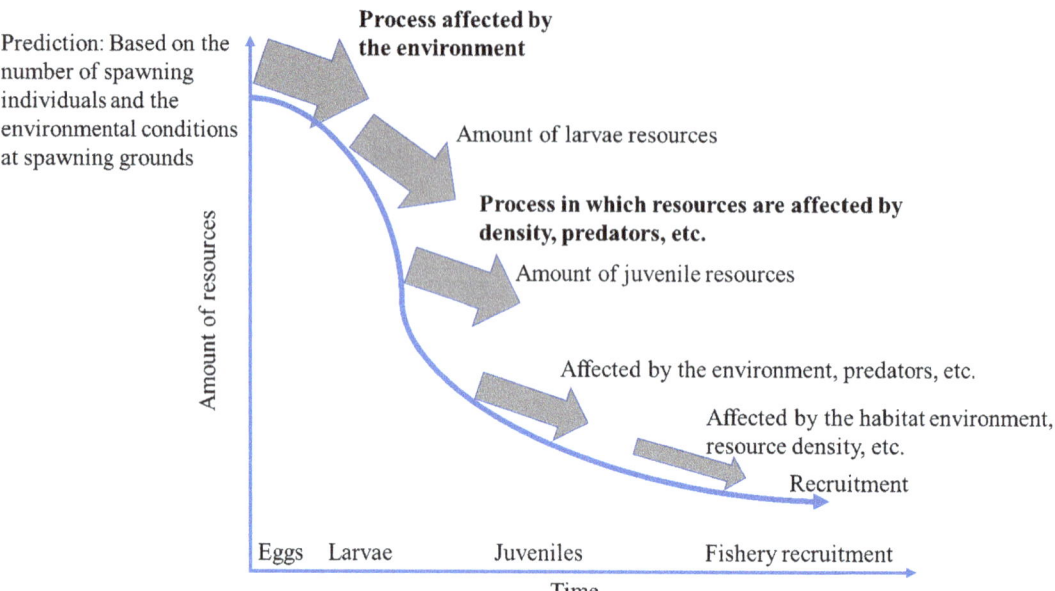

Prediction: Based on the number of spawning individuals and the environmental conditions at spawning grounds

Process affected by the environment

Amount of larvae resources

Process in which resources are affected by density, predators, etc.

Amount of juvenile resources

Affected by the environment, predators, etc.

Affected by the habitat environment, resource density, etc.

Recruitment

Amount of resources

Eggs   Larvae   Juveniles   Fishery recruitment

Time

Fig. 7.9 Schematic diagram of the recruitment process for cephalopods and its affecting factors (Chen 2014, 2016)

temperature at feeding grounds directly affect the formation and spatial distribution of squid fishing grounds (Wang and Chen 2005).

Studies have deemed that the current global climate changes (including temperature) indirectly affect the recruitment of cephalopod by affecting the environmental conditions of spawning grounds; there have been more studies on the relationship between environmental changes at spawning grounds and the replenishment quantities of cephalopods, and some research results have been used to predict resource replenishment quantities. The studies have shown that global climate change has the most substantial effect on the amount of cephalopod resources is the life history stage from hatching to fish larvae (Fig. 7.9), that is, the period of time after spawning because cephalopods at this stage are primarily passively affected by the environment and cannot actively adapt to environmental changes. However, after the larvae develop into fish, cephalopod individuals have comparatively strong swimming ability and can then find a suitable habitat environment by way of migration and actively adapt to environmental changes. However, in the course of reviewing the literature, we sought studies on the

relationship response between environmental changes at spawning grounds and the recruitment of cephalopods (the number of cephalopod adult individuals during fishery exploitation), but there are very few studies on the death and growth of and mechanism of influence on cephalopods during the intermediate stage (movement and growth with the currents). For the sustainable use and scientific management of cephalopod resources, not only do we have to consider the effects of environmental changes on the growth and death of individuals in spawning grounds, but we must also attach importance to their effects on juveniles, larvae, and individuals in different stages of life. Only in this way can the accuracy in predicting marine environmental changes and the recruitment of cephalopod resources be further improved.

References

Arcos DF, Cubillos LA, Núñez SP (2001) The jack mackerel fishery and El Niño 1997–98 effects off Chile. Prog Oceanogr 49(1–4):597–617

Arkhipkin AI, Middleton DAJ, Sirota AM et al (2004) The effect of Falkland current inflows on offshore ontogenetic migrations of the squid *Loligo gahi* on the southern shelf of the Falkland Islands. Estuar Coast Shelf Sci 60(1):11–22

Caballero-Alfonso AM, Ganzedo U, Santana AT et al (2010) The role of climatic variability on the short-term fluctuations of octopus captures at the Canary Islands. Fish Res 102:258–265

Caldeira K, Wickett ME (2003) Anthropogenic carbon and ocean pH. Nature 425:365

Cao J, Chen XJ, Chen Y (2009) Influence of surface oceanographic variability on abundance of the western winter-spring stock of neon flying squid (Ommastrephes bartramii) in the Northwest Pacific Ocean. Mar Ecol Prog Ser 381:119–127

Chen XJ (2014) Fisheries resources and fisheries oceanography. Ocean Press. (in Chinese)

Chen XJ (2016) Theory and method of fisheries forecasting. Ocean Press. (in Chinese)

Chen XJ (2021) Fisheries oceanography. Science Press. (in Chinese)

Chen XJ, Zhao XH, Chen Y (2007) Influence of El Niño/La Niña on the western winter–spring cohort of neon flying squid (Ommastrephes bartramii) in the north-western Pacific Ocean. ICES J Mar Sci 64:1152–1160

Choi K, Lee CL, Hwang K et al (2008) Distribution and migration of Japanese common squid, Todarodes pacificus, in the southwestern part of the east (Japan) sea. Fish Res 91(2):281–290

Dawe EG, Colbourne EB, Drinkwater KF (2000) Environ-mental effects on recruitment of short-finned squid (Illex illecebrosus). Mar Sci 57(2):1002–1013

Dawe EG, Hendrickson LC, Colbourne EB et al (2007) Ocean climate effects on the relative abundance of shortfinned (Illex illecebrosus) and long-finned (Loligo pealeii) squid in the Northwest Atlantic Ocean. Fish Oceanogr 16(4):303–316

FAO (1999) The state of world fisheries and aquaculture—1998. FAO

FAO (2020) The state of world fisheries and aquaculture—2020. FAO

FAO (2022) Global capture production statistics (1950 - 2020). https://www.fao.org/fishery/statistics-query/en/capture

Hernández-López JL, Castro-Hernández JJ, Hernández-García V (2001) Age determined from the daily depo-sition of concentric rings on common octopus (Octo-pus vulgaris) beaks. Fish Bull 99:679–684

IATTC (2001) Status of albacore tuna in the Pacific Ocean. IATTC stock assessment report 1, 255–283

Japan Meteorological Agency (2018). http://www.data.jma.go.jp/gmd/kaiyou/data/shindan/a_1/glb_warm/glb_warm.html

Kimura S, Sugimoto MNAT (1997) Migration of albacore, Thunnus alalunga, in the North Pacific Ocean in rela-tion to large oceanic phenomena. Fish Oceanogr 6: 51–57

Kishi MJ, Nakajima K, Fujii M et al (2009) Environmental factors which affect growth of Japanese common squid, Todarodes pacificus, analyzed by a bioenerget-ics model coupled with a lower trophic ecosystem model. Mar Syst 78(2):278–287

Lehodey P, Bertignac M, Hampton J et al (1997) El Niño southern oscillation and tuna in the western Pacific. Nature 389:715–718

Leitea TS, Haimovici M, Mather J et al (2009) Habitat, distribution, and abundance of the commercial octopus (Octopus insularis) in a tropical oceanic island, Brazil: information for management of an artisanal fishery inside a marine protected area. Fish Res 98:85–91

Lu HJ, Kee KT, Liao CH (1998) On the relationship between El Niño/Southern oscillation and South Pacific albacore. Fish Res 39(1):1–7

Lu HJ, Lee KT, Lin HL et al (2001) Spatio-temporal distribution of yellowfin tuna Thunnus albacares and bigeye tuna Thunnus obesus in the tropical Pacific Ocean in relation to large-scale temperature fluctuation during ENSO episodes. Fish Sci 67(6):1046–1052

Mantua N, Hare S, Zhang Y et al (1997) A Pacific interdecadal climate oscillation with impacts on salmon production. Bull Am Meteorol Soc 78:1069–1079

O'Brien CM, Fox CJ, Planque B et al (2000) Climate variability and North Sea cod. Nature 404:142

Orr JC, Fabry VJ, Aumont O et al (2005) Anthropogenic Ocean acidification over the twenty-first century and its impact on calcifying organisms. Nature 437:681–686

Ottersen G, Stenseth NC (2001) Atlantic climate governs oceanographic and ecological variability in the Barents Sea. Limnol Oceanogr 46(7):1774–1780

Reist JD, Wrona FJ, Prowse TD et al (2006) An overview of effects of climate change on selected Arctic fresh-water and anadromous fishes. Ambio 35:381–387

Shen JH, Han SX, Fan W et al (2004) Resources and fishing ground of Pacific saury (Cololabis Saira) in the Northwest Pacific Ocean. Mar Fish 26(1):61–65. (in Chinese)

Tian YJ, Akamine T, Suda M (2002) Long-term variability in the abundance of Pacific saury in the northwestern Pacific Ocean and climate changes during the last century. Bull Jpn Soc Fish Oceanogr 66:16–25. (in Japanese with English abstract)

Tian YJ, Akamine T, Suda M (2003) Variations in the abundance of Pacific saury (Cololabis saira) from the northwestern Pacific in relation to oceanic-climate changes. Fish Res 60(2–3):439–454

Waluda CM, Rodhouse PG (2006) Remotely sensed mesoscale oceanography of the central eastern Pacific and recruitment variability in Dosidicus gigas. Mar Ecol Prog Ser 310:25–32

Waluda CM, Trathan PN, Rodhouse PG (1999) Influence of oceanographic variability on recruitment in the Illex argentinus (Cephalopoda: Ommastrephidae) fishery in the South Atlantic. Mar Ecol Prog Ser 183:159–167

Wang YG, Chen XJ (2005) Economic oceanic squid resources and fisheries in the world. Ocean press. (in Chinese)

Wyllie-echeverria T, Wooster WS (1998) Year-to-year variations in Bering Sea ice cover and some consequences for fish distributions. Fish Oceanogr 7(2):159–170

Zheng GG, Yang CF (1993) Fishery and prediction of El Niño. Mar Forecasts 10(4):65–70. (in Chinese)